Female Adolescence in
American Scientific Thought,
1830–1930

NEW STUDIES IN AMERICAN INTELLECTUAL

AND CULTURAL HISTORY

Howard Brick, *Series Editor*

Female Adolescence in American Scientific Thought, 1830–1930

CRISTA DELUZIO

The Johns Hopkins University Press

Baltimore

The Johns Hopkins University Press
2715 North Charles Street
Baltimore, Maryland 21218-4363
www.press.jhu.edu

Library of Congress Cataloging-in-Publication Data

DeLuzio, Crista, 1966–
Female adolescence in American scientific thought, 1830–1930 /
Crista DeLuzio.
p. cm. — (New studies in American intellectual and cultural history)
Includes bibliographical references and index.
ISBN-13: 978-0-8018-8699-7 (hardcover : alk. paper)
ISBN-10: 0-8018-8699-6 (hardcover : alk. paper)
1. Teenage girls—United States—History—19th century. 2. Teenage
girls—United States—History—20th century. 3. Research—United
States—History—19th century. 4. Research—United States—History—
20th century. I. Title.
HQ798.D3976 2007
305.235'209730904—dc22   2006103116

A catalog record for this book is available from the British Library.

Contents

Acknowledgments

It is a pleasure and a privilege, at long last, to thank those who have guided and supported me in writing this book.

I began this project while I was a graduate student in the Department of American Civilization at Brown University, and I am grateful to my advisors for the assistance they provided in its initial stages. I thank Richard Meckel for encouraging my interest in the history of childhood early on in my graduate career and for sharing with me his breadth and depth of knowledge in the field. I have benefited over the years from his hard questions and high standards. I so appreciate that these have always been advanced with such patience and good will. Mari Jo Buhle's work in women's and intellectual history has been an inspiration. Her thorough readings of and insightful comments on my work continually prodded me toward more careful analysis and greater intellectual risk. The words of encouragement she offered as this project moved along meant more than she is ever likely to know. Susan Smulyan's insistence on clear, precise thinking and writing were helpful at the outset and at all stages of revising that followed. I am grateful, too, for her attention to my professional development and for modeling so beautifully the ideal of the teacher-student relationship. At Brown, I also benefited from the suggestions and encouragement of the members of my dissertation group. My warmest thanks to Dan Cavicchi, Kathy Franz, Joanne Melish, Janice Okoomian, Kristen Peterson Farmelant, Miriam Reumann, and Mari Yoshihara. I am also grateful to the American Association of University Women Educational Foundation for the fellowship that supported my research and writing during the dissertation phase.

My colleagues in the William P. Clements Department of History at Southern Methodist University provided invaluable guidance as I revised the manuscript for publication. I thank them for the examples of excellence in teaching and scholarship they offer through their own work and for the many ways they assist me as I strive to meet that same standard. With the most sincere appreciation, I thank Jeremy Adams, Peter Bakewell, John Chavez, Dennis Cordell, Edward Countryman, Melissa Dowling, Kenneth Hamilton, Benjamin Johnson, Thomas Knock, Glenn Linden, Alexis McCrossen, John Mears, Donald Niewyk, Daniel Orlovosky, Sherry Smith, David Weber, Kathleen Wellman, and Hal Williams. I am particularly grateful to Department Chair James Hopkins, whose exemplary mix of professionalism and kindness has gone a long way in enhancing my experience as a junior faculty member here. I am also thankful to Mildred Pinkston and Julie Stewart for the consistently efficient and cheerful assistance they provide as I carry out my daily academic duties.

Other colleagues at SMU also make this a stimulating and congenial place to work every day. My thanks to Suzanne Bost, Mark Chancy, David Doyle, Adam Herring, Valerie Hunt, Bruce Levy, Beth Newman, Pamela Patton, Carolyn Sargent, Nina Schwartz, and Rajani Sudan. I am grateful for the encouragement of my work from former SMU colleagues James Breeden, Deborah Cohen, Olga Dror, Michelle Nickerson, Michael Provence, Sarah Schneewind, and Trysh Travis. Billie Stovall in the SMU Interlibrary Loan office deserves my special thanks for her hard work in responding to my many requests for sources.

At the Johns Hopkins University Press, I am grateful to editor Robert J. Brugger for his initial interest in the project and for his diligent commitment to it as he shepherded it through to publication. I also thank Dorothy Ross and an anonymous reader for the many useful suggestions they offered for revision. Andre Barnett's careful copyediting greatly improved the final version of the manuscript.

Many family members and friends cheered me on with enthusiasm and helped me to keep things in perspective. My thanks to Paula and Kevin Appel, all the Burkes, David Criasia, John Crisafulli, Nancy DeLuzio, Tamara DeLuzio, Rosanne and Ralph Francesconi, Cara Harding, Karen Kinsella, Annette and Lee Packard, and Jeanne and Dick Valorie. Tina Nappi continues to remind me through her own remarkable example of the joys to be found in a life well lived and a job well done. I thank my brother, Mark DeLuzio, for his gifts of music and countless other inspirations. My sister, Maria DeLuzio, has championed my efforts for as long as I can remember. I am deeply grateful for all of the ways in which she puts her extraordinary generosity of spirit to the task of easing my worries and embellishing my small victories. My heart is full of thanks for my husband Bob Burke, without whose unwavering confidence and practical support this project would not have been completed. My son Aidan Burke's arrival as I was revising the manuscript rekindled my fascination with developmental thought in the past and present and has also opened up many new and wondrous horizons for thinking and feeling in my life and work. My oldest and deepest debt of gratitude is owed to my parents, Jean and Reno DeLuzio. It is impossible to thank them adequately for all that they have done for me. I hope they will recognize the lessons they taught me about hard work and leaps of faith that are written on every page.

Female Adolescence in
American Scientific Thought,
1830–1930

Introduction

In F. Scott Fitzgerald's short story, "Bernice Bobs Her Hair," the self-proclaimed modern Marjorie flat-out refuses any association with the Victorian conventions of girlhood espoused by the four March sisters in Louisa May Alcott's *Little Women*. "What modern girl," she impetuously demands of the more reserved Bernice, "could live like those inane females?"[1] Instead of wholesome, provincial domesticity, Marjorie vows to embrace worldliness and independence and expects to have a whole lot more fun in the process than the likes of Meg, Jo, Amy, and Beth could ever imagine. Flappers—icons of Fitzgerald's era—promised to replace Alcott's once equally emblematic March girls with a model of youthful femininity more suitable to modern times. By the 1920s, they seemed to be everywhere, found widely represented in literature, advertising, and popular culture, as well as roaming the halls of high schools and colleges, working in factories and offices, shopping in department stores, and frequenting dance halls and movie theaters.[2] Despite the seeming ubiquity of the flapper, however, the transition from the "old" girls of the Victorian era to the "new" girls of the modern age, as contemporary psychologists Phyllis Blanchard and Carolyn Manasses put it, did not occur without considerable social and cultural struggle. Even as Marjorie cavalierly dismissed them as hopelessly "out of style," the virtues of the March girls continued to exist in tension with a more updated set of manners and mores wrought from the rapid and dramatic changes giving shape to American life at the turn of the twentieth century.[3]

Playing a leading role in imparting larger meaning to the transformation of old girls into new were scientists and intellectuals working in the fields of medicine, biology, psychology, and anthropology during the nineteenth and early twentieth centuries. This period marked the inception of the modern scientific study of the child. From within and across these disciplines came descriptions of the child's development into maturity, along with prescriptions for directing that development toward the dual ends of personal happiness and social progress, all based on the authority of science.[4] Operating within this emerging framework of child development studies, scientists eventually came to render their own version of the new girl known as the "adolescent girl." The central intellectual challenge they found, however, was in reconciling the concepts of femininity and adolescence. This book explores the evolving permutations of that challenge from 1830 to 1930 in order to consider the influences of the scientific construction of female adolescence on child development expectations and meanings of gender in the modern age. What sorts of conceptual problems did fashioning a category of the adolescent girl pose to scientists in these years, and how did they go about solving them? What have been the effects of these efforts on ongoing approaches to the study of the child?

The debates during this period over the meanings of adolescence and female adolescence have influenced subsequent thinking about these concepts and categories. Indeed, the dilemma of reconciling femininity and adolescence still resonates for scholars, professionals, and the wider culture in the early twenty-first century. The work of feminist psychologist Carol Gilligan in the 1980s and 1990s initially renewed interest in this topic. Since then, several notable studies have been conducted by the American Association of University Women, and we have seen the publication of bestsellers such as psychologist Mary Pipher's *Reviving Ophelia: Saving the Selves of Adolescent Girls.*[5] This body of work begins with the assertion that girls have not been well served by a psychological tradition in which the model for human development is the white middle-class male. As Gilligan explains, girls face a "crisis of connection" during the teenage years, as their attempts to square the (masculine) adolescent mandate for autonomy with the (feminine) requirements of relationship put them in danger of "drowning or disappearing." She and others propose to correct this problem by promising to ascertain the unique challenges adolescent girls face and to promote educational and cultural changes that would better address girls' particular developmental needs.[6]

Contemporary cultural theorist Barbara Hudson likewise deems the categories of femininity and adolescence to be mutually exclusive. From her analysis of professional and popular variants of the two constructs, she concludes that they are entirely at odds with—even "subversive of"—one another. Adolescence, Hudson argues, is a masculine construct whose behavioral expectations for independence, rebellion, and sexual experimentation fundamentally conflict with the "master discourse" in the girl's life, that of femininity and its requirements for social compliance, enduring relationships, and sexual restraint. Along with Gilligan, Hudson determines that in matters of relationship, especially, "the discourse of adolescence is clearly at variance with the discourse of femininity." Hudson illustrates how teachers and social workers, for example, variably rely on these conflicting discourses to describe, explain, and regulate the adolescent girl's behavior. Such capriciousness, she maintains, works to engender a sense of incompetence, insecurity, and inadequacy in the lives of many girls, conveying to them the sense that "whatever they do, it is always wrong." For Hudson, it is this dilemma, being held accountable by two opposing sets of emotional and behavioral expectations, that constitutes the essence of the female adolescent experience.[7]

The pioneering historical studies of adolescence substantiate Gilligan's and Hudson's assertions that adolescence has long been conceptualized as a masculine construct. The first historians to examine the changing meanings and experiences of adolescence in the United States and western Europe contended that the "invention" of adolescence during the nineteenth and early twentieth centuries was firmly embedded in cultural meanings and social expectations for masculinity and functioned specifically to describe and prescribe the transition from childhood to adulthood for male youth. These historians deliberately positioned boys as the central character in their historical narratives, arguing that a predominating concern with changes in the social experiences of boyhood thoroughly preoccupied the architects of modern adolescence. Class is the primary analytic lens through which these historians view their subject. They maintain that the inventors of adolescence initially described and normalized the experience of white middle-class boyhood. Then they extended those norms across lines of class and ethnicity through the languages of biology and psychology, with the goal of facilitating the social control of minorities and the working class.[8]

According to these historians, girls were at best ignored in and at worst deliberately excluded from the early formulations of adolescence. One scholar does suggestively acknowledge that longstanding cultural anxieties about female sexuality and associations of female puberty with physical, psychological, and social danger in some ways rendered the white middle-class girls of the mid-nineteenth century the "first adolescents." Moreover, qualities that came to characterize the modern adolescent—vulnerability, passivity, and awkwardness—"previously had been associated only with girls." In this analysis, however, female adolescence, as either a locus of cultural meanings or as lived experience, remains largely unexplored.[9] Likewise, another historian who focuses on scholars' neglect of the changing experiences of adolescent girls in the past nonetheless maintains that such changes in girls' lives were not matched by a commensurate "intellectual understanding" of adolescent girlhood and that an "adequate concept" of female adolescence had yet to emerge by the end of the twentieth century.[10]

Even as these historians were writing, others were looking more closely at the ways girls' transition from childhood to adulthood was lived and represented at various moments in the past. Describing the middle-class, working-class, and "delinquent" variants of the female adolescent experience during the nineteenth and early twentieth centuries, this group of scholars offers compelling analysis of the ways in which the categories of class, gender, and age intersected in the lives of particular groups of girls. These historians also examine familial, communal, institutional, political, and cultural responses to the highly visible and oft-rendered problematic presence of the adolescent girl in American society over this one-hundred-year period. Offering up female adolescents as historical subjects in their own right, this body of work endeavors to discern the ways groups of young, single females shaped, and were shaped by, their changing roles in the family, the economy, the framework of social welfare and educational institutions, and the worlds of popular culture in these years.[11]

Efforts to render adolescent girls historically visible have been accompanied by work in cultural studies and contemporary critical developmental psychology that posits that the child and the adolescent are "cultural inventions." The work of Michel Foucault and his method of discourse analysis influenced many of these scholars. This approach sees discourses as composed of ideas and practices that work together to produce knowledge, to organize social relations, to constitute individual subjectivities, and to deploy social power. In examining discourses that have produced the modern child, such scholars subject to critical analysis certain key premises of western science, most notably the presumptions that the child "develops" at all and that developmental norms are universally applicable across groups of social subjects.[12] As psychologist Valerie Walkerdine contends, the "grand metanarrative" of developmental psychology "is premised upon the construction of an object of study, 'the developing child' . . . that . . . is not real, not timeless but produced for particular purposes within very specific historical, social and political conditions." Going further than Gilligan, who seeks to rehabilitate the developmental paradigm for those who have been marginalized by it, Walkerdine insists that psychological studies of the child must now move "beyond developmentalism." "The big story [of developmental psychology]," she claims, "is a European patriarchal story, a story from the centre which describes the periphery in terms of the abnormal, difference as deficiency. I want to explore how this is accomplished and examine how it might be challenged." Despite their differences, both psychologists share a common purpose in seeking to expose the ideologies and power relations of race, class, and gender that have formed the foundation and scaffolding of the scientific study of the child.[13]

In taking another look at the intellectual and cultural formation of "adolescence" during the period of its invention, this study both makes use of recent cross-disciplinary approaches to the study of the child and provides historical perspective on them. How did earlier generations of scientists bring together conceptions of femininity and adolescence to describe and explain the adolescent girl and her development? What range of factors—intellectual, cultural, social, and professional—shaped their conclusions at particular historical moments? What expectations for the adolescent stage of life and female development did they bequeath to the child studies experts in our own time? The intent here is to explore the dynamic interrelationships between ideas about adolescence and femininity and the ways in which these two discourses not only excluded but also intersected with, mutually constituted, and undermined one another at various historical moments. The goal is to trace the multiple and contested articulations that constituted the discourses as they changed over time. In doing so, this analysis follows recent trends in the history of science by resisting dichotomies that associate men with conservative scientific views and women with progressive ones or biological paradigms with the oppression of marginalized groups and cultural paradigms with social equality.[14] Rather, the production of scientific knowledge about adolescence and female development has been more multifaceted than either historians or contemporary social scientists have recognized. Such knowledge admittedly has played a role in perpetuating social injustice. Exploring the nuanced thinking entailed in the transformation of "old" girls into "new" during the nineteenth and early twentieth centuries is essential to understanding that role and the ongoing struggle against its enactments and effects.

In the nineteenth and early twentieth centuries, notions of gender, race, and class figured into the scientific production of adolescence as a "universal," "developmental" category that privileged maleness, whiteness, and middle-class status as its normative characteristics. In this body of scientific thought, often girls were ignored, excluded, or deemed deficient because of their sexual difference. However, the white middle-class girl in particular was not easily dismissed by experts devoted to describing and prescribing the development of the child. Through an exploration of the attention girls sometimes (quite prominently if not necessarily adequately) received, we come to see the ways that ideas about their development helped influence and even challenge the modern concept of adolescence. Indeed, from an examination of ideas about the adolescent girl in the past, we discover that the recent criticisms of Gilligan, Hudson, Walkerdine, and others have a history.

This study offers close readings of works on the topics of adolescence and female development by a range of thinkers, situating them in wider intellectual, cultural, and social contexts. The focus is on what scientists and intellectuals in the United States contributed to initial meanings of the concepts of adolescence in general and female adolescence in particular. Such intellectuals were, however, part of broader transatlantic conversations about development, childhood, and gender. Thus, their engagement with their British and other European counterparts is also an important part of the story.[15] Some of the figures focused on, most notably psychologist G. Stanley Hall and anthropologist Margaret Mead, made major (although certainly not wholly original or undisputed) contributions to their fields of scientific endeavor. Many others were interpreters of other experts' findings and ideas who claimed the mantle of scientific authority in shaping cultural understandings of the adolescent girl and her development. All were popularizers or, more high-mindedly, public intellectuals, who were eager to disseminate and to apply their scientific knowledge to the making of their vision of more evolved human beings and a better society. Such experts garnered so much attention (if not unmediated influence) at certain moments because their scientific knowledge reflected and legitimized broader cultural common sense about the female child, even as they helped to both produce and challenge aspects of that common sense.[16]

Some historians debate the location of the experience and concept of adolescence in history. The term "adolescence" has Latin roots and was used during the Middle Ages and the Renaissance. Young people have always experienced the biological changes of puberty, and descriptions of these changes can be found in the classical texts of medical literature. Several historians have found evidence of rebellious youth cultures and accompanying concern over youth as a dangerous period of life in early modern Europe and Colonial America.[17] This analysis does not discount the possibility that adolescence as a concept or experience existed before the nineteenth century in the United States. Nonetheless, it recognizes that the concept took on its modern connotations during the nineteenth and early twentieth centuries and argues for the important role scientists across several disciplines played in formulating and popularizing the category of adolescence in the wider culture.[18]

In the nineteenth century, a group of unorthodox physicians involved in the antebellum health reform movement initially undertook such a task. They drew from the broader focus on development in British, European, and American thought to begin to delineate a period of life between childhood and adulthood that they referred to as the "age of puberty." Health reformers helped to pave the way for the interpretation of Charles Darwin's work, which predominated later in the century, that conceived all forms of development—whether of individuals, species, societies, and culture—as organic, linear, hierarchical, and purposeful. Such expectations shaped their conception of an age of puberty as both a problematic and an auspicious period of life, to be both managed and enabled by adults enlightened by scientific knowledge about human nature. During the last quarter of the nineteenth century, health reformers were followed by medical doctors, educators, and other scientifically minded intellectuals who sought to define an "epoch of development" between childhood and adulthood within the debate over the merits and detriments of coeducation. As the nineteenth century gave way to the twentieth, the pioneering developmental psychologist G. Stanley Hall emerged as the foremost inventor of the modern concept of adolescence, which he fully explicated in his influential two-volume work *Adolescence*, published in 1904. During the Progressive era, psychologists, sociologists, and reformers attempting to solve the problem of juvenile delinquency reiterated and redefined conceptions of adolescence. By the 1920s, psychologists associated with the flourishing mental hygiene and child guidance movements spearheaded the growth of a scientific study of "normal" child development, which included demarking the characteristics and mandates of "normal" adolescence. Also during the early decades of the twentieth century, anthropologists brought the newly articulated culture concept to bear on both nineteenth-century ethnographic accounts of puberty rites and using evolutionary theory to explain child development among "primitive" and "civilized" peoples.

At each of these particular historical junctures, some scientists seriously considered how the girl (most often white and middle class) would figure into "universal" developmental expectations in general and the adolescent stage of life in particular. They pursued and debated a range of responses to the girl's development: marking, accounting for, and assessing her developmental differences, holding her up as an exemplar of certain developmental norms, and extending to her the same prerogatives conferred by the process of development claimed by her brother. Their conceptualizations of the adolescent girl were shaped by a mix of influences, including ideas about human nature, especially theories about the relative role of and interaction between nature (biology) and nurture (environment or culture) in propelling the child's development and producing sexual and racial difference; their individual and collective biographies, including their gender, race, and class identities and positions; their disciplinary orientations and professional status and aspirations; the activities and experiences of girls and boys; and the first wave of women's rights activism and the emergence of modern feminism. Together, these scientific experts of the nineteenth and early twentieth centuries contributed to a developmental paradigm whose primary subject was the white middle-class male, while also probing some of that paradigm's limits and questioning some of its tenets.

It is, then, a constellation of ideas about adolescence and female development that contemporary child development experts have inherited and with which they continue to wrestle. Significantly, Carol Gilligan alludes to the history of ideas about female adolescence in justifying her attention to the topic in the 1990s. "For over a century," she and her colleague Lyn Mikel Brown write, "the edge of adolescence has been identified as a time of heightened psychological risk for girls . . . This crisis in women's development has been variously attributed to biology or to culture, but its psychological dimensions and its link to trauma have been only recently explored."[19] Yet, positioning herself as newly discovering what was by the late twentieth century an old problem, Gilligan also borrows from the *Handbook of Adolescent Psychology* to note that "[a]dolescent girls have simply not been much studied."[20] In part, this project began as an attempt to provide some historical insight into the current preoccupation with female adolescence. How *was* adolescence formulated as a "crisis" in female development by scientific experts in the past? How did such experts demark the possibilities and limits for the girl's development within the reigning paradigms of biology and culture in scientific thought? How might knowledge of this history inform a critical appraisal and appreciation of approaches to the "crisis" of female adolescence in the present? By focusing on earlier moments in which adolescent girls were the object of scientific and cultural concern, this book illuminates the intellectual origins of the relationship between adolescence and femininity— "subversive" and otherwise—to contribute to a better understanding of efforts to reconcile them in our own time.

CHAPTER ONE

"Laws of Life"

Developing Youth in Antebellum America

In the entry for "nubile" in the 1854 edition of his *Medical Lexicon*, Robley Dunglison cautioned fellow physicians against viewing puberty as a sudden transition to maturity for girls or boys. "Generally, the period of puberty is considered to be the age at which both sexes are *nubile*," he explained. "They are truly nubile, however, only when they are fitted to procreate healthy and vigorous children, and are competent to discharge their duties as parents."[1] Dunglison's definition heightened awareness about what he and other doctors were beginning to notice as a decline in the age of puberty. Although lacking a body of empirical data to support their claims, the observations of these physicians have proved to be correct. In 1780, white middle-class girls reached puberty at about age 17, with boys arriving one year later. Historians estimate that because of rising prosperity, improved nutrition, and lower threats of infectious disease, menarche declined to about age 14 by 1900; boys continued to lag behind girls.[2] For Dunglison and his medical colleagues, this decline threatened social and economic imperatives for later marriages among the middle class, as well as middle-class expectations for male sexual control and female sexual purity. To allay such threats, doctors conceptualized an "age of puberty" (extending up to age 25), along with organic developmental requirements to govern it, that deemed a prolonged period of innocent youth to be mandated by the laws of nature.

Contrary to Dunglison's assumption about the invariable correlation between puberty and nubility, social and economic factors, not changes in the physical body, had long indicated readiness for marriage and the concomitant achievement of maturity.[3] Moreover, in the hierarchical social order of the Colonial period, law and custom clearly denoted youth as a dependent and inferior stage.[4] As with other social groups during the early nineteenth century, the formation of a democratic society and the spread of a market economy threw the subordinate status of youth into question. Social, economic, and cultural changes provided unprecedented opportunities for the young to assert their independence from old forms of adult control. At the same time, these forces gave rise to a constellation of imperatives that were rendering the young immature for longer periods of time. During the first half of the nineteenth century, currents in medicine, education, and religion came together around the newly emerging idea of development to define "youth" as a stage in the life cycle commensurate with these changes. Dunglison's definitions emphasized what postpubescent young people were *not*, that is, physically, sexually, psychologically, or socially mature. Left unanswered were important questions about the quality of this "age," the process by which the immature child grew through it to become the mature adult, and whether—and in what ways—various social groups would be constrained by its limits or be able to lay full claim to its privileges.

Historians have used the terms "semidependence" and "semiautonomy" to characterize the often stressful vacillation of those in their teens and early twenties between attachment to family and community, on the one hand, and personal freedom, on the other.[5] Joseph F. Kett associates this struggle primarily with the white middle-class male and notes, with some regret, that the balance weighed more heavily toward the boy's and young man's increased dependence as the century wore on. Although he contends that the modern concept of adolescence was formulated at the turn of the twentieth century to describe and prescribe the experiences of white middle-class boys, he also suggests that white middle-class girls were the "first adolescents." Declining need for their household labor, new educational opportunities, and widespread cultural anxiety about precocious female sexuality, he argues, came together during the first half of the nineteenth century to render girls the first group of young people to be both free from adult responsibilities and subjected to adult protection and supervision, two qualities that characterize modern adolescence.[6] More recently, Jane H. Hunter's insightful study of Victorian teenage girls' lives has shown that girls, too, were pulled between duty to the family and a longing to cultivate the individual self. However, by century's end, girls achieved more *independence* from their "little women" status and its perpetually circumscribed domestic sphere. By comprehensively examining girls' experiences at school, as part of a peer culture, in the family, and in relation to mass culture, Hunter affirms and expands on Kett's claim that "[i]n defining this modern life-stage [of adolescence], girls led the way."[7]

An emerging discourse of development helped give meaning to and shape the experiences of male and female youths during the first half of the nineteenth century, and its interpretations can be located within the tradition of modern scientific thought on the concept of adolescence. In this period, ideas about child development can be found in novels, domestic advice manuals, pedagogical treatises, and religious tracts.[8] Texts by physicians, with recommendations on caring for infants and young children, a precursor to pediatrics that would emerge in the late nineteenth century, also had growing cultural influence.[9] This chapter focuses on physicians involved in the popular health movement of the 1830s through the 1850s who were particularly self-conscious about articulating developmental principles and norms and explaining "youth" as a stage in the life cycle. As they paved the way for new directions in professional medicine, these doctors departed from orthodox medicine's reliance on heroic therapies and, instead, attended to the role of hygiene and prevention and to the interrelationships among the body, mind, and spirit in shaping health and disease. Hygiene advocates, drawing from and contributing to the broader reform climate of the antebellum years, were optimistic about human nature's tendency toward perfection and the social progress that might follow. Their emphasis on self-improvement led many of them to consider the growth process, and their thinking about it dwelled on many themes that would remain important in the subsequent investigations of child development. Health reformers recognized childhood and youth as distinct periods of life; pondered the relationship between nature and nurture in development; described how physical, mental, and moral development interacted; and prescribed how these should proceed to maximize individual health, happiness, and potential, as well as to ensure the stability and progress of the newly ascendant democratic and capitalist social order.[10]

Health reformers' goals for development—their expectations for what the growing child was progressing toward—were rooted in Enlightenment notions of liberal individualism, particularly as espoused by philosophers John Locke and Jean-Jacques Rousseau. For Locke and Rousseau, reason, autonomy, self-control, and virtue, qualities they exclusively attributed to the elite adult white male, characterized the mature self. According to Locke, youth was the stage of life when the boy tried and tested these qualities and decisively asserted his independence over the parental controls of his childhood. In contrast, Rousseau conceived of growth as a more gradual process, prescribing a longer period of youthful dependence to ensure the proper incubation of the qualities of mature selfhood the boy would eventually assume. Despite these differences, together Locke and Rousseau positioned the boy as the "first adolescent" of developmental thinking, positing that he alone passed through a stage that moved him out of the dependency of childhood and prepared him to enjoy the prerogatives and to assume the responsibilities of the mature individual.

For their part, American health reformers were uneasy about Locke's recommendations for early youthful independence; they instead embraced Rousseau's model of protracted and protected growth throughout the teen years. As a result, the girl came to figure prominently in their thinking about the developmental process. She was foremost in the minds of health reformers who sought to explain the relationship between the development of the body and the mind, especially those who argued that during childhood cultivating physical strength took precedence over "forcing" mental prowess. Girls were also particularly useful for explaining and illustrating the precept of gradual growth that applied to all children because the dangers of their sexual precocity seemed self-evident. However, discerning and explicating the "universal" laws of development in no way precluded health reformers from furthering prevailing cultural expectations for dichotomous sexual difference. Indeed, most conceptualized the "age of puberty" in such a way as to affirm that youth was a stage of life whose greatest privileges exclusively devolved to the boy. Nonetheless, some health reformers challenged this view. Without fundamentally undermining the ideology of separate spheres, they held out expectations for female development that contended that girls were equally entitled to and capable of experiencing a prolonged period of growth that would not only preserve their purity but also allow for the full cultivation of their physical, mental, and moral powers, thereby enabling their active participation in both the private and public realms of democratic society. As health reform advocates outlined them, the natural laws of development governed all young people, boys and girls, working and middle class, in similar fashion. At the same time, their conception of the stage of life that would come to be known as "adolescence" played an important role in the construction and enforcement of categories of social, and especially sexual, difference.

GIRLS AND BOYS GROWING UP, TO 1860

Antebellum conceptions of development and youth emerged out of a web of economic, social, and cultural changes that had been reshaping the experiences and meanings of growing up for girls and boys since the end of the seventeenth century. Institutional age grading was not a prominent feature of Colonial society. Indeed, the structures and rhythms of preindustrial work and play kept the young of all ages in proximity to one another and to adults. Nonetheless, Colonial Americans recognized broad distinctions among the stages of life, and, perhaps most significantly, the superiority of adults over minors was a primary axis organizing relations of power in their hierarchical social order.[11] In Colonial New England,[12] the first stage of life, infancy, was comparatively brief, marked by complete subordination and dependency for girls and boys, although not necessarily lacking in nurturing and love. By age 7, boys donned adult clothing for the first time, and by about age 10, boys transitioned into extended stages of childhood and youth wherein they engaged in productive labor on the farms or in the shops of their fathers or masters, to whom they were bound out as apprentices or indentured servants. At the same age, girls began receiving training from their mothers or other mistresses to prepare for their future domestic roles and responsibilities. During this preparation, young people were expected to acquire greater proficiency with adult tasks and responsibilities, while deferring to the external authority of their parents, masters, and mistresses. In return, adults were obligated to protect and to support the young in their charge, as well as offer clear role models for adult identity, behavior, and occupation.[13]

In this Puritan social order governed by the principles of hierarchy, patriarchy, and mutuality, maturity entailed an assumption of authority, obligation, and deference in relation to others, the particular mix of which was determined by one's rank and gender. Boys' arrival at manhood was signified by some combination of land ownership, proficiency in a trade or profession, marriage (on average at age 25), and the establishment of an independent household. For them, maturity conferred greater freedom, control, and responsibility in relation to minors and women, although even men of the highest positions of social rank continued to owe obedience to God. Girls also anticipated that maturity would bring marriage (usually not before age 20), the creation of a new household, and the assumption of greater domestic responsibilities, including the command of children and servants. Maturity for girls also meant continuing their dependence on adult men, however, as they extended or transferred deference from father to husband.[14]

In part because of the establishment of stable social institutions, particularly the strength of the patriarchal family, youth was experienced and perceived as a relatively smooth period for much of the seventeenth century in the New England colonies, marked by the assurance of one's current status and role and the clarity of the social position and identity one would assume as an adult.[15] Such a condition contrasted with the youth in early modern Europe, where social instability led to the rise in both rebellious youth cultures and widespread anxiety about youth as a perilous period in the life cycle. Puritan theologians most prominently exhorted against the dangers of youth. Drawing on the humoral model of the body from classical medicine, in which puberty in both sexes was perceived as a period of physical disruption, marked by an excess of heat and lust, they characterized boys and girls in their teens as sensual, heedless, willful, and prideful; argued that youth needed discipline and monitoring; and urgently called on the family, the church, and the university (for boys) to fulfill the social control function.[16] At the end of the seventeenth century in New England, rapid population growth, increasing geographical mobility, and commercial development and expansion weakened patriarchal authority and undermined the exceptional harmony in age relations that had characterized the initial Puritan experiment. As court records reveal, some young people chafed against their subordinate status and resisted adult rule. As a result, they were now volubly condemned by religious authorities in the New World, who joined their English counterparts to pronounce against the wickedness of youthful sensuality and to warn of ensuing social chaos if their wickedness were not restrained, primarily through the mechanisms of conversion and church membership.[17]

As historian John Demos explains, beginning in the middle of the eighteenth century and escalating during the decades following the American Revolution, the experiences of youth became even more "disjunctive and problematic," marked by "new elements of [social and psychological] stress."[18] The transformation to a democratic polity and a market economy upset the vertical social order that had enforced the principles of hierarchy, patriarchy, and mutuality in age (as well as gender, class, and race) relations. The adult roles and identities to which youth might aspire, along with the routes by which they might be acquired, proliferated. The changing political and economic order set new standards for the mature self based on the capacities for rationality, autonomy, and self-control, qualities embodied by the adult white male of the emerging middle class.[19] Left unresolved, however, were the limits of individualism and to what extent various social groups would be able to assume its rights, prerogatives, and responsibilities. In this context, young people in their teens and early twenties faced both novel opportunities to exercise independence from adult control and renewed requirements that they depend on adult support and defer to adult authority. While some of these changes united the young around common experiences, how the balance and tension among autonomy, (inter)dependence, and subordination manifested in the lives of youth also diverged along lines of gender, class, race, ethnicity, and geographic location.

The changing social order of the new republic wrought transformations in work; educational, reform, and religious institutions; family life; and peer relations that affected different groups of young people in varying ways. From 1780 to 1840, as modes of capitalist production and exchange penetrated the rural and urban economies of the Northeast, white girls of that region saw their traditional work roles in the home undercut by new manufacturing technologies and the shift from family-oriented production to a market economy. Many joined their mothers and young male siblings in producing goods under the outwork system, weaving cloth or braiding palm leaf for hats at home under parental supervision. Many were also likely to spend some portion of their teenage years working for wages outside the home, in teaching, or, as in the case of the Lowell mill hands, in factory work.[20] Historians disagree about the meaning of such vocational opportunities for girls' lives. At the same time that girls from middling families were leaving home to pursue wage work, their mothers' lives were increasingly associated in ideology, if not always in practice, with the separate sphere of the private home, economic dependence, and the "feminine" values of piety, purity, self-sacrifice, and compassion.[21] The Lowell experience and teaching may have afforded greater variety, mobility, and personal freedom for some girls than their female counterparts in the past or adult women in the present; however, girls in the workplace were expected to follow the same proper feminine behavior required of their mothers. In addition, girls' wages most often remained embedded within a family economy, and they anticipated a return to domestic responsibilities when they married.[22] Moreover, girls' own assessments of the relationship between wage work and domesticity were most likely mixed. Indeed, whether the girl looked forward to or dreaded the prospects of "woman's sphere" following the relative freedom of work depended on whether she saw marriage and motherhood as raising her status or restricting her opportunities.[23]

In a society increasingly stratified by class difference, working-class girls also took jobs that set them apart from their mothers, though with less sanguine consequences for their lives. Native-born white and then immigrant girls who met the growing demand for urban domestic servants performed long hours of menial labor under their employers' scrutinizing supervision. For many girls, almost any other form of wage work was preferable to domestic service and the social subordination, economic insecurity, and risk of sexual abuse that went along with it.[24] Urban outwork and manufacturing jobs, such as those in the New York garment trades, brought risk and hardship to girls trying to make their way alone in expanding urban economies, where freedom from familial over-sight meant the loss of an important source of protection and support.[25] For enslaved girls, the teenage years were marked by increased vulnerability to masters' sexual exploitation and forced labor that remained largely undifferenti-ated from their mothers' work and bore no connection to the capacity for personal independence.[26]

In addition to changes in their work lives, white middle-class girls, in northern states, especially, took advantage of expanding educational opportunities during the early nineteenth century. The free public school systems established in the north in the 1830s attracted such large numbers of girls ages 5 to 12 that, by midcentury, the United States claimed one of the highest female literacy rates in the world. Managers of the Lowell mills encouraged the trend by hiring workers with school certificates.[27] With the influx of immigrant labor after 1840 and the accompanying rationalization of the factory system, white middle-class teenage girls' participation in the manufacturing workforce was undermined, both by labor competition and the attendant decline in respectability associated with this kind of work. With such forms of wage earning denied them and their tradi-tional domestic functions rendered obsolete because of declining birth rates, the availability of immigrant domestic servants, and changes in the technology of housework, these girls increasingly made their way into the new urban public high schools, private boarding schools, and teacher-training regimens at normal schools and female seminaries. For the rest of the nineteenth century, these girls attended school in greater numbers than their middle-class brothers or working-class girls and boys.[28]

Urban middle-class parents sent their daughters to school for many reasons: to occupy them during the years between puberty and marriage; to prepare them to assume the elevated domestic and maternal roles of Victorian woman-hood; to facilitate and to formalize the project of self-improvement and refine-ment at the heart of middle-class self-definition; and to display the family's class status and claim the respectability associated with it. Historians have found that privileged parents were the least likely to acknowledge an economic rationale in supporting a daughter's education. Given the economic volatility of the times, however, unspoken motivations for educating middle-class girls were that they would be better equipped to make a financial contribution to their family of origin and, if needed, to their family when they married (primarily through teaching); in a worst-case scenario, middle-class girls would even be able to support themselves. Whatever parental or institutional intent, though, schooling frequently created inner conflict and familial tensions, as both school structure and curricular content promoted female deference to class-specific gender norms and enabled girls to imagine and experience the self as a rational, controlled, and assertive being.[29]

Lower-class girls sometimes found themselves affiliated with institutions that in no way sanctioned youthful female autonomy. The New York House of Refuge, the first institution for juvenile delinquents in the United States, was founded in 1825, with separate departments for girls and African Americans. Three years later, the House of Reformation in Boston and the Philadelphia House of Refuge were established. In 1856, Massachusetts founded the Lancaster Industrial School, the first reform school for girls. Along with the public schools, these institutions helped to formalize the growing importance of age grading in American society. They also furthered the expanding cultural recognition of childhood and youth as life stages with distinct characteristics and discrete needs, deserving of adult protection but also subject to adult control. Most girls were brought to these institutions, either by reformers or by their parents, because they had committed, or were thought to be likely to commit, a sexual offense. Inclined to see all girls, even misbehaving ones, as essentially innocent, antebellum reformers geared their efforts toward the moral uplift and practical training, rather than the punishment, of female delinquents. Nonetheless, in removing girls from their families and carefully circumscribing their behavior in institutional settings, they also helped to set the limit for claims on the privileges of individualism, a line that poor, neglected, and "deviant" girls were clearly not to transgress.[30]

Religious institutions and practices also occasioned new opportunities for both the erasure and assertion of the self for teenage girls. Girls across the social spectrum embraced the revival spirit of the Second Great Awakening and, along with their mothers, helped to account for the majority of converts in the years from 1790 to 1840, as well as for the majority of church membership after that. Evangelical Christianity promulgated a religion of the heart, equally accessible to all who would surrender themselves to God's love and mercy, which was to be received initially in the form of a spontaneous and emotional conversion. Both proponents and critics of the revivals emphasized the gender and immaturity of converts, thereby fostering an association, whether positive or negative, between this type of volatile, expressive religiosity, femininity, and youthfulness. For some girls and women, conversion was surely an intense and anxiety-provoking experience that also heightened their sense of passivity and dependency. However, many also deliberately chose the evangelical message because they found in it a respect for femininity and recognition of the moral importance of the roles of wives and mothers in family and society. Moreover, many took the message further, out from the home and into the world, and used it to justify pursuing an active life of missionary work or to participate in social and moral reform efforts.[31]

Accompanying and enabling these economic and institutional developments during the early to mid-nineteenth century were changes in the structure and function of family life, which had important implications for the social and psychological lives of teenage girls. With the birth of the democratic republic and the expansion of the market economy, the urban white middle-class family was increasingly defined and experienced as an isolated conjugal unit, set apart from both larger kin and community relations and the workworld.[32] The rise of economic specialization, the spread of the wage system, and the improvement in standards of living rendered the economic labor of middle-class wives both invisible and obsolete. The father was now the family's designated sole economic provider, and the mother was the primary nurturer of children in the private home. As a strategy to protect and promote class status and as a reflection of women's growing domestic influence, middle-class couples consciously limited their fertility, and family size within this group declined over the course of the century. Smaller families made intensive mothering possible, which for girls and boys of all ages meant that their physical health and emotional well being received greater attention. Moreover, such close family relationships were extended over the course of the life cycle. Because girls delayed marriage to work or to attend school, and because they exerted greater control over choosing marriage partners, they remained primarily connected to their parents' households into their early twenties.[33] For girls, the psychological, emotional, and social ramifications of these changes were mixed. The private, affectionate family encouraged female selflessness by fostering intimacy among its members, especially with mothers, and valorizing the feminine qualities of empathy, compassion, and service. It also, however, promoted awareness and sometimes an assertion of the self as a separate, even special, being. In addition, the domestic family mandated prolonged, more careful, and sometimes invasive adult supervision and monitoring of girls' lives, which raised new possibilities for emotional coercion and conflict among family members.[34]

Close connections with mothers and other adult female kin shaped many girls' lives during this period, reflecting and reinforcing the continuities and similarities in the female experience across the life span.[35] Girls' friendships with peers and their participation in a distinct girl culture, however, indicated that female identity was becoming a product of age as well as gender. In female seminaries and the Lowell mills, for example, girls cultivated relationships and sought experiences that not only preserved qualities and values appropriate to adult femininity but also allowed them to fashion the meaning of such values and to experiment with their limits in their own lives. Through their romantic friendships, for example, girls considered how much space intense feelings might consume in their daily lives as girls and, later, as women. In making decisions about spending factory earnings and leisure time, they renegotiated the boundaries between self-sacrifice and self-assertion during this period of their lives.[36] Through their immersion in popular literary culture, as well, girls identified with one another and wrestled with their culture's multiple interpretations of the unique problems and possibilities of becoming a woman. Far from defining girlhood with a monolithic voice, periodical literature and novels of midcentury let girls know what was expected of them at this and subsequent stages of their lives and also opened up possibilities for them to respond to those expectations in voices of their own.[37]

Boys in their teens and early twenties also experienced competing pulls between freedom and subordination as a result of the socioeconomic changes of the late eighteenth to the mid-nineteenth century.[38] Boys' experiences with work, schooling, family life, and peer culture were also differentiated by race, class, and gender. The changes wrought by the market revolution were fundamental in transforming the meaning of youth for boys. By the 1790s, in the urban Northeast, capitalism was beginning to destroy traditional craft production and the apprentice system along with it. With the spread of the outwork system and the rise of industrial technologies, craft labor was being divided into discrete tasks. At the same time, to ensure profitability in the face of market competition and unpredictability, masters replaced long-term, personal, contractual labor relationships with more flexible ones. As free wage labor became more widespread, employers also replaced experienced journeymen with untrained, cheaper juvenile "helpers." In trades from printing to shoemaking, apprentices no longer lived with masters' families or received education or clothing as part of their indenture. Instead youths, or their parents, received cash payments in return for their labor.[39] During the first decades of the dynamic post-Revolutionary economy, some male youths learned to use new technologies more quickly and proficiently than their masters, and in capitalizing on their entrepreneurial spirit, they assumed an early independence as they made their way into the ranks of the new middle class.[40] In the meantime, bound labor was increasingly debased and its subordinate status reinforced, associated as it was with black men in or moving out of slavery and with orphaned and destitute children and youths, who continued to be "placed out" as they had in the Colonial period by those private charities and public agencies assuming responsibility for their welfare.[41]

The continued, if uneven, deterioration of apprenticeship and the mounting prominence of the wage system in the 1820s and 1830s marked a growing class divergence among male youth.[42] Although parents across the social spectrum had long relied on the work of children and youth to maintain or enhance the household economy, some were now so threatened by the vagaries of the new economic relations that their children's wage labor became essential to family survival.[43] This was especially true of Irish and German immigrant parents, arriving in the 1840s and 1850s, who sent their children to work in factories or "sweated" them in cellar and garret shops of their own. While boys working for wages were less constrained by the family claim and were permitted a greater measure of social and sexual freedom than their working-class sisters, the wage system gave rise to new insecurities and exacted new forms of deference and restraint for those on the economic margins. Moreover, enthusiasts of houses of refuge and reform schools made little distinction between poor, neglected, and "troublesome" boys. Boys who came under their care were subjected to stringent modes of adult supervision for their protection and the preservation of social order.[44]

White middle-class boys, whose families could afford to forgo their wages, pursued many paths en route to autonomous manhood, a process that entailed, in the words of Joseph F. Kett, "a jarring mixture of complete freedom and total subordination."[45] The tension had its roots in middle-class male childhood, when boys moved between a domestic culture governed by the values of affection, mutuality, duty, and restraint and a boy culture characterized by aggression, competitiveness, irresponsibility, spontaneity, and independence.[46] When the middle-class boy reached his teenage years, he faced the daunting yet exciting prospect of choosing an occupation. Boys continued to depend on their families for economic and emotional support. However, they also were compelled to rely on themselves, exercising initiative, making decisions, and cultivating abilities beyond the immediate direction and control of others. Thus, during this stage, which extended well into his late twenties, the youth might reside for a time with his family of origin; move to a city to live alone or with kin; continue his studies at a high school, academy, or college; or receive some sort of vocational training in his anticipated career. Whatever route, or combination of routes, he took, the possibilities for both means and ends and the vacillation between dependence and independence such possibilities occasioned frequently generated uncertainty, restlessness, and ambivalence in the growing boy.[47]

The unsettled status of male youth set the stage for familial and social conflict. Parents' financial sacrifices to secure the boy's education could cause tension among family members. Parents were also susceptible to bewilderment as they tried to determine the best way to influence their sons' career and character development.[48] In the broader society, tensions over the relative freedom of middle-class male youth peaked at such colleges and universities as Yale, Brown, and the University of Virginia. School officials tried to rein in youthful independence by publishing detailed conduct books carefully cataloguing rules boys were expected to follow regarding study, dress, recreation, and, especially, respect for authority. Boys caught transgressing these strictures faced warnings or expulsion. Careful regulation of schoolboys dates back to the Colonial era, when the subordination of youth and widespread suspicion about the heedless and irresponsible behavior of boys in groups were assumed. For their part, students earned their reputation as troublesome by carrying out pranks, rioting, fighting, harassing and abusing school officials, and, even in one case, committing murder. Although they were certainly a reaction to the repressive collegiate environment, student rebellions of the early nineteenth century also reflected the new, if partial, experiences of and expectations for freedom that had taken hold in the lives of middle-class boys.[49]

Whether they were at school, in rural communities, or in more anonymous urban settings, teenage boys and young men relied on one another for help in coping with the insecurities of coming of age. Male youth culture incorporated attributes of the boy the youth once was with the man he would become, channeling the impulses of the former toward goals appropriate to the latter. This set boys apart from the adult world and helped to facilitate their movement into it. Voluntary associations, including religious groups, military companies, self-improvement societies, and civic organizations, captured the attention of young men during this period. These organizations, often created by young men, allowed youths to indulge in competitive play with friendly comrades on their own terms, while offering them a chance to obtain knowledge, hone skills, and exercise values central to adult male identity.[50] Intimate friendships with male peers were also were also important during this stage of life, though, unlike female friendships, most lasted only for the duration of youth. Boys were expected to outgrow the dependence and individualized compassion such relationships required. Emotional expressiveness and vulnerability were reserved for marriage, while self-containment, detachment, and restraint characterized relations among men.[51]

The decades following the birth of the American republic thus marked an important transition for youth. As historian C. Dallett Hemphill shows, the transformation to a democratic polity and a market economy led to a shift from a "hierarchical to a horizontal social order" that for young people of the white middle class, in particular, portended a rise in their status. While children were decidedly subordinate to adults in the new republic, even as they were treated with greater benevolence than before, white middle-class youths were integrated into the adult social world. They were presented with opportunities and challenges to acquire the capacity for self-government, rational choice, and individual initiative to ensure the success of the democratic order and the free-market economy, as well as to secure their eventual hegemonic position in both.[52] Even as the status of such youth began to rise, however, it was called into question. What were the limits of social equality in the relationship between adults and youth? What position and posture of youth would best serve the interests of a rising middle class? How important were gender differences in youth to establishing those limits and securing those interests? Young people learned for themselves there were no easy answers. New opportunities for girls to engage in factory work, for boys to select careers, and for youth of both sexes to choose marriage partners entailed a degree of autonomy unknown to youth in the hierarchical, patriarchal society of the Colonial period. At the same time, the emergence of age-graded institutions and the rise of the affectionate family reasserted the (inter)dependent status of youth, even as they both opened up possibilities for the young to craft autonomous cultures from which they might challenge new forms of adult control. While youth had long been recognized for its liminal nature, as a bridge linking two other stages of life, its in-between quality took on new meaning during the early nineteenth century. It referred to not only the transitional function of the life stage but also to the contradictory experiences and ambivalent expectations regarding the capacity for independence that such a transition now entailed.[53]

THE EMERGENCE OF THE DEVELOPMENTAL PARADIGM

The problem of youth's liminality was raised in the explanations of the growth of the child offered by participants in the early-nineteenth-century popular health movement. This group of unorthodox physicians and their lay supporters named and assuaged cultural anxieties about youth as a critical stage in the life cycle. They drew from and built on a broad Euro-Anglo American intellectual context marked by a focus on the meaning and significance of developmental processes. It was this context that prepared the way for Charles Darwin and, paradoxically, for the subsequent widespread acceptance of the "non-Darwinian" view of all forms of development, including child development, as orderly, linear, purposeful, progressive, and even divinely ordained.[54]

The division of human life into successive stages, as well as speculations about phylogenetic evolution, can be traced to ancient times. However, early notions of the occurrence of change in the individual, as well as in the natural and social worlds, were largely viewed as static, as the manifestation of that which had been present all along. Thus, Medieval Christians envisioned a depraved adult existing in full, although miniature, form in the mind of the child, marking differences between the child and the adult in degrees rather than in kind. In the eighteenth-century science of embryology, this view was expressed, albeit with greater nuance, as "preformationism," the idea that complicated organisms unfolded out of preexisting entities in the sperm or egg. In addition, beginning with Aristotle and persisting into the eighteenth century, the concept of the Great Chain of Being described a fixed, hierarchical order with all the beings of the universe occupying unchanging positions within it.[55]

From the eighteenth century and into the mid-nineteenth century, a confluence of influences from philosophy, theology, and science contributed to a new formulation of development whose broadest tenets were dynamic change, relational continuity, and faith in progress. Republican and free-market ideologies challenged the fixed, hierarchical social model, replacing it with a fluid, egalitarian model rooted in the prospects for individual freedom, responsibility, and potential. The teachings of liberal and evangelical Christianities, along with the reform movements they inspired, unified promises for individual and social salvation by proclaiming that perfectible human beings had the power to realize the perfection of God's kingdom on Earth. The dynamic, dialectical theory of history put forth in the ideal philosophy of Georg Hegel and the material philosophy of Karl Marx advanced the idea that any phenomenon must be explained in terms of its role in a continuous historical process. The group of German romantics known as the "nature-philosophers" posited the existence of a fundamental unity among all natural phenomena, deemed progressive, systematic change to be a general process in Nature, and speculated about a correspondence between individual development and the development of species. A dynamic concept of development also emerged as the centerpiece in geological research, most prominently in the work of British geologist Sir Charles Lyell, as well as in early theories of biological evolution, most notably those of French botanist and invertebrate zoologist Jean-Baptiste Lamarck. In embryology, the epigenetic view, wherein undifferentiated organic structures grew sequentially into more complex ones through a process of dynamic change, successfully challenged preformationism.[56] In physiology, the work of British, French, and German scientists on cell formation and activity provided what historian Carolyn Steedman termed "a series of imaginative and figurative paradigms for describing individuals, time and change" that also made important contributions to the organic, teleological model of development that would dominate for the rest of the century.[57]

Meanwhile, new possibilities for postnatal individual development were entertained under the continuing intellectual influences of the enlightened rationalism of John Locke, which emphasized the malleability of the child, and the romanticism of Jean-Jacques Rousseau, which recognized the uniqueness and promise of child nature.[58] Locke's ideas about human nature, the family, and child rearing, as explicated in his *Two Treatises of Government, Essay Concerning Human Understanding,* and, especially, *Some Thoughts Concerning Education,* all published during the 1690s, took hold in the American context during the mid-eighteenth century as Americans struggled to define a paradigm for the self appropriate to a changing political and economic order. As historian Jay Fliegelman explains, "In a new political world in which government was to exist for the governed, the educational paradigm would provide a new model for the exercising of political authority."[59] At the foundation of Locke's philosophy was the premise that human nature was both fundamentally reasonable and inherently self-interested. Countering Calvinist notions of the child as inherently depraved and monarchical assumptions about the fixity of the status of the individual at birth, Locke maintained that children's minds and characters were pliant and that right concepts and attitudes could be instilled in them through an education that fostered the development of the power of reason. His prescriptions emphasized the primacy of nurture and fostered recognition of the difference of child nature. "When I talk of reasoning," he clarified, "I do not intend any other but such as is suited to the child's capacity and apprehension. Nobody can think a boy of three or seven years old should be argued with as a grown man."[60] He also advised individual parents to discover the nature of their own particular child and adjust their educational regimens accordingly. All parents, however, shared the same goal in child rearing: the cultivation of a rational, self-governing, autonomous adult whose self-interest and concern for the common good were completely compatible. In Locke's scheme, that end was best achieved by parents requiring strict obedience from young children to curb their natural, but often excessive, longing for liberty. Ideally, this would be done by appealing to the child's reason rather than relying on corporal punishment. It was also to be accomplished by granting or withholding parental affection. In addition, parents were to exercise careful control over their children's physical habits, inculcating the capacity for self-denial by exposing them to fresh air, cold baths, and simple diets. Under the influence of these child-rearing methods, children would come to obey their parents and, in the process, internalize the normative social values parental strictures enforced, not out of fear or blind submission but because they would understand it was in their best interest.[61]

Once reasoning was established (at about age 12 or 13) and control over individual desires and appetites became habitual, parents would relinquish their external authority, and dependent childhood would give way to independent youth. Locke envisioned youth as that period when the individual experiments with newly acquired capacities for rational choice, self-control, and purposeful freedom.[62] Addressing fathers about how to treat their sons who had made it through childhood, he advised: "The sooner you treat him as a man, the sooner he will begin to be one: and if you admit him into serious discourses sometimes with you, you will insensibly raise his mind above the usual amusements of youth, and those trifling occupations which it is commonly wasted in. For it is easy to observe, that many young men continue longer in the thought and conversation of schoolboys, than otherwise they would, because their parents keep them at that distance, and in that low rank, by all their carriage to them." By treating their teenage sons as equals, fathers would be able to secure their sons' friendship. Paternal authority thus would be established more decisively because it would be founded on expressions of love and esteem, not coercion. Unlike children, whose reason was not fully formed, youth would learn that their happiness depended on wanting for themselves what their parents and society expected of them. Therefore, according to Locke, the competing values of individual freedom and social order would be balanced and secured.[63]

Rousseau's romantic concept of the developing child, as presented in *Emile, or On Education* (1762) influenced some American child-rearing advisors by the 1790s[64] and furthered some Lockean notions while challenging others. Rousseau emphatically asserted the existence of an essential child nature that adults were to discern and education was to follow. He rejected Locke's prescriptions for stern discipline and for thwarting desire in favor of a childhood of self-motivated activity, exploration, and enjoyment. Nonetheless, as the ubiquitous presence of Emile's tutor made clear, Rousseau also proposed that a natural development could only be facilitated by nurture, however carefully adult direction might be disguised. Herein lay the fundamental paradox of all nature-based theories of development. After all, the task of determining the qualities of the child's nature ultimately fell to the child's adult governors. Thus, while Emile feels free to follow his own inclinations, his tutor has already manipulated his environment so that Emile's interests unfold according to a larger, predetermined social purpose.[65]

For Rousseau, as well as for Locke, that ultimate purpose was to transform youth into an autonomous adult, whose concern for the common good would further the stability and progress of civil society. Whereas Locke believed this could be achieved by granting youth a measure of independence early on, Rousseau made the case for a prolonged, youthful dependency. Both Locke and Rousseau deemed the young to be essentially innocent and sought to protect them from the corrupting influences of society. For Locke, though, this meant exposing the youth to the dangers of the environment so that he might apply his powers of reason and self-control to resist and overcome them. Rousseau was more cautious about releasing the youth into the world, in part because he considered the ways that the physical changes of puberty potentially compromised youths' ability to exercise higher powers of the intellect, the will, and the heart. To Rousseau, youth was a "second birth," incited by the maturation of the reproductive organs and the accompanying onset of sexual feeling. He warned that the nascency of the sexual passion had the potential to ignite a "stormy revolution" in feeling, thought, and behavior that could overwhelm the youth, dangerously rendering him sensual, irrational, rebellious, enervated, and passive.[66] However, he was also careful to emphasize that such perils were less determined by nature than by a corrupt education and precocious experience. "Nature's instruction is late and slow," he contended, whereas "men's is almost always premature." Hence, "learned and civilized" people arrived at puberty earlier than "ignorant and barbarous" ones, whose moral "simplicity" enabled them to remain "peaceful and calm" well into the period of their youth.[67] His prescription: that every effort be made to slow the youth's development and follow the gradual course that nature intended. Hereby would Lockean ends still be secured, albeit by quite different means. That is, only by way of a protracted, steady development that did not test the youth's capacities too soon could the newly emerging sexual energy impart the "vigor and force" to the body, mind, will, and heart that would enable the complete development of the free individual who was ready and willing to serve the public interest.[68]

In seeking to reconcile the rights of the individual with the imperatives of the social bond, both Locke and Rousseau highlighted the significance of youth as a stage of life. While Locke regarded the child as morally neutral and Rousseau granted that the child possessed an inherent sense of justice, neither went as far as the eighteenth-century Scottish common-sense philosophers who posited the existence of a moral sense from birth and attributed to the child a natural disinterested benevolence.[69] To both Locke and Rousseau, youth was an important stage of life because the self-centeredness of childhood gave way to the self as a social being. For Locke, goodness throughout life was motivated by self-interest that was discernible by reason. As one's capacity for reason increased with the advancement of childhood, so too did one's capacity for virtue. Rousseau deemed that goodness originated in the instinct for self-love rather than in rationally calculated self-interest. Exceeding Locke in valorizing the youth's acquisition of conscience and capacity for selfless compassion, he described the boy at adolescence as "the most generous, the best, the most loving and lovable of men."[70] Primed by the communal imperatives of the sex instinct, the boy who followed nature's dictate for gradual growth would spend his youth acquiring the feelings of benevolence, empathy, and pity that were to form the foundation of the affectionate family and the democratic state: "Little by little the blood is inflamed, the spirits are produced, the temperament is formed. The wise worker who directs the manufacture takes care to perfect all his instruments before putting them to work. A long restlessness precedes the first desires; a long ignorance puts them off the track. One desires without knowing what. The blood ferments and is agitated; a superabundance of life seeks to extend itself outward . . . One begins to take an interest in those surrounding us; one begins to feel that one is not made to live alone. It is thus that the heart is opened to the human affections and becomes capable of attachment."[71]

With Locke's sons and Rousseau's Emile as the explicit subjects of their educational treatises, these philosophers laid a solid foundation for subsequent developmental thinkers' rendering of the boy as the normative adolescent. Both intended their prescriptions for the rearing of young children to apply to boys and girls alike, contending that there were no salient differences between the sexes during childhood. Youth, however, was a stage of life unique to male development. Whether by way of an early exposure to the world or of a more protected, protracted growth, only boys outgrew the dependence of childhood and achieved the ultimate aim of development: the formation of an autonomous, self-governing individual capable of making a contribution to the common good. Rousseau located the cause of sexual difference in nature, asserting that it was the changes that occurred in the body at puberty that set boys and girls apart and that propelled boys on their way to cultivating the highest powers of the intellect, the will, and the heart. "Up to the nubile age children of the two sexes have nothing apparent to distinguish them," he declared. "Everything is equal: girls are children, boys are children; the same name suffices for beings so much alike . . . And women, since they never lose this same similarity, seem in many respects never to be anything else. But man in general is not made to remain always in childhood. He leaves it at the time prescribed by nature; and this moment of crisis, although rather short, has far-reaching consequences."[72]

Whereas Locke barred girls from the privileges of youth (and by extension, mature adulthood) simply by ignoring them altogether, Rousseau took a different tack. In his account of the education of Emile's future wife, Sophie, he excluded girls by offering up an account of their developmental difference. For Emile, the onset of puberty, if properly managed through channeling the sex instinct toward higher education, self-control, and concern for the common good, inaugurated the capacity for self-determination and social creation—"far reaching consequences," indeed. For Sophie, the emergence of the reproductive powers channeled her into a predetermined, generic female destiny, as wife and mother in the domestic sphere. Rousseau granted Sophie some measure of rationality, autonomy (in the choice of marriage partner), conscience, and moral influence within the private family, which was rooted primarily in her instinctual sexual modesty. He emphasized, however, the ways that her distinct female nature and the education that followed from it rendered her Emile's complement but not his equal in the pursuit of the rights of the individual or the responsibilities of citizenship.[73] Although the approaches of both philosophers to the problem of the teenage girl differed, their expectations for development were much the same: Only boys experienced youth as a stage of life; only they emerged from it poised and permitted to claim the full entitlements of maturity in a democratic social order.

GROWTH AND DEVELOPMENT IN
THE POPULAR HEALTH MOVEMENT

Participants in the popular health movement of the 1830s through the 1850s played a leading role in conveying and construing these many currents of developmental thought for parents, educators, and young people.[74] While diseases plagued many Americans and theories and therapies of eighteenth-century heroic medicine were increasingly under public and professional attack, several disparate groups of unorthodox physicians and their lay supporters sought to restore Americans' health through programs for better hygiene. Within medicine, the turn to hygiene was propelled by research from the Paris, or Clinical, school, which stressed the empirical evaluation of disease and treatment. It was also furthered by advancements in the science of physiology. Both regular and unorthodox early-nineteenth-century physicians increasingly relied on a growing body of scientific knowledge about the function and organization of the body's tissues, organs, and systems, whose workings were perceived to be directed by predictable laws of nature. Of interest to physiologists in this period was the process of bodily growth and development.[75] Such physiological knowledge was not monolithic. Academic physiologists and medical practitioners disagreed about the functioning of respiration and digestion, the performance of the nervous system, and whether materialism or vitalism explained the origins of life forces. Health reformers occupied a range of positions in these debates, with some quite closely aligned with orthodox medicine and others more decidedly on the margin of medical theory and practice. For their part, most orthodox physicians dismissed all types of health reform as inferior "quackery," even as they also sought to enhance their prestige by attempting to appropriate popular physiology for their own domain. Health reform thus constituted a challenge, a complement, and a vanguard to the medical establishment of the early nineteenth century.[76]

Outside of medicine, what historian James Whorton calls "hygienic religion" thrived on the meld of optimistic, romantic, individualistic, and Christian perfectionist orientations of antebellum American culture. Health reformers were drawn from the ranks of the urban white middle class that led the way in creating the values of that culture, and they played an important role in fashioning and propagating its tenets. In particular, it was their use of scientific authority to confirm and extend the tenets of liberalized Christianity, including a code of Christian morality, that made health reform so appealing to middle-class audiences. Like other antebellum reform movements, health reform was firmly rooted in the emerging denominations of evangelical and liberal Protestantism. Despite important differences among them, all the variants of liberalized Christianity positioned themselves against orthodox Calvinist beliefs in original sin, predestination, and a judgmental God, embracing instead the inherent goodness of humankind, the role of free will in achieving salvation, and a vision of God as benevolent Creator. Espousing a "natural theology," health reformers believed that all of nature, including human nature, revealed God's will and his beneficent promises to humankind as corporeal, spiritual, and social beings. Health reformers, in concert with their fellow reformers who crusaded for temperance, moral reform, and abolitionism—and with whom they often collaborated—forcefully linked individual improvement and salvation with social regeneration and progress. Their particular contribution to the antebellum reform impulse lay in their insistence that the scientific study and understanding of natural laws were vital for accessing the religious truths and moral principles that were to guide human beings toward perfection, in both this life and the next.[77]

Whether promoting homeopathy, hydropathy, or vegetarianism, antebellum health reform advocates maintained that physical, spiritual, and social perfection could be achieved from a knowledge of and willingness to apply certain fundamental laws of physiology and health. As Sylvester Graham, the Presbyterian minister turned self-educated physiologist and icon of the health reform movement, explained, "The constitutional nature of man, is established upon principles, which, when strictly obeyed, will always secure his highest good and happiness:—and every disease, and every suffering which human nature bears, result from the violation of the constitutional laws of our nature."[78] In hundreds of instructional tracts and lectures, such as those given under the auspices of the American Physiological Society, founded in Boston in 1837, health reformers explicated what they called the "laws of life" for eager audiences. For all health reform advocates, the effective comprehension and application of the laws of human nature depended on the acceptance of three basic principles. First was the belief in the beneficence, and also the primacy, of nature. God had created nature—human and otherwise—for the purposes of pleasure and enjoyment. Health and vitality were every person's birthright. Disease resulted when human beings wittingly or unwittingly rejected their divine inheritance through their failure to understand and to obey nature's precepts for proper diet, dress, exercise, and sexual hygiene. Following from the belief in nature's absolute goodness was the correlation health reformers drew between physical and moral laws and among the physical, intellectual, and moral aspects of human nature. Reviving and recasting ancient beliefs about the interrelatedness of the body, mind, and soul, health reformers argued that sickness in the body both reflected and produced sickness of the mind and soul and vice versa. Likewise, physical vigor, mental well being, and moral virtue were related as causes and effects of one another. Finally, all versions of the health reform creed preached the importance of self-help and prevention. All individuals were responsible for learning the laws of nature and for adjusting their habits and environment accordingly. Adults were to pursue self-knowledge for themselves and cultivate it in their children. Doctors were to watch for indications of abusive habits in their patients and to educate them about the changes they needed to make before diseases struck.[79]

If the will of God intended for all human beings to be healthy and happy, his will also meant for them to grow and develop, particularly, though not necessarily exclusively, during the early stages of life. Indeed, health and happiness throughout life depended on following the laws of development during childhood and youth. Therefore, it was not enough for the would-be health conscious to grasp generic laws of human nature. How nature changed over time had to be noticed. "In judging . . . the propriety, advantages, or evils of exercise, food, and clothing," asserted Andrew Combe, "we must take into consideration not only the kind of exercise, the kind of food, and the kind of clothing, but also the age, health, and kind of constitution of the individual who uses them, and adapt each to the degree in which it is required."[80] Amariah Brigham, whose *Remarks on the Influences of Mental Cultivation and Mental Excitement Upon Health* had a major influence on school reform in the 1830s, agreed. All educational programs, he argued, "should be formed, not from a partial view of [the child's] nature, but from a knowledge of his moral, intellectual, and physical powers, and of their development."[81] According to William Andrus Alcott, the physician and educational reformer who published numerous popular guides on health and character development (and the relationship between them) for young men and women, such powers had merit on their own terms: "I wish to see [youth] so educated that they will not only be what they should be, when they come to adult age, but also what they should be now. They have or should have a character to acquire *now*; a reputation to secure and maintain *now*; and a sphere of personal usefulness and happiness to occupy *now*."[82] The first president of the American Physiological Society, Alcott had earlier served as an assistant to William Channing Woodbridge, the editor of the *American Annals of Education* and the most important popularizer of the ideas of Swiss pedagogue and Rousseau disciple, Johann Pestalozzi, in the United States.[83] In keeping with Rousseau and Pestalozzi, Combe, Brigham, and Alcott promoted the notion of childhood and youth as distinct periods of life with particular characteristics, discrete needs, and even unique contributions to make to society.

So, too, did Orson Fowler, hygiene enthusiast and one of the leading American popularizers of the widely embraced science of phrenology. Phrenology was a simpler and more empirical version of the faculty psychology espoused by the Scottish common-sense philosophers, particularly Thomas Reid and Dugald Stewart. Its practitioners enumerated some thirty-seven mental faculties and located each of these in a specific area of the brain. They claimed that they could read the "bumps" on a person's skull to determine the relative strength and weakness of these traits so as to offer a prescription for directing the individual's capabilities and character.[84] Fowler and others conceived of phrenology as a developmental science. That is, in mapping the site of the range of mental capacities and character traits from combativeness to cautiousness to agreeableness onto different parts of the cerebral cortex, phrenologists also associated their appearance and relative influence with the stages of the life cycle. "Man is not brought forth, like the fabled Minerva from the brain of Jupiter, in the full possession of every physical power and mental faculty," Fowler explained, "but a helpless infant, yet grows by slow but sure gradation in strength and stature to ultimate maturity." First to develop, according to phrenologists, were the lower "animal propensities," the physical drives for food, sex, and survival, whose "organs" were located at the center and back of the skull. Next to develop, and located in the front of the head, were the more advanced powers of reason and perception. Emerging last from the area at the top of the brain were the moral and religious sensibilities, the highest and most important capacities of human nature. The structure of the brain thus provided important clues into the nature of the child at each stage of development and offered an infallible guide for the deployment of children's education. Alcott's regard for the "now" notwithstanding, all health reformers joined with phrenologists in casting the process of development in such decidedly teleological and progressive terms. Nature's universal tendency was toward perfection, proclaimed the ever-optimistic Fowler, who declared improvement to be "the practical watch-word of the age." Childhood and youth were not, therefore, the most enjoyable stages of life, for happiness increased in direct proportion to the augmentation of intellectual and moral excellence.[85] Thus, while promoting the recognition and even the appreciation of the different qualities marking the stages of life, health reformers also nonetheless essentialized a clear hierarchy of value among them.

Health reformers such as Brigham, Alcott, and Fowler understood the capacities, traits, and requirements marking the stages of development to be determined not only by the imperatives of organic growth but also fundamentally assisted by the processes of "cultivation" and education, as well as by the exercise of the individual will. In this way, these developmental thinkers posited something of an interaction between the forces of nature and nurture. This enabled them to counter Locke's "materialism" by asserting that children possessed a God-given nature, their minds endowed with inherent capacities to know and to love eternal truths. Thus Brigham found it imperative to declare that he did not deny the existence of the "immortal and immaterial mind," while he also recognized that the brain must be cared for as the organ through which the mind operated.[86] At the same time, health reformers' appreciation of a relationship between nurture and nature also allowed them to check what they perceived to be the excesses of Rousseau's romantic naturalism with Locke's conception of the self as educable, autonomous, and self-governing. Indeed, their particular reading of Rousseau, which largely ignored the philosopher's own focus on education and self-determination in fashioning the individual into a social being, revealed a deep ambivalence toward the tenets of romanticism. On the one hand, they embraced romanticism's sanguine view of nature, its veneration of the innocent child, its valorization of the unique individual, and its optimistic allegiance to the process of becoming. On the other hand, they remained suspicious of its rejection of civilization, registered a profound distrust of its celebration of the passions, and were reluctant to acknowledge an embodied, emotional self beyond the reaches of rational self-control.[87] For Elizabeth Blackwell, as for other health reformers ascribing to an ethos of Christian perfectionism, the ultimate goal of development was divine, not primitively natural, perfection. Thus, Adam and Eve in the garden, not the noble savage in the wilderness, constituted her "ideal of the Human race," which amounted to a harmonious blend of "beauty and strength, lofty intelligence, powerful action, and purity of soul."[88] Those prescribing more secular developmental outcomes were nonetheless also careful to set limits on the romantic conception of the self. "Rousseau has advocated with much speciousness and sophistry this unthinking, savage, or what he calls the natural condition of our species," William Sweetser proclaimed. "[However,] [t]he tendency of man is obviously to civilization and mental progress, whence the highest moral and intellectual advancement of which he is capable, is the only natural state that can be predicated of him."[89]

Blackwell offered up a systematic explanation for realizing this progressive tendency. In her 1852 work *The Laws of Life*, she carefully distinguished between "organic life," in which nature determined both the means and the ends of growth, and "related life," in which education and individual will played crucial roles in establishing and achieving the higher aims of human existence. After all, she asserted, the aim of development was for the body and the mind not only to become functional but also beautiful. The object of each human being was not merely to live but to live well. "[N]ot our life, but the purpose of our life," she emphasized, "is under our own control." Moreover, the balance between organic life and related life shifted with age. In infancy and early childhood, when nature's intentions were entirely clear, the only responsibility of the child's caregiver was to furnish the optimum environment for nature to do her work. After age 7, when nature did not speak in such a clear voice, however, favorable conditions for organic growth still had to be provided, as did the noble aims toward which physical, mental, and spiritual development ought to be directed and the training to enable the individual to act independently to achieve those ends.[90]

In attempting to reconcile the forces of nature and nurture, Blackwell laid out four principles of development for children and their caregivers to follow. First, the "law of exercise" dictated that all the faculties had to be used for them to grow. Second, the "law of order in exercise" deemed the growth of the various faculties to be conditioned by "periodicity," with different aspects of human nature predominating during different stages of the life cycle. Most important, mothers and educators must understand that the development of the mind was to follow the development of the body. Childhood and youth were thus to be ruled by the "sovereignty of the body," with the maturity of the mind and soul to be achieved later after physical education was complete. Third, the "law of balance of exercise," or the "law of compound movement," qualified the law of periodicity by maintaining that at no stage of life could the body, mind, and soul be wholly separated, that indeed at every age the requirements and possibilities of the whole being had to be attended to. Fourth, the "law of use in exercise" proposed that every being possessed a *"special* purpose" and a *"universal use,"* which if conformed to, was guaranteed to lead to the individual's utmost health and happiness. "The perfection of our human nature, in its double capacity of body and soul, ready for strong and healthy action," Blackwell summarized, "can only be attained by the *gradual unfolding* of this nature, according to the Divine *order* of growth. This order requires that the material development shall precede the spiritual growth; that during youth the mind shall grow *through the physical organization*; that our education of the mind shall always be subordinate to our education of the body, until the body has completed its growth."[91]

Failure to abide by these laws of development resulted in two possible "evils" that threatened the child's movement toward divine perfection: the problem of imperfect or arrested development and the problem of precocious or early development. Both of these offenses against the developmental process violated what Fowler deemed to be nature's universal motto of "on time."[92] They also both posited the body as a closed system with a limited amount of energy. In this assumption, health reformers were influenced by the groundbreaking work in the physical sciences, most notably the formulation of the First Law of Thermodynamics, or the law of conservation of energy, which was most cogently articulated by German physicist and physiologist Hermann von Helmholtz in the late 1840s. When applied to organic matter, the law dictated that bodies possessed a finite quantity of "vital force," or nervous and muscular energy, that had to be carefully apportioned and balanced to achieve growth and maintain health. A body expending energy on one task, such as strenuous physical activity, would not have any energy to spare for another task, such as high-level thinking.[93] In keeping with this law, Blackwell explained that the first problem of imperfect development arose when adults neglected to provide children with the proper conditions that would aid appropriate and timely development. A missed opportunity to cultivate the body during childhood and youth inevitably led to perpetual weakness and permanent suffering because energies later in life were to be devoted to other capacities. "Precocity" was even more dangerous. Blackwell warned that the "reserve of vital force" possessed by children and youth was never to be misdirected toward objects of later existence. If mental cultivation were pursued at a moment when the body was nature's main concern, the child would grow disproportionately in one direction at the expense of the other. Ultimately, both mental and physical development would be arrested, with neither reaching its fullest potential.[94]

Of particular concern to antebellum health reformers were infant schools, those experimental educational institutions founded in the 1820s and 1830s that presumed to give children an advantage by cultivating their intellects at an early age. The fate of the child prodigies produced by such a regimen was perceived to be dire indeed. "This green-house method of forcing premature development," Fowler starkly cautioned, "weakens all their powers while alive and hastens death." Health reformers also decried the teaching methods prevalent in the common schools, which promoted the cultivation of reason too soon through early reading and rote memorization. Rather, children learned best by observing and focusing on areas in which they have a "natural aptitude." "That mind will likely to attain the greatest perfection," Sweetser counseled, "whose powers are disclosed gradually, and in due correspondence with the advancement of the other functions of the constitution."[95]

Even more detrimental than the menace of intellectual overpressure was young people's involvement with dangerous temptations that health reformers perceived to be rampant in an industrializing, urbanizing society, such as stimulating food and drink, wearing fashionable dress, reading sensational literature, and socializing with disreputable associates. Exposure to these perils was exacerbated by girls and boys working and attending school, prostitution in urban areas, proximity to working-class and foreign servants in the home, and commercial publishing. According to health reformers, these activities unduly stimulated the emotions, weakened the nervous system, and compromised healthy development of the body and mind. Particularly troublesome was the potential for licentious thoughts and behaviors. Although health reformers held diverse views about sex, most advocated a single standard of morality—self-control in men, purity in women, and innocence in children.[96] For Christian physiologists such as Blackwell, Alcott, and Graham, who celebrated the beneficence of nature while linking physical and moral cause and effect, immoral habits did not originate with the body but would surely produce damning somatic results. Thus, the original instincts of the body were not impure but could be made so by unchaste thoughts acted on by a weak or immature will. Bodily indulgences, in turn, led to more immoral thoughts, which led to more indulgences.[97] One of the worst offenses in this regard was the practice of masturbation, which, warned health reformers, wasted the energy of the bodily economy and frightfully compromised the developmental process. When untreated, masturbation inevitably led to sickness, to insanity, and to untimely death.[98]

If the laws of development were rightly followed, however, health reform advocates promised that normal growth would proceed in a consistent, balanced, and favorable fashion. Harmony and ease, not conflict and difficulty, were thus the watchwords of this body of early-nineteenth-century developmental thought. "All nature's operations are gradual," Fowler explained. "The sun does not burst suddenly upon our earth, nor go down instantaneously, but rises and sets gradually, besides being preceded and succeeded by slowly-increasing and diminishing twilight."[99] Such analogies from the natural world offered scientific justification for the perspective of one of the most influential texts on religious education of the antebellum period, minister Horace Bushnell's *Christian Nurture,* first published in 1847. Bushnell denounced the evangelical practice of sudden, difficult conversion from depravity to goodness in favor of a process of moral training in which "the child is to grow up a Christian, and never know himself as being otherwise." The traumatic conversion from sin to piety had a long history in Protestantism and was expected to occur during the late teens and early twenties as the moment of harrowing choice between the heedlessness of youth and the righteousness of Christian adulthood. More recently, during the revivals of the Second Great Awakening, conversions occurred earlier, in the early- and mid-teenage years. Bushnell argued that dramatic conversion need not occur at any age, that goodness was the product instead of consistent parental nurturing of the moral sensibilities from infancy onward. Since children were neither inherently depraved nor essentially moral, he maintained, they would certainly struggle between good and evil. The gradual inculcation into goodness would mitigate such conflicts, however, and make their proper resolution habitual, rather than a cause for suffering or concern for parent or child.[100] Physicians in the health reform movement affirmed that as the spirit developed, the body and mind would grow. Thus, according to Sweetser, a slow and regular course of development kept the young "along in the beaten track of existence," away from the "burs and briars of life," and was most conducive to physical growth, moral improvement, intellectual advancement, and individual happiness.[101] Likewise Alcott, who expected the young to strive toward excellence to develop all aspects of their natures to "a high pitch," nonetheless prescribed that this be done by living according to the motto, "Make haste slowly."[102]

To the ideal of harmonious, gradual growth, the "age of puberty" proved both the rule and the exception. In asserting the precept of protracted, steady growth, Blackwell deemed "earthquakes and revolutions" to be "destructive, not creative" forces. Like the other major life transformations that preceded them (birth, first dentition, and second dentition), the changes that occurred at puberty were to be accomplished in a "slow and complete manner," "the result of long-continued action, working silently but constantly in the right direction."[103] That acknowledged, Blackwell nonetheless also conceded that growth was not an entirely uniform process but was punctuated by "special periods of excitement" when the body and mind were both particularly vulnerable before nature's requirements and uniquely susceptible to influences from the external environment. Puberty stood out as the most important of the successive periods of rapid growth because the new functions acquired at this age were crucial both for the development of the individual and the progress of the "race." It was also a critical age because the "evil" of sexual precocity, which bred both physical weakness and moral degeneracy, posed its greatest threat. Even here, though, Blackwell's tone was ultimately reassuring. Like Rousseau, she recognized that nature potentially bequeathed to youth some measure of "storm and stress" but prolonging "natural" dependence and innocence also provided the antidote to the dangers puberty posed: "[T]he physical education of the body, its perfectly healthy development, *delays the period of puberty*, and . . . a true education in which all the *bodily powers* were strengthened as well as the mental and moral ones, would be the most effectual means of outrooting this evil." As long as society, parents, and youth did not seek to rush what nature intended to be a period of gradual and benign growth, Blackwell promised, the transformations wrought during puberty would proceed unremarkably and optimally and provide an important source for the vital energy that was essential for the child's sustained progressive development into rational, autonomous, and moral adulthood.[104]

In rejecting Locke's recommendations for early youthful independence in favor of Rousseau's prolonged dependency, health reformers articulated wider

cultural concerns about the unsettled status of youth during the early decades of
the nineteenth century and then attempted to temper such anxieties by rendering the traditional social subordination of youth a natural imperative. Their
veneration of youthful sexual innocence, which was to extend well through the
teenage years, also expressed a resistance to the forces of social change. In this
way, health reformers' developmental principles, especially the dictate of gradual
growth, performed a certain amount of conservative cultural work by offering
up new explanations of and justifications for the preservation of the age hierarchies of an older order.[105] By emphasizing the advantages that accrued to those
youths whose growth proceeded in gradual fashion, however, health reformers
also affirmed a new model of the self that they recognized was essential for the
success of a new social order. Thus, they functioned as what historian Steven
Mintz refers to as both "moralizers and modernizers."[106] Health reformers expected that scientifically enlightened mothers and educators would become
cognizant of the organic demands for and benefits of children's gradual unfolding. By following this natural law, such adults would be able to prevent "dangerous" youth from self-destructing or wreaking havoc on the social order as
well as shelter "innocent" youth from the worst uncertainties and corrupt influences of the era, while preparing those young people in their charge to benefit
from the promises for self-determination and individual advancement held out
by such dynamic times. This was a tall order and one replete with many unresolved intellectual and practical contradictions. Not the least of these was to
whom the privileges of mature selfhood secured through the development process ultimately belonged. For Locke and Rousseau, these privileges were to be
realized exclusively by the elite male, in whom they were tested and incubated
during the period of youth. Many health reformers' views of child development
and youth reinforced this expectation explicitly or implicitly; some also, however,
opened up the possibility for alternative interpretations.

On the one hand, the laws of development that health reformers espoused
were derived in conjunction with the changing imperatives of white middle-class
work and family life. These "laws" offered both a description of and a prescription for the sort of growing-up experiences children and youth from this select
social group needed to negotiate the process to their advantage. The precept
of gradual growth, in particular, supported the expanded educational and training regimens that were increasingly crucial to securing middle-class status for
nineteenth-century boys and girls. It also helped to protect the relations within
the private, affectionate family from succumbing to the centrifugal forces of
individualism.[107] On the other hand, health reformers' particular blend of organi-

cism and environmentalism led them to perceive and insist on the "universality" of developmental laws across class lines. Indeed, for Blackwell, it was the working-class family that best honored the gradual-growth dictate and provided the normative model for others to follow. As evidence, she noted that youth in the laboring classes arrived at puberty later than those in the middle and upper classes. Because, she maintained, honest, moderate farm or factory labor enabled the imperatives of organic development more so than the pernicious indulgences of wealthy urbanites, which inevitably bred the ill moral and physical effects of precocity.[108] Others, less sanguine about the inherent healthfulness of working-class life, relied on the law of order in exercise to make the case against child labor. Fowler asserted that because nature demanded that bodies attend almost exclusively to the requirements for rest and exercise throughout the period of their growth, limits had to be placed on children's and youths' participation in factory work. Likewise, Combe wanted legislators to be properly educated in the laws of "the constitution of the human body" so that they would see "the utter impossibility of combining [factory labor during childhood and youth] with . . . that moral and intellectual cultivation which is so imperatively required."[109] Whether health reformers saw working-class children and youth as exemplars of developmental laws or as possessing the same inherent rights as their middle-class counterparts to be shaped by them, their explicit proposition was that the requirements of age were to trump the exigencies of class, even as their class-based assumptions always implicitly informed those age requirements.

Health reformers also grappled with the relationship between the laws of life and the meaning and significance of sexual difference. Women were active creators, avid disseminators, and eager recipients of popular scientific knowledge about health, and they joined their fellow male hygiene enthusiasts in occupying a spectrum of positions regarding views about gender roles and identities.[110] Ranging from endorsements of women's premier moral authority in the home to radical calls for their equal status in society, this spectrum of views was also represented within and across other reform movements, from temperance to abolitionism to women's rights.[111] Many health reformers, men and women alike, offered up and embraced the cause of hygiene as an avenue by which middle-class women could exalt in their new role as primary caretakers of the family. Such a prospect was heartily endorsed by the members of the American Physiological Society—almost one-third of whom were women—in a resolution passed at their second annual meeting: "That woman in her character as wife and mother is only second to the Deity in the influence that she exerts on the physical, the intellectual, and the moral interests of the human race, and that her education should be adapted to qualify her in the highest degree to cherish those interests in the wisest and best manner."[112] All health reformers deemed it the responsibility of women to ensure the physical, mental, and moral health of the men and children in their families. Women needed to be rigorously educated about the laws of physiology and health and trained in the scientific management of the household. Women also needed to attend more conscientiously to their own health, which health reformers and orthodox physicians alike recognized as particularly dire throughout the antebellum period. In varying degrees, health reformers advocated for improving adult middle-class women's health by adopting exercise regimens, abolishing restrictive and decorative dress, and adhering to sexual restraint within marriage. Whatever their particular perspective on these remedies, all of those who propagated and heeded the call of health reform took part in elevating the status and enhancing the influence of women within the domestic sphere.[113]

Some health reformers went further than this, more boldly extending the bounds of woman's sphere into the public realm, although they most often did so by appealing to the logic of sexual difference. Indeed, in writing, in speaking, and in teaching about health, and, for some, in seeking medical training and entrance into the medical profession, women health reformers attested to women's capacity to claim social roles for themselves beyond the purview of domesticity. Elizabeth Blackwell was one prominent example of those health reformers who sought to endorse women's domestic role as well as enlarge it. Born in 1821 in England to progressive parents who were active in the temperance, women's rights, and abolitionist movements, Blackwell received an education equal to her brothers, and, from an early age, she was encouraged to cultivate an independent mind and sense of social responsibility. The Blackwell family moved to the United States in 1832, first to New York, then New Jersey, and finally Cincinnati, where her father unsuccessfully pursued opportunities in the sugar-refining business. Her father's death in 1838 left the family emotionally bereft and financially destitute. Uninspired by the teaching she took on to help support the family, dreading the prospect of marriage, and increasingly drawn to transcendental and Swedenborgian ideas, Blackwell turned to medicine as a moral calling. Determined to overcome resistance to her pursuit of medical training, and winning some admirers and supporters in the process, Blackwell became the first woman to receive a medical degree in the United States, from Geneva College in New York in 1849. Following further study in England and Paris, she returned to New York hoping to launch her medical career. She again faced obstacles from colleagues and a public hostile to female doctors. Blackwell turned to writing

lectures about hygiene, which were published as *The Laws of Life* in 1852. She
went on to become a pioneer advocate for women's health and medical education in the United States and England, as founder of the New York Infirmary for
Women and Children and the Women's Medical College of the New York Infirmary. When she returned permanently to London in 1869, Blackwell strongly
supported establishing similar institutions in the United Kingdom.[114] According
to Blackwell, it was women's superior capacity for compassion that made them
such ideal healers, both as mothers in the home and as medical professionals. In
bringing their heightened moral sense to the task of caring for their own bodies
and for those of others, women also nursed the goodness of the soul and thereby
held the potential to transform the practice of medicine and bring about the
reformation of the larger social world. Blackwell thus went further than the
American Physiological Society's recognition of women's maternal roles in also
singling out for praise those "spiritual mothers of the race" who were "often
more truly incarnations of the grand maternal life, than those who are tech-
nically mothers in the lower physical sense."[115] For Blackwell, women's innate
moral superiority not only necessitated their greater authority in the domestic
sphere but also justified their expanded participation in the medical profession
and other social reform projects.

　　Multifaceted views about gender figured into health reformers' descriptions
and prescriptions for the development of the child. Echoing Locke and Rousseau,
health reformers emphasized the similarities, rather than the differences, between the sexes in childhood. In doing so, they challenged characterizations
within orthodox medicine that depicted the female body as essentially weak and
debilitated.[116] Indeed, for many health reformers, the primary motive for alerting
mothers and teachers to the imperatives of organic development in the first place
was to expose the violations of such principles by girls and those who cared for
them, as well as to offer remedies for the current epidemic of ill health among
female youth. Girls thus often became the more prominent subjects of developmental thinking among health reformers, a trend that would continue in the
history of ideas about child development. Thus, Brigham saw the tendency to
cultivate the mind at the expense of the body to be "more particularly true as
regards females" and argued that more careful attention had to be paid to the
girl's physical education during childhood. The reason the female body was so
often neglected, Combe explained, was that orthodox physicians, teachers, and
parents were operating under the mistaken assumption that the laws of development differed according to the sex of the child. "[S]uch is the dominion of
prejudice and habit," he declared, "that, with these results [of girls' poor health]

meeting our observation in every quarter, we continue to make as great a distinction in the physical education of the two sexes in early life, as if they belonged to different orders of beings, and were constructed on such opposite principles that what was to benefit the one must necessarily hurt the other." "No time is lost [in the girl's education]," chastised J. Wilson, "in impressing her young mind with the great idea that is to govern her whole after-life-that *she is not a boy,* and not even a *child,* but a 'little woman'—that she must be prim, demure, and cautious in all her movements, 'like mamma'—that to run and romp is 'unladylike,' and to kick up her heels an indelible reproach on her embryo womanhood." Blackwell's *The Laws of Life,* significantly subtitled *"with special reference to the physical education of girls,"* intended to correct the misconception of "the great idea" by insisting that developmental principles be uniformly applied to girls and boys. For all children, she contended, the only way to achieve the ideal of slow, steady, balanced growth, along with the physical, mental, and moral benefits that flowed from it, was to keep early education focused on physical development and to avoid any sort of excessive study or undue emotional excitement that might compromise the body's energies and threaten its progression toward its divinely ordained potential and purpose.[117]

This insistence on the uniform application of the laws of development to girls and boys alike dovetailed with a broader Victorian conception of androgynous childhood, signified in the homogeneously feminized clothing of infants and young children, which both confirmed the dependent status of children as a group and held them up as paragons of sexual innocence. Such a construct linked the establishment of sexual difference with reproductive capacity and (hetero)sexual desire and disassociated all of these from the province of childhood.[118] Even so, it also ultimately functioned in conjunction with, rather than in subversion to, the ideology of separate spheres. As historian Karin Calvert finds in her examination of the material culture of early childhood in this period, "An androgynous image of children was acceptable to Victorian parents only so long as the nature and destiny of each sex seemed unalterable and secure . . . All of this meant that parents were faced with the somewhat conflicting need both to reassure themselves of their children's asexual innocence and to receive clear indications that any child would be able to fill its proper place in society."[119] In the same way, health reformers' disavowal of the assumption that girls and boys occupied "different orders of beings" during childhood worked to reinforce, rather than undermine, prevailing notions about the complementary characters and divergent destinies of the male and female sex and about the importance of sexual difference to social order and progress. Indeed, their main point about sexual difference was that it was organically developmental—that it appropriately emerged at a particular moment in the life cycle but was essential, nonetheless, ordained by God and Nature and static in the ideal qualities that were attributed to femininity and masculinity.[120]

In the health reformers' conceptions of development, puberty was a particularly important time in the life of girls and boys because the bodily differences that determined their future identities, roles, and responsibilities were manifested. For boys, however, the changes that occurred during puberty established not only a uniform gender identity but also secured a dynamic individual identity. Health reformers' conceptualizations of the "age of puberty" in the male thus affirmed Locke and Rousseau's expectation that youth was a critical stage of life for boys that was unique to their development because their passage through it secured for them alone this highest privilege of mature selfhood. Two European medical texts with American editions that extensively described the phenomenon of male puberty and its implications for development were British physician William Acton's *Functions and Disorders of the Reproductive Organs* and French physician M. Lallemand's *A Practical Treatise on the Causes, Symptoms and Treatment of Spermatorrhoea*. For both Acton and Lallemand, male puberty was marked by the body's production of semen and the accompanying onset of feelings of sexual desire. Both recognized that the inauguration of such capabilities could not help but influence the boy's body and mind, although whether for good or for ill was the point of some disagreement between them. "Nothing can prevent the genital organs, at the time of their development, from reacting on the economy and giving rise to new sensations and ideas," Lallemand maintained. Whereas he conceded that such reactions might have detrimental effects on the boy, Acton instead emphasized that the secretion of semen was the source of an increase in vital force for the boy at puberty, which in turn was responsible for sustaining the progress in physical, mental, and moral development that made possible his realization of mature adulthood. The key to accessing this vitality, he explained, was that excepting those occasional involuntary nocturnal emissions that provided the continent individual with a natural release of plethora, the semen that began to be produced at puberty was to be "reabsorbed into the animal economy," thereby "[augmenting] in an astonishing degree the corporeal and mental forces." He went on to describe this process with much awe and enthusiasm: "This new . . . powerful vital stimulant-animates, warms the whole economy, places it in a state of exaltation and orgasm; renders it in some sort more capable of thinking and acting with ascendance—with a superiority, as we equally observe among animals in the rutting season." According to Acton, nature intended youth to be marked by "robust health and absence of care." It was only when boys violated nature's laws by consciously rejecting incontinence that the period of their youth took on qualities too often accepted as inevitable in the popular imagination: debility, sadness, sensitivity, restlessness, agitation, and apathy. Such a state of "sexual suffering" was, Acton averred, often much exaggerated, "if not invented" for the purpose of justifying immoral behavior.[121]

Complete continence throughout youth accrued to the individual boy physical strength and mental power because it allowed his vital energy to be used for its proper function of building up his growing frame and cultivating his intellect. In addition, it ensured the robust reproduction of the human race, for only boys whose reproductive organs were fully mature could give birth to healthy offspring. Most important, it propelled the boy's moral development because resisting temptation strengthened the boy's moral fiber and made possible his capacity for self-government and moral autonomy later in life. As Acton asserted, the continence that he advised was not "mere ignorance." Rather, "[t]rue continence is complete control over the passions, exercised by one who knows what they are, and who, but for his steady will, not only could, but would indulge them." If there was anything difficult or unpleasant about youth, this was surely it, although the rewards of emerging victorious from such a struggle were great indeed. "Grant that continence is a *trial*, a sore trial, a bitter trial, if you will," Acton elaborated. "But what is the use or object of a trial but to *try*, to test, to elicit, strengthen, and brace, whatever of sterling, whatever of valuable, there is in the thing tried?"[122] Following Rousseau, Acton posited that out of the boy's repeated refusal to gratify the selfish sexual passion during youth emerged the reproductive instinct and, finally, the social sense and hereby the highest moral capacities of the man for love, compassion, and justice were born.

American health reformers likewise characterized male pubertal development as naturally entailing a vigorous body, an energetic mind, and a rigorous moral sense (although Graham preferred to focus on the role of the nervous system as the source of such vitality, as opposed to the semen, the importance of which he deemed to be "exceedingly overrated"). The progression through youth into manhood, Samuel Woodward explained, "requires all the energy of the system, greatly increased as it is at this period of life, which, if undisturbed, will bring about a vigorous and healthy condition of both the mental and physical powers."[123] In his *Familiar Letters to Young Men*, Alcott provided a depiction of male puberty for young men that was both less and more sanguine than this. "There is a period in every young man's history," he acknowledged, "when dangers of every kind thicken around him, and seem to threaten inevitable destruction . . . And such is the violence, to most, of the storms that assail at this critical period, that we are not to wonder if thousands and millions of our race are left to suffer, under their influence and their own experience, a most fatal and terrible shipwreck." Such dangers were the product of the inharmonious nature of the teenage boy's body and the society in which he lived. Alcott's point, though, was both to name the problem and to challenge its inevitability: "If it should be argued that young men, such as I am addressing, are not to be expected to have those well-balanced natures which farther education and a more extended experience would be apt to develope (sic)—that indeed they cannot have them—I should meet the argument by a flat and positive denial. Your character should be as harmonious at four as at sixteen, and at sixteen as at sixty. It should be in harmony at every age of moral accountability, and in all circumstances." Moreover, insofar as struggles against inharmonious conditions, both internal and external, ensued, it was this stress that made men great.[124] According to Alcott, strength and harmony of body, mind, and character were every boy's birthright; at puberty, the boy was poised and challenged to exercise his manly demeanor to determine his destiny and shape his world—to become *"whatever* [he] *will resolve to be."*[125]

Physicians in the emerging disciplines of obstetrics and gynecology established a stark contrast between the possibilities for the boy's physical, mental, and moral progress occasioned by the onset of puberty and the limits placed on the girl's development by the maturation of the reproductive organs. The girl not only failed to secrete semen, the impetus for the boy's physical, mental, and moral growth, but the first menstruations actually impeded her development by pitting her body's extensive demands for vital energy against her educational prospects, placing both of these beyond the control of her individual will. This rendered her "age of puberty" inherently volatile, enervating, passive, and unhealthy and made it impossible for her to achieve the ends of mature adulthood promised by Locke and Rousseau.[126] Of far less concern to these doctors than girls and boys being treated too differently during childhood was that pubertal girls and boys were treated too much alike. For them, puberty in the girl was, exclusively, the moment when sexual difference was established and secured and understanding how that happened and what it meant for both individual and society depended on a thorough scientific explanation of the female body and mind during this most critical development. As shown in Chapter 2, conservative doctors' exclusion of girls from the privileges of youth by providing an account of their developmental differences and deficiencies reached its height during the second half of the nineteenth century. It was intensified by the rising authority and mutual influence of reproductive medicine and evolutionary science and made ever more imperative by the growing incursions by girls and women into the public sphere.

Meanwhile, health reformers provided an important counterpoint to the rising conservative medical discourse about female puberty. They tended to describe girls' experience with puberty in broadly similar terms to that of boys— as potentially dangerous but not inexorably so. Expressing their confidence that health maintained during childhood and youth would more than suffice to see the girl through the physical and moral demands of this important stage, they were intent on establishing and communicating the organic developmental laws that would enable mothers, teachers, and girls to preserve that health, rather than documenting female developmental failure and lack. Indeed, in his *The Young Woman's Book of Health*, Alcott went so far as to apologize for focusing on the "dangers and pitfalls" of this stage in girls' lives, lest he "either lead them to injudicious dosing and drugging, or make them moping and melancholy." As with his prescriptions for male development, his larger intention was not to resign girls to succumbing to the inharmonious conditions marking the age of puberty but to encourage them to manage internal and external pressures to emotionally and morally auspicious ends. "Seeing, as [the girl] must by what I have here written, to how great an extent God has placed her happiness and her misery within the range of her own choice," Alcott proposed, "will she not be led the more earnestly to secure the one and avoid the other?"[127]

Nonetheless, Alcott and his fellow health reformers both assumed and propagated notions of sexual difference. One way they did this was by deeming the imperative of gradual growth to be even greater for girls than for boys because "natural" female purity was more threatened by the dangers of sexual precocity. Most health reformers were advocates of a single standard of sexual morality and emphasized the importance of chastity for both young men and women. Lust was, however, also grudgingly recognized as an attribute of male nature, and the young man's effort to control it was one factor that propelled his moral development and provided the foundation for his self-rule and authority over others later in life. In contrast, premature sexual expression by girls served no such purpose, for it violated the essential "passionlessness" of their nature and therefore portended only the direst physical, moral, and social consequences.[128] The implications of health reformers' assumptions of female innocence for understandings of the girl's moral development were mixed. On the one hand, the girl, in whom the sex instinct was always the reproductive instinct, was perceived as inherently more moral than the boy. Her innate love and compassion served as the vital source of her moral authority in the family and, to some extent, in society. On the other hand, because the girl never had to struggle to control the sex instinct, she was also seen as incapable of achieving the same degree of moral rigor as her brother and therefore unable to exercise the self-control and broad social responsibility expected of and enjoyed by him. Joining other antebellum reformers in accentuating women's moral superiority and recognizing the importance of "female values" to personal salvation and social improvement, health reformers tended toward the former interpretation of the girl's moral development. In either case, though, female sexual desire was patently denied and deemed to require more stringent external controls than boys' sexual desire to ensure the protection of both the girl and the larger community.

Furthermore, although health reformers did not depict female development as a process marked by debility and lack, they did not necessarily expect that the girl would move through it to become "whatever [she] will resolve to be." Thus, they appealed to Blackwell's principles of use, order, and balance to claim for female as well as male children the right to a healthy, vigorous, and active development. At the same time, they also construed her fourth law of development, which stated that every being reached its "highest welfare" and greatest happiness by conforming to its "*special* purpose," to mean that the final aim of the female child's growth was the achievement of sexual difference, which entailed, exclusively, the identities and roles of wife and mother realized in the private home. As Brigham explained, it was because girls were endowed with a distinctively sensitive nature that their physical education had to be attended to in equal measure to boys'. Girls who were given the freedom to romp and play and did not adopt the manners and mores of adult women too soon had the best chance that their more active imaginations and more intense emotions would not be "rendered excessive" but instead would give rise to the "finer sensibilities" in which women were superior to men.[129] Alcott, who hoped to convince young women that such traits as rationality, originality, perseverance, and decisiveness were not the exclusive province of their brothers, nonetheless saw their main task in life to be putting these qualities to their own distinctive use as a gendered class: "It is quite time that woman should understand her power and her strength, and govern herself accordingly. It is quite time for her to stand upright in her native, heaven-born dignity, and show to the world—and to angels, even, as well as to men—for what woman was made and wherein consists her true excellence." Fowler put the fine point on Alcott's vague injunctions. Girls needed strong and fully developed bodies to fulfill their unique and sole destiny as mothers of the human race.[130]

Elizabeth Blackwell did not disagree. However, she also envisioned a more expansive potential for female development, applying her developmental thinking to enlarging the girl's capacity as individual and female, albeit still largely within the context of woman's sphere. For Blackwell, as for her fellow health reformers, gradual growth was perhaps best illustrated by the imperative of protecting the sexual innocence of the girl. She went further than this, however, to argue that girls could and ought to pass through a prolonged stage of youth that not only preserved their purity but also enabled their physical, mental, and moral powers to unfold and be cultivated to their fullest potential. Nature would take care of some of this; the right kind of education and the girl herself would do the rest. Thus, Blackwell declared that the first problem with female education, which was recognized by other health reformers as well, was that the girl's body was not adequately cultivated during childhood and youth. Overlooked or uncritically accepted by others, however, was the additional problem that the education of the girl's mind was thought to be complete when she reached 16, just the age, Blackwell proclaimed, at which it should begin. Whereas bodies completed most of their crucial growth in the years up through the end of puberty, the mind had the capacity to continue to develop throughout life. Too often, she lamented, girls spent the years of their young womanhood engaged in the frivolous pursuit of the pleasures of partygoing or novel reading, or, worse still, succumbed to early marriages out of misplaced yearnings for excitement. Instead, the years from 16 to 25 were to be spent in the pursuit of higher mental cultivation that would prepare the girl for the "active duties" of all of the roles of her adult life—wife and mother, most surely, but also "member of society" and "human being." The laws of development dictated that the growth of the body was to *precede* the growth of the mind, Blackwell pointedly maintained, not to *supersede* it.[131]

Furthermore, per the law of compound movement, some sort of balance among the various realms of development was to be sought at all stages of the girl's and woman's life. During girlhood, the emphasis on physical growth would be tempered by some attention to the mind and spirit. During youth, the prominence of mental cultivation would be enabled by maintaining sound bodily health. During adulthood, woman's focus on the family would be enriched by her ongoing engagement with the wider world in mind, in body, and in spirit.[132] Blackwell did not fundamentally challenge the masculine norms of youthful development established by the Enlightenment philosophers and reinforced by nineteenth-century physicians of both conservative and unorthodox ilk. Instead, she maintained that rationality, autonomy, and social responsibility were as much the product of the girl's development as of the boy's. Along with her fellow health reformers, she did celebrate the special contribution women made to the family and society through their superior moral capacity. She went further than most, however, in conceiving of the range of possibilities for individual women to express this tendency and in illuminating its harmonious relationship to other developmental imperatives throughout the female life cycle.

As interpreters of Locke and Rousseau in light of the changing experiences of early-nineteenth-century American youth, health reformers made lasting cultural contributions to thinking about child development, not the least of which was to help lay some important groundwork for the modern concept of adolescence. Within the context of the growing authority of medical science, they raised the specter of youth as a problematic stage of life and highlighted the importance of puberty to physical, mental, and moral development. They also attempted to assuage cultural anxieties about the dangers of youth, especially the dangers of puberty, with prescriptions for gradual development that reinforced the immaturity, innocence, and dependent status of youth. At the same time, they promised the young the opportunity to achieve new standards set for the mature self marked by autonomy, rationality, and moral responsibility. That realizing such standards was the product of natural growth, aided both by the right kind of education and individual wherewithal, held out the possibility that the privileges and responsibilities of mature selfhood derived from the process of development might be universally parceled out across lines of class and gender. Girls of the white middle class in particular, who in their social behaviors were at the forefront of carving out a new phase of life for youth, were defended in this discourse and even appealed to as exemplars of certain inviolable developmental laws. They were, however, contained and constrained by it as well, as health reformers also applied the discourse of development to explaining and reenforcing categories of social difference.

CHAPTER TWO

"Persistence" versus "Periodicity"

From Puberty to Adolescence in the
Late-Nineteenth-Century Debate over Coeducation

During the last three decades of the nineteenth century, a cultural fervor ensued over the merits and detriments of "identical coeducation." Coeducation referred to either educating boys and girls together or, if they were educated separately, using the same methods and purposes. It became a lightning rod in the late Victorian era because of a constellation of anxieties about gender roles and identities and their relationship to social change and "racial" progress. The essence of this debate was not new. Indeed, since the founding of the republic, the issue of female education—whether to educate girls, by what means, and to what ends—had been repeatedly engaged by intellectuals, pedagogues, theologians, and reformers. What *was* new in this latest incarnation of the ongoing debate was the influence of the current climate of developmental thinking based on the authority of biological science. So noticed Mary Putnam Jacobi, one of the key participants in the late-nineteenth-century debate over coeducation. In attempting to clarify the effects of evolutionary thought on contemporary conceptions of womanhood, she wrote: "A remarkable change has taken place in the tone of habitual remark on the capacities and incapacities of women. Formerly, they were denied the privileges of an intellectual education, on the ground that their natures were too exclusively animal to require it. Today, the same education is still withheld, but on the new plea that their animal nature is too imperfectly developed to enable them to avail themselves of it."[1]

Jacobi's observation amounted to an indictment of the dominant scientific and "habitual" interpretation of evolutionary theory that rendered woman biologically inferior to man. According to this reading of Darwin, woman was so determined by her bodily forms and functions as to be "too exclusively animal" to be capable of exercising the highest powers of reason, self-control, and self-assertion possessed by the occupant of the pinnacle of the evolutionary hierarchy, the western, white middle-class—or "civilized"—male. In this scheme in which perfection was determined by biological fitness, however, woman was now also depicted as imperfectly animal, with deficiencies in her phylogenetic and ontogenetic development accounting for her subordinate place on the evolutionary ladder. As a scientist steeped in developmental thought, Jacobi weighed in on the coeducation debate with an account of development in the individual human female that strove to sever her supposed closer connections to animals on the evolutionary continuum *and* to attribute an equivalent degree of biological fitness to her essential animal nature. Challenging the prevailing trend in the biological as well as the emerging psychological sciences of the time, Jacobi refused to see the female mind as overdetermined—and therefore limited—by the female body. Her refutation, however, was likewise grounded in the authority of biological science, in alternative conceptions of female physiology and of the changes that occurred in it over the course of the life cycle.

Such uses of developmental thinking in the debate over coeducation contributed to the further emergence of scientific and cultural conceptualizations of adolescence at the end of the nineteenth century. The work that sparked the intense debate over coeducation in the United States was the 1873 treatise by prominent physician Edward H. Clarke entitled *Sex in Education*, a widely popular book that was reprinted seventeen times in thirteen years.[2] Historians have previously placed Clarke's work in the context of the history of women's education, determining how educational policies, programs, and practices, as well as young women's attitudes, were shaped (or not) by Clarke's directives.[3] This chapter will focus on Clarke's concerns with the physical, mental, and emotional growth of the child and locates the work within the emerging intellectual tradition of child development studies. Here, I will attend to the major influences on Clarke's ideas—to evolutionary science, particularly the developmental theories of Herbert Spencer, and to the reigning understanding of the menstrual cycle in reproductive medicine. I will then turn to the outpouring of responses to *Sex in Education* for the ways in which Clarke's conception of what he termed, interchangeably, the "epoch of development" and the "sexual epoch" was challenged and revised by his opponents in the coeducation debate.[4] Jacobi's work figures notably here as the most scientifically authoritative alternative to Clarke's understanding of puberty and adolescence for a late-nineteenth-century audience.

The convergence of evolutionary science and reproductive medicine in

Clarke's work countered and temporarily eclipsed the potentially enabling accounts of female development proffered by the health reformers of the antebellum period. In the descriptions of the developmental process advanced by the
opponents of coeducation, male puberty, when it was discussed at all, was reasserted as a time of steady, protracted, harmonious growth, with its gradual and
prolonged nature deemed to be the source of the superior development of the
boy's body, mind, and character. Girls' bodies, in contrast, grew too fast, too
soon, and with great demands placed on them. This precluded their advanced
intellectual and moral development and turned puberty into an experience of
pronounced somatic vulnerability and emotional volatility. Women's rights advocates, educators, and some doctors protested these accounts and advanced their
own alternative explanations of female development. Significantly, one such
alternative suggested that insofar as puberty *was* characterized by vulnerability
and volatility, these were not qualities of gender, but of age (as well as race and
class), and so were shared by all civilized girls and boys undergoing the stage of
life that some were now beginning to refer to as "adolescence."

The late-nineteenth-century coeducation debate was prompted by the growing claims of white middle-class young women to the male prerogatives of
higher education and to the life beyond the home that educational opportunity
portended. During the first half of the nineteenth century, economic, social, and
cultural changes gave rise to a dramatic influx of white middle-class northeastern
girls, in particular, into newly established public and private primary and secondary schools. Conservatives warned that educating girls would endanger the
natural separation between man and woman's spheres. More persuasive and
more vocal, though, were the proponents of female education, among them the
pioneering founders of female seminaries Catharine Beecher, Mary Lyon, and
Emma Willard, who assured that education would not undermine the identity
and purpose of the true woman but would rather strengthen girls' ability to
perform their future roles as wives and mothers. According to this justification,
girls needed to be rigorously educated so that they would be prepared to teach
children, in the home and in the classroom, the republican principles and Christian values that were the foundation of the greatness of the new American
nation.[5] The rhetorical links between female education and domesticity were
strong and certainly had some effect on institutional policies and practices, as
well as on the ways in which girls experienced and made sense of their education.
As Jane H. Hunter illustrates, however, conservative fears of female education
were not entirely unfounded, as schooling "laid down self-expectations and

experiences which would profoundly unfit [girls] for domestic subordination."[6] The unintended consequences of female education notwithstanding, given the range of middle-class family needs that could be met by sending girls to school and girls' own enthusiasm for student life, the question by midcentury became not whether girls should be educated but how to educate them.[7]

In 1850, most public elementary schools, or "common schools," the majority of which were one-room schools in rural communities, were coeducational. Such arrangements were made most often in the interest of efficiency and cost. In most locales, it was simply not possible or practical to fund separate institutions for girls and boys. Coeducation in common schools was also acceptable because it mirrored the "natural" mixed sex relationships that abided in the family. In addition, given the limited mission of the common school—to teach children basic literacy skills and to instill in them a set of "universal" moral values and civic virtues—and in keeping with Victorian acceptance of androgynous childhood, many educators and parents remained largely untroubled by prepubertal girls and boys learning a similar curriculum together and in the same fashion. For all of these reasons, coeducation was practiced in, but not much discussed in relation to, rural common schools.[8]

However, coeducation in urban public schools, particularly high schools became a matter of dispute in a number of cities at midcentury, including Boston; Washington, DC; San Francisco; St. Louis; and Savannah. Urban public schools were larger and more diverse in terms of class and ethnicity than rural schools, although in most secondary schools, sons, and especially daughters of the middle class, predominated. Urban school were also more bureaucratized and more standardized than rural schools, creating a context for school administrators to formulate uniform and centralized policies about coeducation, which set the stage for the debate. The dialogue about the detriments and merits of coeducation took place among school officials in local communities, where policy decisions were made, as well as in broader professional forums, including state teachers' associations, the reports of state superintendents, and the meetings and publications of the National Educational Association. Opponents of coeducation voiced many arguments against the practice. They cautioned that pure and delicate teenage girls would be corrupted by too much contact with the opposite sex, particularly with indecorous working-class and immigrant boys. They feared that male teachers would be forced to discipline boys and girls according to similarly harsh methods (this despite a predomination of female teachers and a shift away from the uses of corporal punishment). Gesturing to increasing evidence that girls were outperforming boys in schools of all kinds, they argued that boys' lower academic standards would impede girls' mental advancement. And, they worried that educating girls with their brothers and in similar fashion would make them "like boys," unable or unwilling to fulfill their "natural" domestic functions that were so important for sustaining and advancing "civilized" society.[9]

In addition to propagating practical arguments for economy and bureaucratic efficiency, some coeducation advocates responded by idealistically defending the lofty goals of public education. Public schools were meant to provide all children with equal education; mixing classes and sexes in the classroom was deemed to be central to their mission. They also maintained that coeducation held out moral and pedagogical benefits for both sexes. Girls and boys educated together would moderate the excesses of masculinity and femininity in each sex, creating more balanced characters as youths matured. In addition, regular academic engagement between girls and boys would reduce sexual tension and enhance the respect of each sex for the higher qualities and characteristics of the other. And, the different learning styles of each sex—girls more receptive and concrete thinkers, boys more critical and abstract—would facilitate the intellectual progress of both. Following the Seneca Falls Declaration of Sentiments of 1848, which denounced women's lack of access to higher education, a small group of women's rights advocates voiced the argument, as others had since the founding of the republic, that equal education was essential for women to claim the full political, social, and economic rights to which they were entitled as citizens in a democracy. The school administrators who voiced support of coeducation did not go that far, however. Like most of the supporters of female education before them, they contended that girls ought to be educated with and in the same manner as boys so as to foster the highest potential of their sexual difference.[10]

By the 1870s, coeducation in secondary schools was not settled, even as the practice was becoming firmly entrenched. The debate took on even greater urgency in the post-Civil War decades, however, as young women entered what had been the almost exclusively male preserve of higher education. The first experiments in collegiate coeducation were carried out in the antebellum period at Oberlin and Antioch colleges, founded by evangelical Christians in 1833 and 1853, respectively. At both institutions, male and female students were separated in their extracurricular activities and social encounters, policies that some women students protested. From the perspective of the schools' founders, separatism signaled respect for women's differences, and their presence on campus was meant to elevate young men to the same standard of piety and social behavior as their sisters. Even so, presidents of both colleges in the late 1850s expressed doubts that such a goal had been achieved at their own institutions or that it could be at other establishments less religiously inclined than theirs.[11] During the Gilded Age, a number of public and private colleges and universities began admitting women, but not without struggle. According to historian Lynn D. Gordon, "Collegiate coeducation usually did not come about naturally or because of American democratic traditions." Rather, "[a]ccess to higher education . . . became a reality when women themselves—mothers, civic leaders, potential students, or women's rights advocates—pressured state and university officials to open the doors of colleges and universities." A number of prestigious women's colleges were founded during this time, mostly in the East and the South. Modeled on female seminaries, which emphasized the distinctly feminine contributions educated women were to make to society, Vassar, Smith, Wellesley, Sophie Newcomb, and Agnes Scott colleges also established the highest standards of academic rigor for their female students. As with coeducational colleges and universities, these institutions came under immediate criticism from those who argued that their seminary orientation overly constrained female students and from those who believed their intellectual focus threatened young women's femininity.[12]

The perception that female education, whether coeducational or single sex, either portended great possibility or peril for girls and society was heightened during the Gilded Age in the face of tremendous flux in the Victorian gender, class, and racial order. During the last decades of the nineteenth century, forces of industrialization, immigration, and urbanization were forging the United States into a modern nation. Class and race relations were marked by intense conflict as members of the working class, various ethnic groups, and African Americans struggled to claim equal rights and opportunity against the nearly insurmountable odds of economic exploitation, nativism, and racial discrimination.[13] At the same time, middle-class white adult women, many of whom had secondary or some college education, challenged the limits on their autonomy the ideology of separate spheres placed on them. Instead they pursued professional careers, participated in the fledgling campaign for woman suffrage, and organized and joined clubs devoted to self-improvement and social reform. Although education, the engagement in public activities, and women's rights intersected in varied and complex ways in individual middle-class white women's lives, these were perceived as indelibly linked in the minds of those who raised and pondered the late-nineteenth-century "woman question."[14] As important were the connections such women made between their status as civilized women, with a superior racial capacity for moral agency, and their right and responsibility to assume public authority over racial and ethnic minorities and the working class. In this way, white middle-class women's gender transgressions opposed patriarchy, while also serving to buttress class, racial, and religious hierarchies in a nation striving for territorial expansion, commercial dominance, and cultural ascendancy at home and abroad.[15]

At the same time, the shift to a corporate, consumer-oriented capitalism, and the serious economic depressions that accompanied it in the 1870s and 1890s, was making it difficult for white middle-class men to exercise the manly virtues of autonomy and self-control that were the hallmarks of maturity and the prerequisites for full citizenship rights in the early republic. These threats to white middle-class manhood were exacerbated by the contentions for political and economic power by working-class, immigrant, and African American men, as well as by white middle-class women. In response, as historian Gail Bederman demonstrates, white middle-class men "remade manhood" in the years between 1880 and 1917, assuming a new set of powerful, "primitive" "masculine" traits that sought at once to resist the enervating effects on their manhood by the forces of modern life and "to ensure the continued millennial advancement of white civilization."[16] In the face of such momentous change, the stakes of the girl's education were deemed to be higher than ever before. To make the case about whether this was for good or ill, both opponents and proponents of identical coeducation marshaled the resources of the increasingly authoritative biological sciences. In the process, they constructed new knowledge about the development of the child and helped to formulate the modern category of adolescence.

CLARKE'S EPOCH OF SEXUAL DEVELOPMENT

Edward Clarke first opined against young women joining the collegiate student body in a speech before the New England Women's Club in Boston in 1872. Clarke, a member of Harvard's Board of Overseers and a former member of its medical faculty, was invited to speak by women's rights advocate Julia Ward Howe, under the assumption that he would endorse women's right to higher education. Indeed, three years earlier, Clarke had suggested that if women proved capable of training themselves to be physicians, they should be allowed to do so. To the dismay of many of his listeners at the New England Women's Club, Clarke now clarified his position on the matter, asserting that woman's distinctive physiology, the center of which was her reproductive function, prevented her from succeeding in any program of higher education, medical, or otherwise. He explained that women were innately unfit for the rigors of higher education, while they were supremely equipped for the roles of wife and mother, made possible by their unique biological endowment. Club members engaged Clarke in heated discussion following the address. Somewhat taken aback by his audience's response, he decided to explicate his argument more fully in a book, published the next year as *Sex in Education,* which was rather disingenuously subtitled, *A Fair Chance for the Girls.* This was followed in 1874 by a sequel dealing with similar themes, *The Building of a Brain.*[17]

That the womb and the ovaries dominated the life of the woman, determining her physical capacity, mental ability, emotional temperament, and consequently her social role and responsibilities, was a commonplace belief in nineteenth-century medicine.[18] Met with the growing incursions by white middle-class girls and women into the public realms of schooling, work, and politics in the post–Civil War years, Clarke's contribution to the conservative scientific justification for separate spheres renewed claims about female destiny by recasting that destiny in relation to the processes of individual development. According to Clarke, it was the extraordinary demands of the *growth* of the female reproductive system during the teenaged years that necessarily interfered with female educational opportunity. His argument drew from the latest innovations in evolutionary science and reproductive medicine. He began, though, with a long-extant model of the female life span that divided the life of the woman into three eras: childhood, from birth to the early teens; maturity, from the early teens to age 45; and old age, from age 45 until death. Such a model prescribed the process of female maturation as marked by the continuous unfolding of a singular destiny, with younger females lacking a distinct identity of their own and instead occupying the position of "little women." However, Clarke also altered this older formula by revisioning, and granting enormous significance to, the period of transition from childhood to maturity, which spanned from age 14 to 20.[19] Like the antebellum health reformers, Clarke was intent on disassociating the phenomenon of puberty from the achievement of maturity. He then went on to endow puberty as the key determinant of the personal attributes of the girl, the mandates of girlhood, and the girl's future physical and psychological health. As a result, he made a case for a distinct and extended period in the girl's life in which she was neither a child nor a woman, a period when the occurrence of sudden and critical physiological changes gave rise to sexual difference, circumscribed the capacity for mental activity, prompted the emergence of a peculiar emotional temperament, and determined the quality of the stages of the life cycle to follow.

As were most scientists and intellectuals of his day, Clarke was influenced by the theories of biological and social evolution of Charles Darwin and Herbert Spencer, particularly their explanations of the cause of sexual difference in the development of the human species and its importance for the advancement of western civilization.[20] In *The Descent of Man,* published just before Clarke's work in 1871, Darwin explained how the two fundamental phylogenetic principles he formulated in *On the Origin of Species* (1859), natural selection and sexual selection, applied to the evolution of human beings and gave rise to sexual difference. The human male, he posited, possessed superior physical strength, intelligence, endurance, inventiveness, and courage because he led the way in the struggle for survival and, early in the evolution of the species, was also engaged in competition with rivals for mates. In contrast, the female of the species, characterized by a slower metabolic rate and a natural dependency on the male due to the demands of human motherhood, was unable to benefit fully from the forces of natural selection. Consequently, she failed to develop the highest forms of mental acumen and therefore displayed less intellectual variability than the male. More emotional and intuitive in her ways of knowing and being, the human female also possessed greater tenderness and benevolence because of her role as the primary caretaker of children. The laws of heredity, as Darwin explicated them, worked in conjunction with these principles. Thus, traits that the individual acquired early in life, such as physical size, were the product of natural selection and were transmitted, although not developed, in both sexes. Traits arriving later in life, most especially mental abilities, were the product of sexual selection and were imparted only to offspring of the same sex. Spencer's widely discussed article, "Psychology of the Sexes," published in 1873 in *Popular Science Monthly,* provided similar analysis of the cause of sexual difference in human evolution. More popular than Darwin, historian Louise Michele Newman asserts, "Spencer gave his version [of evolutionary theory] a sharply teleological cast, asserting that evolutionary changes occurred so as to produce social and racial progress, leading to ever higher and more advanced civilizations and races." A whole array of evolutionary thinkers followed Darwin and Spencer to link sexual difference with evolutionary progress, contending that sexual divergence was both sign and cause of the superiority of the white race.[21]

If Clarke accepted these analyses of the constitution of sexual difference on the level of the species, the particulars of the establishment of sexual difference in individual human development, and indeed of individual human development itself, remained open for further scientific explanation. Darwin's model of species development as gradual, continual, and adaptive posed fundamental and enduring questions for the modern scientific study of child development—questions about the role of history in the life of an organism, the relationship between an organism and its environment, and the connection between the evolution of species and individual development. How "revolutionary" his effect on child development studies was, however, remains a matter of some debate.[22] Scholars of developmental psychology Roger A. Dixon and Richard M. Lerner both recognize and qualify Darwin's contributions to their field. It was, they offer, "the intellectual climate of historical or evolutionary thinking—a climate that . . . both preceded Darwin and gained impetus from him and that, in some ways, was epitomized by him—[that] is an originative core of developmental psychology."[23] Darwin was influenced by the work of two scientists whose ideas linked individual development with the evolution of species or ontogeny with phylogeny.[24] The first was Karl Ernst von Baer's findings from comparative embryology that embryos of different species resembled one another in their earliest stages, becoming more differentiated as they developed. The second was Jean-Baptiste Lamarck's theory of acquired characteristics, which posited that traits attained during the individual's lifetime could be passed on genetically to the next generation.[25] He also showed some specific interest in ontogeny when in 1838 he raised questions in his notebooks about the emotional development of infants. The following year he started a baby diary to record observations of his infant son, which was published in 1877 in the journal *Mind* as "A Biographical Sketch of an Infant." Nonetheless, Darwin's primary scientific concern remained changes in species over long periods of time and not changes over time in the life of the individual.[26]

Beyond Darwin, it was the highly popular theories of "individual evolution" articulated by Spencer that directly influenced the ideas about child development and the concepts of puberty and adolescence that were deployed by Clarke and contested by his opponents in the coeducation debate. Spencer's "doctrine of evolution" was founded on the law of conservation of energy, which he referred to as the "persistence of force." Proceeding from the physical laws of matter and motion, and indebted as well to von Baer's formula for change, he defined evolution as the natural process by which all phenomena—individual organisms, species, societies, the cosmos—changed from a condition of homogeneity to a condition of heterogeneity and differentiation. Although such change was not inevitable and could be assisted or impeded by external forces, nature intended it to be progressive, a process of moving the developing entity toward a superior state of "individuation," "integration," and "equilibration." For Spencer, coiner of the phrase, the "survival of the fittest," and leading architect of Social Darwinism, the most advanced human beings in the most advanced social state would be free from all external restraints, while their supremely evolved moral sense would ensure their voluntary cooperation with one another and generate peace and harmony in the whole. Thus, evolution marked the "gradual advance towards harmony between man's mental nature and the conditions of his existence," the ultimate end of which could only be "the establishment of the greatest perfection and the most complete happiness."[27]

Key to Spencer's appeal was the synthetic nature of his philosophy. Evolution, he contended, occurred simultaneously among the "several orders of existences," not as "many metamorphoses similarly carried on" but as "a single metamorphosis universally progressing."[28] That such change occurred at all was perhaps no better illustrated than in the transformations that took place in the individual human being from conception to adulthood: "In an individual development, we have compressed into a comparatively infinitesimal space, a series of metamorphoses equally vast with those which the hypothesis of evolution assumes to have taken place during those immeasurable epochs that the Earth's crust will tell us of . . . If a single cell, under appropriate conditions, becomes a man in the space of a few years; there can surely be no difficulty in understanding how, under appropriate conditions, a cell may, in the course of untold millions of years, give origin to the human race."[29] Ontogeny was important in Spencer's scheme because it followed and illustrated the universal laws of biology and physics governing all natural and social phenomena. Given his acceptance of Lamarck's theory of acquired characteristics, individual evolution was also important because of its dynamic relationship to both phylogeny and progressive social change.[30]

Spencer dealt most pointedly with themes related to child development in his 1860 work *Education: Intellectual, Moral, and Physical,* which was his first book published in the United States and certainly among the most read of his publications.[31] Asserting that "the development of children in mind and body rigorously obeys certain [natural] laws," he recognized that he was building on the organic developmental theories of early-nineteenth-century educational reformers, most notably those of Swiss pedagogue Johann Pestalozzi, a follower of Rousseau.[32] He also criticized Pestalozzi for failing to systemize his thought and condemned him and his followers for "the numerous crudities and inconsistencies" in the educational methods by which the Pestalozian doctrine was carried out. "These general notions" of development and education held by Pestalozzi and others, Spencer averred, "must be developed in detail,—must be transformed into a multitude of specific propositions, before we can be said to possess that *science* on which the *art* of education must be based." That science was, of course, evolution, whose general principles had to be grasped and followed by parents and teachers if "serious physical and mental defects" in the child's development were to be avoided and "a perfect maturity" achieved. Rhetorically wondering why, if there existed in the organism itself "a prompter to the right species of activity at the right time," children should not be left "*wholly* to the discipline of nature," Spencer recognized and offered a resolution to the paradox of the organic developmental paradigm. Development dictated the course of education, but education was also necessary, particularly in complex organisms dependent on parents for long-term survival, to maintain "the *conditions* requisite to growth," and even, if carried out without any coercion, to facilitate actively the fundamentally spontaneous process of individual evolution.[33]

Conforming to evolutionary principles in intellectual education meant that lessons should "proceed from the simple to the complex" and "start from the concrete and end in the abstract" because, Spencer explained, the mind, like all phenomena, grew progressively "from the homogeneous to the heterogeneous."[34] Drawing on the recapitulation tradition in evolutionary thought, which held that the development of the individual passed, in linear, progressive fashion, through all the lower stages of life of its ancestors, he also contended that the education of the "civilized" child should follow the same "order in which the human race has mastered its various kinds of knowledge." For Spencer, such an order was revealed in the mental capacities of living adult peoples of primitive races, whose simple environmental conditions had arrested their intellects at lower stages of development.[35] Following evolutionary principles in mental education also meant that learning should be propelled from children's own self-directed inclinations, as free as possible from the management of others. Children "should be *told* as little as possible and induced to *discover* as much as possible," he argued, for self-instruction was the primary impetus for human progress. If this occurred, the final requirement of intellectual education would be met—that it be an enjoyable and exciting endeavor for the child. This would, in turn, provide impetus for more advanced learning, which was the source of the further advancement of civilization.[36]

Spencer's descriptions of moral development and prescriptions for moral instruction followed these same evolutionary principles. Parents should not expect a great deal of moral goodness from young children, he advised, since such youngsters were as yet in a "barbarous" phase of character development, when tendencies toward selfishness, cruelty, and dishonesty abided. Nor should parents try to force their children's virtue too soon, for the higher moral faculties, as the higher intellectual ones, were acquired late in the evolutionary process and could be reached only by "a slow growth," which would be impeded by parents foolishly trying too soon to "make their children what they should be."[37] In matters of discipline, he recommended that parents be neither too lenient nor too strict and that they allow children to experience the "natural" consequences of their indiscretions. This was the only way to foster in them a rational sense of "pure justice" and cultivate in them the highest moral capacities of self-governance and voluntary cooperation with others. It also was the only system of moral governance that preserved the equanimity in the tempers of both parent and child and that encouraged "a much happier, and a more influential state of feeling" between them.[38] Given that parents themselves were morally imperfect (which manifested itself in their children genetically and in their unduly harsh modes of family governance) and given that the society in which children were being educated to enter was as yet morally imperfect, Spencer recognized that such an approach could be carried out only by increments but should be endeavored nonetheless: "It . . . follows that the dictates of abstract rectitude will, in practice, inevitably be subordinated by the present state of human nature—by the imperfections alike of children, of parents, and of society; and can only be better fulfilled as the general character becomes better."[39]

Spencer followed his essays on intellectual and moral education with an exposition on physical education, in which he made clear that the development of the body was the foundation upon which the perfection of the mind and character depended. "[T]he first requisite to success in life," he insisted, "is to be a good animal; and to be a nation of good animals is the first condition to national prosperity." With this as his premise, Spencer laid out one of the most important laws of child development enumerated in *Education*—the law of growth itself. The law of conservation of energy dictated that the body possessed a limited amount of vital energy and that overexertion in the development of one part would result in a depletion of energy available for the development of other parts. Of great concern to Spencer was that the early forcing of mental prowess might undermine children's physical fitness "for the struggle before them."[40] Precocity and a potential conflict between the development of the child's mental and physical powers had received much attention in earlier nineteenth-century developmental discourse and had figured into antebellum critiques of teaching methods and calls for physical education. Although Spencer acknowledged that the danger of precocity was becoming more recognized by parents and teachers, he again sounded these same alarms, reminding his readers that Nature was, indeed, a "strict accountant," who must respond to any forcing of the young child's mental proficiency by leaving "some of her more important work undone."[41]

Spencer went further than previous developmental thinkers, however, in positing the existence of an "antagonism" between growth, or "increase of size," and development, or "increase of structure." The law, he asserted, "is, that great activity in either of these processes involves retardation or arrest of the other."[42] Thus, the simple caterpillar was exclusively devoted to augmenting its bulk during its early stage of existence, turning into the complex butterfly only when, once in the chrysalis, growth ceased and development began. In the higher animals, Spencer conceded, these two processes seemed to be carried out simultaneously. Significantly, though, the workings of the law could still be observed in the differential development of the sexes. "A girl develops in body and mind rapidly," he wrote, "and ceases to grow comparatively early. A boy's bodily and mental development is slower, and his growth greater. At the age when the one is mature, finished, and having all faculties in full play, the other, whose vital energies have been more directed towards increase of size, is relatively incomplete in structure; and shows it in a comparative awkwardness, bodily and mental."[43] Spencer made explicit the implications of this developmental difference in The Principles of Biology (1866) and, more succinctly and more accessibly, in the article "Psychology of the Sexes." Here, he posited another antagonism, this one between "Individuation" and "Genesis," by which he meant "growth, development, [and] activity," on the one hand, and reproduction, on the other. The girl's early growth spurt was necessary, he explained, so that she would have some vital energy remaining to meet the high "cost of reproduction" entailed in her role in the production of offspring. The price she paid for this, however, was that both her physical growth and her mental development were normally arrested at a level below that of the slower, more steadily and continuously growing boy, whose contribution to genesis exacted no significant expenditure of vital force and thus did not compromise his process of individuation. Alternatively, the girl who insisted on the pursuit of a "high-pressure education" was likely to experience an abnormal "deficiency of reproductive power" in any number of guises, including sterility, early menopause, and the inability to nurse offspring.[44] In Spencer's formulation, then, precocity, once understood as a condition visited on young children under the influence of misguided educators who failed to see the importance of physical education, was now determined to be an essential—and detrimental—attribute of female development. Smaller and weaker in body than her more slowly growing brother, the girl also never attained the complex mental and moral capacities for abstract reasoning and impartial justice that for Spencer were the latest—and greatest—"products of human evolution."[45]

In describing girls as normatively precocious, Spencer denied them passage through youth, a stage of life that he deemed the exclusive province of civilized boys. Spencer's conceptualization of youth echoed and furthered the ambivalence of earlier developmental thinkers. The civilized boy's prolonged development, which rendered him uncomfortably awkward, immature, and dependent throughout his teenage years, was nonetheless both cause and effect of his superior evolutionary advancement. It was this dependence that made education and discipline necessary, ministrations that Spencer preferred to be executed by parents rather than potentially coercive social institutions. While recognizing the imperatives of adult guidance over the direction of youth, however, Spencer was also careful to specify the desired developmental outcome of the "active and elaborate course of culture" to which the boy was to be subjected. The primary goal of mental and moral education, he dictated, was to raise "free men," who, in their capacities for intellectual innovation and self-governance would make possible the continued evolution of civilization.[46] Such a goal required the exercise of as much self-direction and self-rule during youth as possible. "The association between filial subservience and barbarism . . . and the fact that filial subservience declines with the advance of civilization, suggest that such subservience is bad," declared Spencer. "Whilst, on the other hand, a non-coercive treatment . . . must . . . accustom the child to that condition of freedom in which its after-life is to be passed." Treating youth with greater latitude also promoted greater happiness in the young and genuine sympathy between fathers and sons, both of which forestalled destructive youthful rebellion and encouraged boys in the purposeful use of their freedom.[47]

To this Lockean conception of youth, Spencer added his own evolutionary optimism. Youth required some measure of restraint in the current, transitional social state because they still possessed some of "the dispositions needed for savage life," which were adaptive in an as yet imperfect society. The more the advancing social state called forth the exercise of freedom in the young, the less necessary control of any kind would become, and in turn the more capable the young would be of furthering social advancement. In a fully civilized society, then, education itself would become obsolete. At that point, "[t]he young human being will no longer be an exception in nature—will not as now tend to grow into unfitness for the requirements of after-life, but will spontaneously unfold itself into that ideal manhood, whose every impulse coincides with the dictates of moral law."[48] In neither the current social state nor the perfect one he imagined, however, did Spencer allow for the possibility of the girl's full individuation. More mature and less developed than boys, girls were, paradoxically, less dependent on adult guidance and regulation because they came of age sooner than boys, and far more so, given that they were biologically incapable of benefiting from the extended development and education that would promote their ultimate freedom.

Clarke employed Spencer's theories of "individual evolution" to make his case against identical coeducation, although he downplayed female developmental deficiency in favor of the positive aspects of female difference. From the outset, he positioned himself against what he decried as a "new gospel of female development" that insisted on establishing the identity between the sexes. "The notion is practically found everywhere," he lamented, "that girls and boys are one, and that boys make the one." As did Spencer, Clarke ascribed to the view that as children, girls and boys were indeed "very nearly alike," and that it was that sameness that allowed them to exist in an appropriate state of "innocent *abandon* that is ignorant of sex." Late-nineteenth-century evolutionary theory bolstered, and was bolstered by, the longstanding notion of the late acquisition of sexual difference in ontogeny by maintaining that sexual divergence increased as human evolution progressed—the more advanced a culture, the greater the distinction between the male and female spheres of life. As in the biological and cultural evolution of the species, the thinking went, so in the development of the individual, and vice versa. According to Clarke, the careful observer could, of course, recognize even in the girl baby the "divine instinct of motherhood" in her gentle treatment of a doll "that her tottling brother looks coldly upon." Likewise, that same infant boy could be seen breaking "the thin disguise of his gown and sleeves by dropping the distaff, and grasping the sword." Nonetheless, as the antebellum health reformers before him, he recognized that there were dangers to the physical, mental, and moral development of the child when sexual difference was allowed to manifest too soon. All of that accounted for and understood, however, Clarke's primary contention was that not nearly enough attention was being paid to the epoch of sexual development—the critical stage of life when, because of the biological changes that occurred with puberty, the sexes unmistakably diverged in their development and, consequently, in the requirements for their education.[49]

Imparting dichotomous gendered meanings to concepts Spencer used to describe the physical laws of matter and motion, Clarke characterized the pubertal development in the male as governed by "persistence" and described female pubertal development as dominated by "periodicity."[50] Before and during the teenage years, he maintained, both boys and girls were governed by the physiological principle that growth only occurred when more of the body's cells were created than were destroyed. Accordingly, for both sexes, mental and physical work had to be apportioned and accompanied by sufficient sleep and proper nutrition, so that repair exceeded waste, with a margin left for "general and sexual development." In the teenage boy, however, sexual development was subsumed under the process of general development, with the male body and all of its parts growing "steadily, gradually, and equally, from birth to maturity." Reiterating the formula for prolonged, harmonious growth that health reformers thought enabled the development of the boy into the autonomous man, Clarke's analysis nonetheless stands out for its rendering entirely inconsequential the effects of male puberty on the boy's growth process. Indeed, repeatedly throughout *Sex in Education*, he insisted that the momentous changes he described as occurring in the development of the pubertal girl had but "feeble counterpart . . . in the male organization." Because Nature did not fundamentally transform the teenage boy's mind or body, nor require anything special from him at this time of his life, Clarke concluded, he best developed "health and strength, blood and nerve, intellect and life, by a regular, uninterrupted, and sustained course of [mental and physical] work."[51]

Female development, Clarke argued, followed the male norm of slow, persistent, harmonious growth up until puberty. Then, because of "the larger size, more complicated relations, and more important functions, of the female reproductive apparatus," sexual development came to dominate the course of general development, and the girl's "engine within an engine" began a (normatively) regularly recurring reliance on the body's limited vital forces. Whereas health reformers such as Blackwell and Fowler had used the concept of "periodicity" to refer to the proper timing of the emergence of a particular developmental capacity, Clarke's terminology was meant to signify the predictable and unmanageable alterations in energy and mood characterizing the female's body and mind with each menstrual cycle. Here, he selectively drew on the theory of "vital periodicity" established by English neurophysiologist Thomas Laycock, who, in a series of important papers published in *The Lancet* between 1842 and 1844, asserted that all physiological processes, including reproductive ones, were governed by regular temporal cycles. Clarke then applied the law of conservation of energy to the epoch at which the periodical reproductive function was first established. He asserted that since the body was unable to perform two functions well simultaneously nervous energy used for mental cultivation during this critical age diverted much-needed nervous energy away from the reproductive development process, thereby endangering the reproductive system, the nervous system, or both. For boys, pubertal development was a continuation along the same even, reliable, imperturbable, and linear path of growth they had always known. For girls, it was marked by patterned conflict: a continual contest among the "constant demands of force" for general growth, the "equally constant demands of force" for education, and the "periodical demands of force" for the establishment and maintenance of the reproductive function.[52]

That the female expended vital nerve force during the menses that would precipitate a conflict in the pubertal girl had been a mainstay in reproductive medicine since the mid-1840s when the ovarian theory of menstruation achieved prominence. This was the basis for Spencer's claim, and Clarke's endorsement, that the "cost of reproduction" was higher for the female. Before the eighteenth century, when what historian Thomas Laqueur terms a "one-sex model" of human sexuality held sway, menstruation was seen as a generally harmless purging of plethora, similar to other forms of bodily self-management in men and women. Ovulation, with its analog in male ejaculation, was understood as occurring only during intercourse and was linked directly to both the experience of pleasurable orgasm and the achievement of conception. With the rise of a "biology of [sexual] incommensurability," ovulation still retained its connections with intercourse, orgasm, and conception, although some doctors, trying to reconcile this with expectations for female passionlessness, supposed that women experienced these collective phenomena "without sensation." Then, in 1843, Theodor von Bischoff discovered that dogs ovulated "spontaneously," that is, independent of the act of coitus and during the regularly occurring estrus, or the rut. Despite the absence of any direct evidence about human ovulation, prominent naturalists and physicians quickly extended Bischoff's findings to human females, and the result was a new and highly influential theory of ovulation and menstruation.[53] Ovarian theory disassociated ovulation from intercourse and fecundation, correlating it instead with menstruation and the excitement of the nervous system. Keeping with the Victorian conception of the body as a mechanical system driven by nerve force, ovarian theorists supposed that it was the accumulated effect of irritation to the nerves of the ovary caused by the periodical enlargement and rupture of the ovarian follicles that produced the congestion of blood in the uterus and the ensuing menstruation. Rendered as an inner drama wholly unique to the female sex, the process was deemed to be analogous to the experience of heat in animals, with the sexual pleasure and autonomous passion once associated with ovulation giving way to the perception of a generalized "animal madness" afflicting the woman with each menstrual cycle.[54]

Moreover, if each menstruation was likely to induce in the woman a heightened nervousness and precipitate an inner conflict over limited vital forces, the undue demands of the first menstruation(s) tended to hyperbolize such effects in the teenaged girl. On this point, Clarke quoted the eminent British psychiatrist Henry Maudsley: "In the great mental revolution caused by the development of the [girl's] sexual system at puberty, we have the most striking example of the intimate and essential sympathy between the brain, as a mental organ, and other organs of the body."[55] Ovarian theory described female puberty as commencing with the *rapid* development of the ovaries and the *sudden* assumption of the pattern of the discharge of ova, accompanied by the initial menstrual hemorrhages. While the phenomenon of female puberty was premised on the qualities of haste and abruptness, however, the onset of these functions was also understood to predate their regular establishment by several years. And, as Clarke asserted, like all developmental mandates, "unless the reproductive mechanism is built and put in good working order at that time, it is never perfectly accomplished afterwards." The teenage girl was thus perceived as experiencing an extended period of pronounced physical and mental vulnerability, during which she was more prone to weakness and disease in the present, more at risk of compromising her future health, and more susceptible, in exaggerated form, to the feminine temperamental attributes of sensitivity, irrationality, melancholy, inchoate sexual feeling, restlessness, and self-absorption.[56]

Like all those who ascribed to ovarian theory, Clarke recognized the importance of the sudden and rapid female growth spurt in shaping the teenage girl's body, mind, and temperament. Unlike Spencer, though, who saw the girl's precocity as a cause of female developmental inferiority, he was more interested in the problem of how to secure "the best kind of growth during this period, and the best development at the end of it." Woman, for Clarke, was not to be looked on as "as a nondescript animal, with greater or less capacity for assimilation to man," but as a distinctive being who was capable of a complete and perfect development of her own kind, in her own way. Clarke's pubertal girl was thus neither a lesser boy nor precisely a little woman, in that she possessed a nature of her own that demanded both explanation by scientists and understanding, respect, and accommodation by the girl and those adults entrusted with her care.[57] Such ends were defied by identical coeducation, Clarke declared, which "put up the same goal, at the same distance" for both sexes, and required them "to run their race for it side by side on the same road, in daily competition with each other, and with equal expenditure of force at all times." In a section of his book entitled "Chiefly Clinical," he provided evidence of the invidious effects of such a practice: a series of examples of teenage girls and young women suffering from general paleness and weakness, menstrual disorders, dyspepsia, sterility, hysteria, and nervous exhaustion, all consequences of the neglect or ignorance of the demands of the female organization encouraged by the pressures of identical coeducation. To forestall such suffering, Clarke proposed a *"special and appropriate"* educational regimen for the girl that provided for a moderate academic program for three-quarters of each month, which incorporated plenty of opportunity for exercise and rest, and required absolute remission from mental work during the week of the catamenial period. Education, he carefully explained to his readers, did not refer to "intellectual or mental training alone," capacities at which girls had, admittedly, shown themselves to be quite proficient, but "the drawing out and development of every part of the system," which "necessarily include[d] the whole manner of life, physical and psychical, during the educational period." Thus did the periodicity of the girl's nature require a similarly periodical routine to ensure its full and proper emergence.[58]

In Clarke's formulation, it was only the western, white middle-class girl who warranted such assiduous protection during the teenage years. Only she was capable of experiencing the "best kind" of development and education that constituted an extended period of female youth with challenges, privileges, and responsibilities all its own, equal in value, if not in kind, to the stage of youth Spencer prescribed for the civilized boy. By virtue of both inferior heredity and less complex social environments, Clarke explained, the girls of the "Orient," of the European peasantry, and of the American working class possessed desirable physical strength but not the refined physical, mental, and emotional natures that portended both great risk and significant possibility for their civilized counterparts.[59] On the problem of the greater developmental peril faced by civilized girls, many fellow physicians joined Clarke in declaring that much was normatively, and uniquely, dangerous, difficult, volatile, vulnerable, and therefore limiting about their stage of pubertal development, even under the best of circumstances. He went further than many of his contemporaries in medicine, however, in also describing girls' youthful female nature as potent, beautiful, and important, the source of their individual freedom and power, as well as their indispensable contribution to society and the race.[60]

With this latter point, all those opposed to coeducation could also agree. The primary responsibility of white middle-class women was to propagate healthy boys and girls in inheritance of those evolutionarily superior mental and moral traits that would propel the continued advancement of western civilization. Civilized girls who did not attend to the requirements of their organization were not only committing the "slow suicide" of themselves as individuals, Clarke warned, but "race suicide" as well, threatening the "non-survival of the fittest" and the future reproduction of the race "from its inferior classes." As evidence for such a claim, late-nineteenth-century opponents of coeducation pointed to census reports that showed a decline in white women's fertility rates and a corresponding increase in the birth rates of immigrants and African Americans. By 1900, marriage rates among those who attended college were less than 50 percent and of those educated women who did marry, 20 percent did not have children.[61] Were civilized girls to honor their difference and allow for its complete unfolding under the guidance of an education in which the development of the body was paramount, Clarke insisted, they would be uniquely poised to experience and facilitate the creation of a more advanced womanliness and, with it, an ever-improving society. "[I]f it were possible to marry the Oriental care of woman's organization to the Western liberty and culture of her brain," he mused, "there would be a new birth and loftier type of womanly grace and force."[62]

That the epoch of sexual development for the civilized girl (in contrast to both her civilized brothers and her uncivilized female counterparts) represented a period of "new birth," honed out of physical and psychological stress and productive of individual and racial regeneration, was echoed in numerous medical texts of the period. Thus declared the authors of *The Practical Home Physician,* who assumed the racial and class markers of the civilized girl and boy they sought to compare:

> The girl therefore demands and must receive other treatment than her brother; for him, sexual development is a more gradual and less integral process; on which does not materially change the bent of his inclinations, the direction of his pursuits, nor his physical habits, which intrudes itself upon his attention by no imperious calls; change, indeed, of which he is often long unconscious. For her, it is an introduction into a new world . . . it is the transformation of the caterpillar into the butterfly. For her there is no possibility of ignoring the change at hand, body, mind and soul unite in calling her attention to the duty of the hour; the strange, uneasy, perhaps painful bodily sensations, the mental languor and indisposition for accustomed pursuits, the indefinable longings and emotions, indicate as plainly to others also the dawn of the new existence.

For all of the arduousness and wonder of the girl's transformation, though, these physicians, like most of the antebellum health reformers before them, also established clear parameters for the "new world" she was about to enter. "[T]he chief aim of the girl is to become a woman," they succinctly declared, asserting that the development of the girl, however complex and fully realized it might be, was complete with the achievement of sexual difference.[63]

No image more ubiquitously encapsulated the limits and possibilities of civilized girlhood entertained by Clarke and his fellow late-nineteenth-century physicians than that of the "budding girl." As expressed in the nomenclature of the "kindergarten," the more generic image of the child as seed, with the environment as garden and the adult as gardener, had been represented in multiple discourses of child health, education, and welfare since the early nineteenth century.[64] Taken together, these two images performed a certain amount of similar cultural work. They both conveyed a new interpretation of development that recognized childhood and youth as distinct stages of life with their own qualities and needs—just as the seed and the bud were not miniature trees, plants, or flowers, so neither was the child a miniature adult. They also both communicated the dictates of the organic paradigm of development. As with the seed and the bud, all children's growth was the product of innate biological imperatives, which emerged according to the precepts of a uniform pattern and moved the individual toward a preordained state of full maturity. The child's education was to follow, guide, and assist these imperatives but was not to thwart and did not determine them.

That the girl approaching puberty required an additional natural metaphor to give meaning to her maturation, however, suggested that she was somehow overdetermined by this developmental paradigm.[65] Indeed, late-nineteenth-century medical discourse held up the pubertal girl as the exemplar of organic development, as well as the primary signifier of the terms of its paradox. Thus, the teenage girl's body, with its dramatic and all-encompassing physical changes, constituted as visible and easily decipherable sign as any of nature's plan for the growing child. Concomitantly, it also served as the primary locus of anxiety over the need for the right kind of intervention required to ensure the proper procurement of that plan. In contrast, beyond establishing a persistent trajectory of growth for the boy, nature revealed few definite plans for his development. Moreover, per Herbert Spencer, while adult monitoring of the boy's education was certainly important in guiding him along in his course, so too always was his own self-directed initiative, discovery, and experimentation. In this vein, John Harvey Kellogg offered an effusive rendition of the budding girl metaphor. "Real girls are like the opening buds of beautiful flowers," he wrote in his popular medical guidebook, *Plain Facts for Old and Young* (1888). "The beauty and fragrance of the full-blossomed rose scarcely exceeds the delicate loveliness of the swelling bud which shows between the section of its bursting calyx the crimson petals tightly folded beneath. So the true girl possesses in her sphere as high a degree of attractive beauty as she can hope to attain in after years, though of a different character."[66] In addition to his rendering of the pubescent girl as an aesthetic and erotic object (a tendency in which he was not alone), Kellogg asked that the girl be recognized as differing qualitatively from her mother and that the "sphere" of her youth be protected and tended to with a spirit dedicated to the fullest emergence of her being. At the same time, he insisted that the "real" girl's development was a matter not of dynamic adaptation and progressive movement toward unknown potential, but of mere unfolding, revealing, and realizing the female destiny that was already, always, and exclusively the essence of her nature.

In many ways, then, the "budding girl" constructed by Clarke and others did not closely resemble the figure of the "youth" that had come before or the "adolescent" that would follow, figures that at least in part embraced the possibility for relaxed expectations, expanded opportunities, and assertions of independence during this developmental stage. Hereby formulated as a part of a tradition that accentuated the significance of biological sexual difference, the teenage girl would thus continue to require explanations all her own in the scientific literature on child development to follow, with such explanations often serving as justifications for constraints on her opportunities and behavior. In other respects, though, the "budding girl" anticipated, and even enabled, the modern construct of adolescence in quite significant ways. The conception of adolescence as a critical life stage; the grounding of adolescent development in bodily changes occurring at puberty; the links emphasized between physical, mental, and emotional development; and the connection between adolescent development, individual adult well being, and the advancement of society and the race were all central components of G. Stanley Hall's theory of adolescence. These components received their most elaborate articulation in the late nineteenth century in discussions about the pubertal girl. Indeed, one of Hall's tasks would be to reconcile what he would call a "feminized" stage of development with the claims of the *boy* to the dramatic transformation and consequential "new existence" that he depicted as so vital to individual development and the advancement of modern civilization.[67] Before he did so, however, several advocates of coeducation contended that girls and boys continued to be "very nearly alike" throughout their teenage years, with girls capable of far more physically and mentally than Spencer or Clarke allowed and, as importantly, with boys

resembling their sisters in some of their physiological functions, their temperamental characteristics, and their requirements for education.

"THE QUESTION OF REST": THE RESPONSE TO CLARKE

Immediately following the publication of *Sex in Education* numerous women's
rights advocates, educators, and medical experts issued pointed critiques challenging Clarke on practical, methodological, and theoretical grounds. Some
critics, most notably those public school leaders whose responsibility it was to
implement theories such as Clarke's, dismissed his prescribed educational regimen for girls as utterly unworkable in the face of the realities of school life. Many
also charged him with doing bad science, condemning him for relying on scanty,
exaggerated, and even faulty evidence in the presentation of his clinical case
studies, all in the interest of supporting his own prejudices about the limits of
woman's sphere.[68]

At the forefront of these latter critics was Mary Putnam Jacobi, who along
with Elizabeth Blackwell, was one of the foremost women pioneers in the
medical profession and was, during the late nineteenth century, the most prominent female physician in the United States. Born in London in 1842, the oldest
child of the prominent publisher George Palmer Putnam, Mary Putnam was
amply schooled in the prescriptions for Victorian femininity that Clarke sought
to defend, while also being exposed to alternatives beyond them. Her parents
ascribed to the traditional division of labor dictated by the ideology of separate
spheres, with her father providing economic support for the family and her
mother presiding over the domestic realm. Her devout Baptist paternal grandmother, who helped to raise the nine Putnam children, taught and exemplified
the virtues of female piety, nurturance, and self-sacrifice, the core values of
nineteenth-century womanhood. Putnam's interactions with her father's world
of publishing, however, also offered her a more cosmopolitan worldview, introducing her to many of the most renowned and popular literary figures of the day,
including several women writers, and prompting her engagement with such
vital social and political problems as slavery and women's rights. As an intellectually precocious teenager with literary ambitions, Putnam attended the public,
progressive Twelfth Street School for girls in New York City and set her sights on
work outside the home. Graduating in 1859, she embarked on a teaching career
and continued writing nonfiction essays and short stories, the first of which was
published in the *Atlantic Monthly* when she was 17 years old.[69]

During her teenage years, Putnam also struggled to make sense of and imbibe the beliefs and values of her grandmother's evangelical religion and expressed hope that she would one day be saved. She had her first conversion experience just before her sixteenth birthday, following her father's brief turn to evangelical revivalism as an emotional ballast against the financial difficulties he faced resulting from the Panic of 1857. Putnam's religious devotion was, however, short-lived, as well. Additional family tragedies, the untimely death of the minister who brought her and other family members into the church, and, most notably, the beginnings of her scientific training, taken up first with private tutors and then through classes at the New York College of Pharmacy, led her during the early 1860s to question and reject many of the religious doctrines she had only begun to embrace. She challenged Calvinist beliefs on many grounds. She was particularly critical of the church's teachings about the inferiority of women and the institution's greater concern with salvation in the next world instead of the improvement of this one. It was, she ventured, the methods and insights of positive science that would reveal more accurate truths about woman's nature and proffer viable solutions to social problems, including woman's subordination to man.[70]

Putnam continued to formulate connections between science and women's rights during the years of her medical training, first at the Female Medical College of Pennsylvania and then at the École de Médecine in Paris. In both the United States and France, women seeking medical education faced formidable obstacles from within the medical establishment and suspicion and hostility from the culture at large. Although Putnam benefited from the efforts of liberal educational reformers in both countries endeavoring to counter this trend, her own considerable struggles to obtain her degrees spurred her to a lifelong commitment to winning women equal access to medical education and improving their status in the medical profession. In addition, the research Putnam conducted for her theses at both institutions, on the function of the human spleen and the relationship between cellular degeneration and cellular nutrition, laid some of the groundwork for her research into the physiological problems related to female health and development that would occupy her later in her career. Putnam's efforts as a medical student to forge links between laboratory research and clinical practice placed her at the forefront of scientific medicine. As a student in France, Putnam's exposure to and enthusiasm for philosophical positivism, which eschewed metaphysics to emphasize knowledge derived from empirically observable phenomena, as well as the principles of socialism, republicanism, and women's rights, led her to believe that science could serve as a key tool for facilitating social change.[71]

The integration of scientism and political activism, especially on behalf of women's rights, became the touchstone of Putnam's subsequent life's work. When she returned to New York in 1871, she entered into private practice, began teaching at the Woman's Medical College of the New York Infirmary for Women and Children, continued to conduct laboratory studies and experimental research, and pursued opportunities to publish her findings as well as her views on many scientific and social issues. In 1873, she married the German-Jewish physician and socialist Abraham Jacobi, who directed his own commitments to science and social reform to forging the specialty of pediatrics and securing a public commitment to improving the health and welfare of children. In the 1890s, Mary Putnam Jacobi joined the ranks of the Gilded Age's women's reform network, bringing her scientific viewpoint and expertise to bear as a prominent voice in the campaign for suffrage in New York State and as a founding member of and leader in the Consumer's League of New York City.[72] Jacobi's devotion to science earned the respect of both men in the medical field, who were increasingly wedded to a model of scientific medicine, and an educated public willing to accept scientific authority. However, Jacobi was frequently at odds with other female physicians, most notably her friend and fellow medical pioneer Elizabeth Blackwell. Both Blackwell and Jacobi strove to improve medical education for women and to enhance their opportunities and respect within the profession. Both also applied their scientific knowledge to challenge perceptions of women's limited capabilities and to argue for their capacity to make vital contributions to social change. For Jacobi, though, Blackwell's Christian perfectionist view of health and disease and her emphasis on female doctors' distinctive and superior powers of morality and compassion threatened to compromise the achievement of women's equality in the medical profession and the society at large. Thus, when the occasion arose shortly after her return from Paris to counter medical and evolutionary orthodoxy about female development, Jacobi reiterated many of Blackwell's earlier conclusions; only hers were arrived at and buttressed by the language and methods of what she deemed to be a far more rigorous, and therefore far more politically efficacious, scientific paradigm.[73]

Jacobi directly responded to Clarke's treatise as a contributor to the 1874 collection, *The Education of American Girls*, in which she exposed the conservative social agenda driving the anti-coeducation argument. "The singular avidity with which the press and the public have seized upon the theme discussed in Dr.

Clarke's book . . . is a proof that this appeals to many interests besides those of scientific truth," she wrote. "The public cares little about science, except in so far as its conclusions can be made to intervene in behalf of some moral, religious, or social controversy." This was followed by her own effort to establish the "scientific truth" about woman's nature: her prizewinning work entitled *The Question of Rest for Women during Menstruation*, which explained "the real succession of phenomena in the menstrual process" and its relation to "the other processes of the [bodily] economy."[74] Given that most of Clarke's critics were not physicians or research scientists, it is not surprising that some sought to shift the ground on which knowledge about woman was to be derived. "The question comes down then to the one point," pondered coeducation advocate Eliza Duffey, "which know most about the capabilities and disabilities of the female sex, the doctors or the women themselves?"[75] Even so, for Duffey and others, the protest against Clarke was most effectively waged by countering one set of scientific "truths" about woman's nature and development with another. It was here that Jacobi's response to Clarke was especially helpful because Jacobi married her "common sense" about women from her own experience with the authority of science, which the late-nineteenth-century "public" did, indeed, care about, and often for the reasons Jacobi suggested.[76]

Like all of Clarke's critics, Jacobi endeavored to remove the stigma attached to female puberty and to reformulate some of the qualities that were thought to constitute teenage girlhood, without, necessarily, rejecting all notions of essential sexual difference or the importance of white middle-class girls and women to the advancement of western civilization. At the same time, she and others also reinforced some of the attributes of the teenage years ventured in Clarke's account of the epoch of sexual development in the girl, suggesting that such qualities of body and mind had to be attended to when explaining and guiding the boy's development as well.

Jacobi led the way in undermining the argument that rigorous mental work axiomatically threatened healthy female development by providing alternative explanations of both female growth and ovulation and menstruation. In formulating these explanations, she relied on and offered different interpretations of the same physiological paradigms accepted by ovarian theorists. She also introduced new empirical evidence about women's experiences with menstruation. This evidence was based on statistical analysis of 268 women's responses to questions about such factors as their experiences with menstrual pain, the amount of exercise they engaged in, and their family and educational histories, as well as on measurements of the changes that took place in a group of six women throughout their menstrual cycles in the excretion of urea, pulse rates and temperature, and degrees of arterial tension. From her examination of this collection of data, Jacobi concluded that girls and women did not normatively require mental and bodily rest during menstruation and that rather than marking the nadir of vital energy in the female, "the menstrual period represents the climax in the development of a surplus of nutritive force and material."[77]

Jacobi accounted for this by offering up a "theory of supplemental nutrition," which challenged Spencer's and Clarke's accounts of the limitations that the growth spurt placed on female development. Accepting the notion of the body as a closed system with limited energies, she posited that the force required for the establishment of the reproductive system came not from "a spontaneous diminution in the functions of certain nerve centers" but, rather, from a reduction of force in the "motor apparatus." This system, she granted, obviously grew greater in mass in the male but was not, as was so often assumed, consequently inferior in its "structural development" or "functional activity" in the female. "Any general inferiority in the entire mass of the central nervous organs in women, as compared with those of men, . . . principally relates to the organs of motility," she asserted, "the anatomical conditions of sensibility and of thought remaining the same." That established, she proposed that, because the girl's muscles and bones finished growing sooner than the boy's, the girl had a surplus of organic nourishment sent from all parts of the body to the reproductive organs at the onset of puberty. Contrary to Spencer's "useless, and even untrue" generalization about women's aborted development, then, rapid and arrested growth of "general development" was not the *price* the girl had to pay for her reproductive development. It was, instead, the natural added *benefit* that allowed her reproductive functions to become established and all without undue taxation on either her physical or mental powers. Furthermore, the girl's experience with the first menstruations set the pattern for the duration of the woman's reproductive life, with periodical reproductive demands continually compensated for by "a gradual deviation from muscular nutrition resulting in an accumulation in the blood vessels of nutritive fluid refused by the muscles." It was, Jacobi declared, the girl's, and later the woman's, capacity to produce this reserve of nutritional material that constituted "the essential peculiarity of the female sex."[78] With a view of nature as kind, wise, and fundamentally just, as opposed to antagonistic and strenuous, she asserted that structured into the rigors of female development was a unique provision that endowed girls and women with the capacity for health and vitality.[79] While Clarke had also claimed to put his faith in nature's intentions, Jacobi accused, he drastically misconstrued those intentions as limiting, rather than enabling, the scope of the developing girl's potential for physical and mental activity as she prepared to enjoy the opportunities and assume the responsibilities of adult life.

If Jacobi's theory of supplemental nutrition attested to the greater degree of biological fitness of the female animal, her critique of ovarian theory sought to effect a theoretical separation of the animal from the human and the body from the mind. Although she did not dispute the trends in modern science that alleged continuities and relationships among these realms of life, she did protest the excessive attention on woman as the exemplar of somatic determinism.[80] To that end, her account of female reproductive physiology severed the links between both ovulation and menstruation and sex and reproduction. She severely objected to unsubstantiated inferences about menstruation in the human female from the experience of heat in the lower animals. She also argued that while menstruation and ovulation had been shown to be "parallel facts," their "causal dependence" was by no means proved in the scientific record. Her own review of that record identified very few autopsy cases that found ruptured ovarian follicles in women who died during menstruation and noted the many autopsy findings of ruptured follicles in the absence of menstruation and of menstruation with no ruptured follicles. Also supporting her contention were the "tolerably numerous cases" of conception occurring, not just before or after menstruation as ovarian theorists surmised but midway through the menstrual cycle. According to Jacobi, ovulation was, indeed, a "spontaneous" phenomenon, as ovarian theorists claimed but rather than occurring periodically and precipitating menstruation, the consecutive maturation and rupture of the vesicles containing the ova "strictly resemble[d] the successive growth of buds on a bough," a purely random, and rather unremarkable, occurrence, tied only statistically to the menstrual flow.[81]

After she dismissed the links between ovulation and menstruation, Jacobi proceeded to contest ovarian theory's core assumption that the reproductive function in the female involved "a peculiar expenditure of nerve force, which was so much of dead loss to the individual life of the woman." Such a claim, she argued, was based on a mistaken conflation of two, wholly distinct aspects of the female reproductive system, the sexual and the reproductive. Reproduction, she explained, was essentially a nutritive process that bore no fundamental connection to the sexual instinct. Unlike the male, for whom the expulsion of the reproductive cell coincided with sexual pleasure and the experience of intercourse, in the female, the ova dehisced independent of coitus, with the sexual instinct remaining unawakened. For Jacobi, this was the significance of the spontaneous occurrence of ovulation propounded by ovarian theory. It was also another factor that determined the essential difference between man and woman. "The theory of *spontaneous* ovulation," she asserted, "means precisely that in the female the essential part of reproduction can be effected without any sexual act. The superior contribution to the nutritive element of reproduction made by the female is balanced by an inferior dependence upon the animal or sexual element; in other words, she is sexually inferior."[82] As nutritive phenomena, Jacobi concluded, ovulation and menstruation had no more effect on mental activity or the nervous system than any of the other bodily nutritive functions, such as respiration, circulation, or digestion. Rather than constituting "an extraordinary exception among physiological phenomena" that dramatically interrupted ordinary physiological life, the periodicity of the menstrual flow, like all other rhythmic bodily functions, marked "the simple climax of a series of consecutive processes perfectly continuous with one another."[83] Jacobi allowed that insofar as the reproductive system required additional vital energy for its proper functioning, the other difference in the female body worth remarking—its capacity to replenish quietly and constantly a supplemental nutritional reserve—guaranteed that the need would be met without affecting the woman's intellectual capacity or her general nervous stability.

The first menstruations were conceived by Jacobi as instigating no great crisis in the life of the girl just as each adult menstruation did not upset the equilibrium of the nervous system. To make such a case, she illustrated the physiological continuity between childhood and maturity in the female. Again challenging a basic presumption of ovarian theory, she explained that Graafian vesicles and ova indistinguishable from those of the adult female composed the ovaries of girl children after age 2. During childhood, the vesicles gradually and successively developed, eventually atrophying without rupture. In the three or four years preceding the onset of menstruation, the ovaries increased in size gradually, a process that never exceeded the girl's capacity to meet the added nutritive requirements. The establishment of puberty was not, then, characterized by sudden or rapid growth of the reproductive organs or by the assumption of entirely new functions but merely by the greater vascularization of the Graafian vesicle, accompanied by "the secretion of fluid into its cavity, leading to the rupture of its walls and the escape of the ovum."[84] Although the commencement of the menstrual flow also marked the onset of puberty, it was the changes in the ovaries, conceived as such, that defined female pubertal physiology for Jacobi and that indeed provided the physiological basis for a period of adolescence. She explained: "The period of [the uterus's] prominent activity does not come until after the action of the ovaries has been completely established; that is, the period of maternity is, or should be, consecutive to the period of adolescence, and the work of gestation only entered upon when the work of ovulation has long been thoroughly accomplished." As Clarke had asserted, Jacobi also maintained that the pubertal girl was not yet in full possession of her reproductive powers. For her, though, the first menstruations, rather than occasioning immediate danger, constituted a more or less benign rehearsal, in rudimentary form, of a crucial nutritive function that would be carried out at a future epoch in the life of the woman. That nature determined female development so was more of a "caprice" than a threat and warranted far less anxiety than was currently being propelled in the direction of the growing girl. "[T]o impose on the girl the precautions necessary to the mother," Jacobi admonished, "is one way to enfeeble and prematurely age her."[85]

Many of Clarke's other critics joined Jacobi in depicting girls as "good animals," albeit decidedly immature ones, insisting that they possessed the natural strength, essential vitality, and native intelligence during childhood and youth that provided an optimum foundation for future physical and mental health. Yes, they conceded, nature made some undue demands on the girl at puberty, but it also overcompensated for the added difficulties by endowing females with a "surplus vitality" present from birth. This extra energy bestowed on girl infants a greater capacity to thrive, as well as conferred on pubertal girls the rapid growth spurt, which was designed as such to meet the particular requirements of the female reproductive system.[86] Still, this optimistic characterization of the girl's nature did not explain why so many civilized girls and women suffered from ill health. Indeed, most women Jacobi questioned reported suffering from some sort of "pain, discomfort, or weakness" during their menstrual periods.[87] To explain this, many of Clarke's critics returned to the group of dangers subsumed under "precocity," which they defined, not as Spencer did, as an essentially limiting attribute of the girl's development, but as the antebellum health reformers had, as artificial influences acting on the girl that belied and frustrated nature's true intentions for her development. Whereas Clarke was concerned that the pursuit of education during the teenage years dangerously postponed the girl's attention to her reproductive function until it was "too late," then, his opponents were preoccupied with the developmental problems that ensued when the ways of childhood were "too early" relinquished for adult manners and mores. Thus, girls were vulnerable to weakness and disease because they succumbed to the trappings of adult femininity that abided in civilized society: a sophisticated diet of rich or spicy foods, fashionable dress, unwholesome novel reading, frequent socializing, no outdoor play and exercise, and, worst of all, sexual passion. It was the emotional excitement induced by such rushed female development that produced the ominous consequences that Clarke associated primarily with rigorous education. Sickly constitutions and menstrual problems resulted, to be sure, as did vapid and sensuous personalities, nervous temperaments, and dull intellects. " 'Critical' fudge!" exclaimed Duffey in decrying Clarke's conception of the girl's teenage years. "Let nature have fair play, and she is perfectly capable of managing the child without repressing physical manifestations of activity or checking mental ones . . . The headache, the dyspepsia, the nervous paroxysms, are so many protests of nature against the compressment and confinement—in direct words, against the unnaturalness of her life."[88]

Management of the female child was not, of course, to be left entirely to the play of nature, however "fair" this may have been. Indeed, the point of all of the responses to *Sex in Education* was to argue that civilized girls were capable and in need of a rigorous intellectual and moral education comparable to what their brothers received. This would both counter the pernicious environmental pressures that preyed on the female child in modern society and enable her "animal nature" to be trained toward what Anna Brackett called her "second and better nature." "The very essence of civilization, of morality, and of religion, consists in the overruling and directing of the merely natural," Brackett insisted. "By nature, man is not man at all. Only in so far as by force of spirit he overcomes, rules, and directs the nature in him can he lay claim to manhood."[89] Whatever she intended by it, Brackett's use of the generic "man" here would have recalled for her readers Spencer's and Clarke's view that it was the civilized male who was most capable of exercising the rationality, self-control, and self-determination that were the driving forces behind individual, racial, and social evolution.

Other advocates of coeducation more directly countered such notions of male advantage. Like Jacobi, they deemed that woman's "animal nature" was "inferior" to that of man, with the gap between her "barbarous" childhood and "civilized" adulthood therefore narrower and more quickly closed. Endowed with the strength, health, and intellectual capacity to support its wide-reaching influence, woman's "second nature" was superiorly oriented toward the qualities of altruism, compassion, and service. Moreover, it was these attributes that would provide an important counterpoint to men's destructive selfishness, competitiveness, and greed in the modern world and play a leading role in revitalizing the race at this stage of its evolution. Jacobi, too, conceded that moral differences between men and women were more marked than intellectual ones, although she hedged on whether these were either entirely innate or exclusive to woman's nature. Even so, she advised, the failure to recognize and to cultivate "feminine" morality in the education of the girl "would be an injury to society, that requires, not uniformity, but increasing complexity, by means of increasing variety of character among its members."[90]

In their outcry against precocity, then, Clarke's opponents expressed their distaste for what they perceived to be the dominant feminine ideal in modern society—weak, superficial, vain, dependent—as well as registered their hopes for an advancing civilization in which women of the white middle class would have a larger role to play.[91] The problem was not that the girl received too much education, Duffey exhorted, but that she received too little, and the wrong kind, in that it encouraged her toward the "follies and dissipations of fashionable society" from earliest childhood, turning her into "but a pitiful caricature of what she should be." Clarke's program for special education would only make things worse, she averred, because by placing exclusive emphasis on the development of the reproductive system, "woman will be woman no longer, but an exaggerated female, weak and wanting in all other functions and faculties, and abnormally developed in the peculiarly feminine parts of her organization."[92] Several of Clarke's critics joined Duffey in deriding these "sexual monsters" for their inability to perform their roles as wives and mothers, deeming them physically, mentally, and morally unfit to transmit the attainments and values of civilization to the next generation.[93] Others envisioned the identity of "woman" as encompassing the functions of the female, certainly, and the roles of wife and mother, most likely, but also the possibilities of the individual, the "citizen," and the "human being."[94] It was, Julia Ward Howe insisted, no wonder that girls professed only to be concerned with fashion or that they became ill after leaving school. Unlike boys, who had "the healthful hope held out to them of being able to pursue their own objects," girls and young women faced only "the dispiriting prospect of a secondary and derivative existence."[95] Jacobi weighed in here, too, with an equal measure of critique and sympathy: "A healthy objectivity is one of the greatest desiderata for modern women. To knock the nonsense out of them, to direct attention from self, to substitute a cosmic horizon for that of their own feelings, who does not know the importance of this for thousands of hysterical women? and equally the impossibility of attaining it?" For the more optimistic Mrs. Horace Mann, the degree of civilization was directly correlated to the number of unmarried women in a society. It was, she contended, those women who cultivated independent lives—developed and educated both into and beyond their "femaleness"—who would "bear the noble fruits of culture, benevolence, and devotion to human improvement" that were the well-springs for ongoing racial and social progress.[96]

Clarke's opponents hereby made the case for increased educational opportunity for girls during their teenage years and for a wider range of possibilities for female identity in adulthood. Their emphasis on "keeping girls girls," however, also established its own limits on the maturing girl's capacities for autonomous action and self-determination.[97] After all, Duffey's question about who was to be the authority on female potential supposed that adult women had as much to say about the matter as male doctors but did not consider what the developing girl might know, imagine, or seek to express about her current or future self. Insofar as white middle-class girls *were* engaging in the sorts of "precocious" behaviors that Clarke's critics found so objectionable, they were depicted as surrendering to dangerous external influences that were beyond their control, at worst, or, at best, as misguidedly pursuing their natural inclination for activity and independence. American girls "will rule themselves," Brackett reluctantly conceded and proudly proclaimed, "and it therefore behooves us to see that they are so educated that they shall do so wisely."[98] Duffey, too, recognized and applauded the girl's penchant and capacity for autonomy, certainly a departure from Spencer's and Clarke's characterization of her dependence and passivity, while also cautioning that this tendency had to be channeled through the proper education of the body, mind, and will toward her "better nature," the substance and limits of which were determined by the values and mores of adults of the white middle class. Given the prevalence of all of the "crying evils of our age," she chastised, Clarke was remiss for not realizing that "there is an energy in girlhood that will not be repressed, that will *always* indulge in some sort of activity." "A girl at this age never does anything by halves," she continued. "It is the time for a life-choice with her, and what is chosen is chosen. She will throw all her energies into flirtation and turn out a coquette of the first water. Or, if she have not strength of character to do any of these wholesale things, she will fritter away her time in small nothings, and so fix herself in the habit of inactivity that when her years of enforced rest are passed she cannot be aroused to effort of any sort." Alternatively, Duffey anticipated, the girl who chose to expend her energy in exercising her superior capacity for love and compassion marshaled a power that was "sufficient almost for the regeneration of the world."[99] Whatever route the girl took (and it was patently clear where Duffey's preference lay), the more subtle and novel contribution here was that for girls, too, the transition from childhood to adulthood was to be marked by a tension between dependence and independence. The gradual development and sustained education that were to take place during this stage of life were at once to foster the autonomy and self-directed energy in the female child that coeducation supporters viewed as her birthright and as vital to the advancement of civilization *and* to provide for adult protection and supervision so that the girl would be sure to grow up to become "what she should be."

As their identification of this tension suggests, coeducation advocates' characterization of the teenage girl as strong, active, healthy, intellectually capable, and in possession of a heightened moral capacity did not mean that she did not also face a unique set of physical, mental, or emotional challenges. For Clarke's critics, though, such difficulties were as much attributable to her age as to her sex. With this claim, they made the case that if, on balance, female puberty required less attention than it received, the male epoch of development warranted far more. It was, again, Jacobi who provided the initial medical justification for this assertion. Returning to the ancient one-sex model of the body, she boldly asserted that the organs of reproduction were anatomically and physiologically equivalent in the two sexes. "An excess of bulk in one direction," she slyly noted, "is compensated by an inferiority in another, so that the sum total is the same." Moreover, "the period of [the development of the reproductive organs], the influence of such development on the entire nutrition of the body, the irregularities of nutritive or of cerebro-spinal action that may be caused by irregularities in such development" were also entirely analogous. Whereas girls required a "trifling" amount of nutrition for the "extra development" of the egg and more nutrition for the production of the menstrual blood (which did not immediately cost them anything), boys spent more nutrition on the new and more abundant development of sperm and less on the production of semen. Such comparisons were useful to Jacobi both to counter declarations about the girl's physical and mental weakness and to draw attention to the age-specific potential dangers faced by girls *and* boys at puberty. Thus, despite her highly sanguine depiction of female development, she nonetheless also warned that during adolescence, both boys and girls were as likely to suffer physically and psychically from the injurious effects of competitive schooling, lack of sufficient exercise, and ignorance of basic physical care. They were also both likely to experience a degree of "morbid emotional excitement" that was natural to their age, simply because all young people had yet to cultivate the kind of sophisticated mental power and moral judgment to always temper their emotions properly.[100]

Jacobi's recommendation for handling such psychological challenges and for cultivating maximum physical health for both girls and boys was to secure the predominance of the intellect and the will over the emotions and the instincts. This was to be achieved by providing all adolescents with "a larger, wider, slower, and more complete intellectual education" than was currently the norm for either sex, by accompanying it with a systematic program of physical exercise, and by making individual adjustments to meet particular mental, moral, and physical needs.[101] Ideally, though, girls' education during the early teenage years was to be separate from boys', even as it was to be intellectually identical. This was counter to the view of some of Clarke's other critics, who saw integrated education as morally uplifting for both sexes and single-sex schools as pernicious "breeding grounds" for the "secret vice."[102] In Jacobi's determination, the most compelling reason for separate education was to prolong the "unconsciousness of sex" during those years when the physical, mental, and moral organizations of boys and girls were still imperfectly developed and unable to manage effectively the emotional excitement aroused by the close contact between them that co-education required. To claim that woman was "sexually inferior" to man was thus not necessarily to discount the importance of safeguards that would protect female purity, especially during teenage girlhood. Still, Jacobi was also careful to add that such precautionary measures ought to abide only during the early teenage years and in no way justified denying girls age 18 or over access to identical and integrated higher educational opportunities.[103]

Fellow physician Elizabeth Blackwell drew similar conclusions to Jacobi's about the development of girls and boys during the teenage years, although these were derived from her markedly different Christian-physiological perspective. Blackwell's *The Laws of Life* pioneered in laying out general principles of child development for mothers and teachers in the antebellum period, as well as in claiming for the girl the same moral, intellectual, and physical benefits of gradual, steady development deemed to be normative for the boy. In her 1884 work, *The Human Element in Sex*, she was as interested in revealing the ways in which the boy's pubertal development seemed to mimic that of the much more frequently commented on pubertal girl. Like Jacobi, she declared that the organs that produced the two essential secretions for reproduction—the ova and the sperm—were "strictly analogous in the two sexes." Departing from earlier medical and religious admonitions against all forms of sexual incontinence in young men, Blackwell compared the nocturnal emissions that occurred spontaneously, and "even with a certain degree of periodicity," in the teenage boy with the "natural healthy" functions of ovulation and menstruation in the girl. For Blackwell, in both cases, the body beneficently discharged the unused sexual secretions with the divine purpose of realizing the "perfection of human growth": "Thus in the female the constant formation of ova is subordinated to the needs of individual freedom, and to the power of mental self-government, by the function of menstruation . . . In the male the slower secretion of sperm is adapted to the same individual freedom and power of self-control by the natural function of sperm-emission." The benevolent intentions of the "Creative Power" notwithstanding, however, the "sudden appearance" of the functions of menstruation and sperm emission was sure to occasion "fright" in the teenage girl and boy alike, and these functions certainly had the tendency to become diseased when "unduly stimulated" by the brain and nervous system. Thus were parents to be ever vigilant during this critical stage of development: "It is of vital importance to the parent to know that such action [occurring at puberty] is as natural and healthy in the growing lad as in the growing girl, but that in both it is a time requiring guidance, both moral and physical."[104] As in her earlier work, Blackwell's late-nineteenth-century text recognized physical differences between the sexes, celebrated women's distinctive moral and spiritual power, and advocated for the special contributions women could make to private and public life. Here, though, it was the girl who stood out as the normative model for pubertal development; knowing what happened to her body and mind provided a way to understand, to talk about, and to respond to the boy's epoch of development as well.

Clarke's other critics also took this dual tack. On the one hand, they were determined to establish that the epoch of sexual development was not so debilitating that it impeded the girl from taking advantage of educational opportunities or exercising her capacity for "individual freedom." On the other hand, they were willing to allow that important biological and psychological changes *did* occur during this transitional developmental stage but only insofar as these were attributable to boys as well as girls. Duffey asserted that the "outer characteristics" of impending maturity were "as striking" in the boy as in the girl and identified these as the change of voice, the appearance of facial hair, and the seminal secretion, "the normal and proper production and retention" of which determined the young man's future health just as surely as the regular establishment of the menstrual cycle determined the girl's future health. These corresponding changes in the boy's body called for a similar response to the changes visited on the girl, which, Duffey regretted, was not currently forthcoming. "Just as great a care, just as watchful precautions, are required in the one sex as in the other," she avowed, "though to the shame of humanity be it spoken, the needs of the young man are overlooked, while his sister attracts all the attention."[105] Boys suffered equally, too, from precocious behaviors and especially from an educational system that was in dire need of reform for the benefit of all children. The debate over the education of the girl hereby gave voice to calls for schools that would attend to the physical, mental, moral, and social development of children on the basis of their discrete needs—children of "quick," "average," and "dull" intellects and of weak and strong constitutions receiving opportunities that matched their distinctive deficits and capabilities. "When development begins," explained Caroline Dall, " special treatment is required; not according to the sex so much as according to the individual . . . That school or family is an absolute failure which does not allow a margin large enough and loose enough for all possible contingencies, as regards boys or girls."[106] Duffey contended that more attention to providing the boy with a balanced education that attended to more than the cultivation of his intellect or to his prospects for individual success would restore to him his "obligations to posterity" and his "responsibilities of the coming race." Countering Clarke's portrayal of the cold, "tottling brother" with the example of her own son's clasping and caressing of his sister's doll, she posited the existence of a "divine instinct of fatherhood," which, she lamented, the current system of education would continue to destroy until there were but "few really worthy, good and loving fathers in the land."[107]

At the same time that some of Clarke's critics drew attention to those biological and psychological imperatives of age that cut across gender lines, others took note of the changes that were occurring in working-class girls' lives in this same period and ventured to universalize those same imperatives across lines of class and ethnicity as well. With the growth of cities and the expansion of industrial capitalism, working-class girls, many of them immigrants, were taking jobs in factories, department stores, and offices that afforded them new opportunities for social independence and sexual autonomy and subjected them to various forms of economic and sexual exploitation. This same dynamic abided in the free time of these girls as well, as they sought out the many new forms of commercialized leisure that strove to appeal to a youthful, heterosocial audience.[108] Many of the same advocates of coeducation who condemned precocity as the primary culprit in the poor health and nervousness of white middle-class girls also charged the life of "common toiling" as destroying the essential vitality of working-class and immigrant girls and of robbing them of a gradual development in which they, too, were to be sheltered from the temptations and free from the responsibilities of the adult world. Far more taxing on the girl's physical energies than even the most rigorous schoolwork, rebuked Marie Elmore, were the long hours and poor conditions faced by female factory operatives, clerks, and seamstresses. And yet, she rhetorically demanded, "Has Dr. Clarke written a book on 'Sex in Manufacturing Establishments'?"[109]

Following the initial brouhaha over coeducation, working-class and immigrant girls came under even closer scrutiny by moral reformers, who feared the challenges the changes in their lives posed to the Victorian gender and sexual order. More wont to see such girls as victims of class, gender, and age exploitation, rather than as willfully rebellious, reformers involved in the late-nineteenth-century campaigns to raise the age of consent, eliminate prostitution, and regulate leisure affirmed that the teenage years were vulnerable ones for all girls, whose imperiled bodies required the special protection of adults. Indeed, one of the arguments of those advocating to raise the age of consent (a campaign in which the voices of physicians also played an important role) was that puberty *not* be relegated, as some would have had it, as the "natural position" at which the girl could legally agree "to this form of degradation."[110] Working-class girls just growing up, declared Emily Blackwell (Elizabeth's sister) in her testimony on the issue, were the most helpless class in society and deserved the same opportunity as their more privileged sisters to develop physically, mentally, and morally before being held responsible for making decisions about sexual activity.[111] Not everyone agreed. As one opponent of age-of-consent legislation opined in an article entitled "The Age of Consent from a Physio-Psychological Standpoint," for girls of the working class and girls of color, puberty conferred maturity. To fail to recognize this was potentially to subject an innocent boy to "the hands of a lecherous, sensual Negro woman, who for the sake of blackmail or revenge would not hesitate to bring criminal action even though she had been a prostitute since her eleventh year!"[112] For this writer, as well as for the opponents and, in some ways the proponents of coeducation, insofar as an epoch of development bestowed on the female child protections for her growing body, allowances for her immature mind, and opportunities to claim some capacity for independence in the pursuit of an intellectual education, these privileges were to be enjoyed by civilized girls alone.

The rhetorical onslaught against coeducation had no effect on the predominant trend in gender practices in education that had been unfolding since the early nineteenth century. From 1870 to 1900, public secondary education remained "overwhelmingly coeducational," with teenage girls accounting for 57 percent of the students and 65 percent of the graduates of high schools in 1890. In addition, young women continued to take advantage of expanding opportunities for higher education. By 1900, 36 percent of all students enrolled in colleges and universities were young women.[113] Even so, as historian Margaret A. Lowe demonstrates, concerns about the developing female body continued to loom large throughout the period, particularly on college campuses. In response to the fears incited by Clarke and in keeping with the recommendations offered by his critics, college administrators at single-sex and coeducational schools established health education programs and instituted practices that encouraged and, in many cases, required girls to take care of themselves by eating well, exercising, getting sufficient rest, and observing proper hygiene. They measured how their "student bodies" were doing by conducting mandatory physical and medical exams on girls at specific points during their college careers. Moreover, they prominently reported on their success at maintaining and even improving female students' health, allegedly at no compromise to their femininity, in annual reports, promotional materials, and public speeches. In personal letters and student publications, girls also revealed a heightened attention to their physical health. While some registered anxiety that they may indeed be putting their health at risk by studying too hard, far more expressed confidence, pride, and even pleasure in their bodily well being and played active roles in defining the meaning of good health and sustaining it in conjunction with their educational endeavors and goals. Meanwhile, following Jacobi's lead, numerous scientific studies were conducted by individual physicians and organizations, such as the newly founded Association of Collegiate Alumnae, that corroborated the re-assurance about education and female health being offered by college officials and their students.[114]

Along with these influences on the girl's educational career, the coeducation debate also left an important legacy to the intellectual tradition of child develop-ment studies. Taken together, the accounts of the detractors and defenders of coeducation constituted the most comprehensive and multifaceted examinations and explanations of female adolescent development before the publication of Hall's text. The points of contention and convergence over the nature of female puberty; the temperament of the teenage girl; the relationship of sexual differ-ence to physical, mental, moral, and emotional development; and the relative influences of nature and nurture in the manifestation of femininity across the life cycle all worked to construct the adolescent girl as a meaningful object worthy of future scientific investigation. In addition, from the collection of portrayals of female development that composed the coeducation debate, the girl emerged as a prominent model of the modern adolescent, helping to establish those links between puberty and adolescence, biology and psychology, and indi-vidual growth and social progress that the developmental tradition's concerns with the growing teenage child would both reinforce and question throughout the twentieth century.

CHAPTER THREE

From "Budding Girl" to "Flapper Americana Novissima"

G. Stanley Hall's Psychology of Female Adolescence

In a 1909 article for *Appleton's Magazine* entitled "The Budding Girl," pioneering developmental psychologist G. Stanley Hall rendered his own interpretation of this familiar metaphor. In doing so, he claimed the adolescent girl as a subject worthy of modern science's attention:

> Girls with hair demurely braided down their backs and skirts just beginning to lengthen toward their ankles *are* buds that should not blossom for some time, but should be kept as long as possible in that green stage; or, to change from a floral to a faunal trope, they are only squabs and not yet doves, maturing pupae and not yet butterflies, and this calf or filly stage should be prolonged by every artifice. She is no longer a little girl, but by no means yet a young woman, nor is she a cross between or a mixture of the two, but a something quite unique and apart, because this is a stage not at all explained by anything in the pedigree of the race, which the history of the individual otherwise tends to repeat. Even among those primitive races where girls are not wed at pubescence, there never was a backfisch stage. That is one reason why she is now the most intricate and baffling problem perhaps that science has ever yet attacked.[1]

Hall began an "attack" on the problem of explaining the girl's development five years earlier with the publication of his book *Adolescence*. As several scholars have noted, it was this work that discursively marked the inception of the "era of adolescence." During the first decades of the twentieth century, the combined effects of public institution building and private market forces, along with the influence of the young themselves, more clearly than ever before demarcated adolescence as a unique and important period of life, perceived and experienced as critically related to the fostering of individual potential and happiness, as well as the preservation of social stability and the advancement of modern civilization.[2] A synthesis of the entire range of the components of human development —physical, mental, emotional, moral, sexual, and social—*Adolescence* represented the first systematic effort in the nascent discipline of psychology to depict adolescence as a distinct life stage, characterized by a particular set of biological imperatives, a corresponding psychological organization, and a consequent constellation of predictable social behaviors. While many of the particulars of Hall's developmental scheme, namely, his reliance on recapitulation theory as its driving mechanism, were challenged by some of his contemporaries and by subsequent practitioners in the field, his work nonetheless played a major role in establishing a twentieth-century science of child development, as well as in inciting wider scientific and cultural concern for adolescence as an individual life stage and a collective social category.[3]

What Hall termed his "genetic" psychology categorized the young according to age-related attributes emerging out of a process of organic maturation. As with previous organic explanations of development, such a biological orientation professed the universality of developmental categories, while it upheld current notions of inherent difference and hierarchy between various groups of social subjects. Thus, as historian Gail Bederman shows, Hall's developmental theory relied on an evolutionary explanation of the biology of racial difference to prescribe a phase of savage masculinity for all boys during childhood, which only civilized ones outgrew during adolescence. "Adolescence was singled out [in Hall's developmental theory] as a crucial point at which an individual (and a race) leaped to a developed, superior, Western selfhood or remained arrested in a savage state," cultural and educational theorist Nancy Lesko further asserts. "Adolescence marked the race and gender divide to be crossed by each boy-man if he were to [be]come civilized and if he were to help the advancement of civilization." Historians Joseph F. Kett and John R. Gillis likewise contend that Hall's work functioned simultaneously to elide and reinforce differences of class and culture by looking to a white middle-class male experience as the model for growth and from there normalizing a set of shared biological, psychological, and behavioral characteristics and dictating common mandates for the treatment of all boys ages 14 to 24.[4] As these scholars argue, the primary focus of Hall's developmental theory was on boys, boyhood, and masculinity. While these scholars notice that girls were left out of many of Hall's descriptions of and prescriptions for adolescent development, however, they do not pursue the reasons or the ways in which Hall also attempted to reconcile ideas about girls, femininity, and adolescence.[5] This chapter explores the interplay among these ideas to map the role of their relationship in shaping conceptions of female development, as well as the modern category of adolescence, within the context of Hall's pioneering contributions to the emerging field of developmental psychology.

The chapter begins with a look at Hall's treatment of boyhood in his developmental theory and at the social context in which a heightened anxiety about imperiled masculinity prompted a conception of adolescence as structured by a set of attributes and needs instrumental to the preservation of the status of the white middle-class male. It then turns to Hall's psychology of female adolescence and his efforts to understand and explain the adolescent girl as a figure in her own right. The late-nineteenth-century discourse about the developing girl figured prominently in Hall's conception of adolescent girlhood, as well as in his more general construct of modern adolescence. Hall characterized adolescence as a "generalized or even feminized stage of psychic development."[6] In doing so, he was able to hold up the girl as the exemplar of certain fundamental qualities of adolescence, without ever threatening, and in fact even reinforcing, the role of the civilized boy as the driving force behind the continued progressive processes of biological and cultural evolution. In addition, even as it was firmly embedded in nineteenth-century paradigms of gender and race, Hall's psychology of female adolescence also presaged several "modern" expectations for the adolescent girl's body, mind, and behavior. Hall helped to break new intellectual ground by positing that the establishment of gender identity at adolescence gave rise to a psychological conflict that was central to the girl's development, by raising the specter of a normal female adolescent sexuality, and by sanctioning a number of other behaviors formerly deemed "precocious" and beyond the pale of Victorian girlhood. To be sure, Hall's treatment of these themes was informed by the new science of sexology, particularly by the ideas of British sexologist Havelock Ellis, and by the psychoanalytic theories of Sigmund Freud, whom Hall famously hosted at Clark University for his only visit to the United States in 1909.[7] The changes modern girls were experiencing and facilitating in their own lives at the turn of the century also influenced Hall's thinking. Of greatest interest and concern to Hall in this latter regard were white middle-class girls' pursuit of equal opportunities in secondary and higher education, as well as their involvement with leisure activities and experimentation with modes of self-expression as part of a distinctive peer culture. In engaging with and responding to such intellectual and social changes, Hall posed several "intricate and baffling problem[s]" about the girl's development for his fellow scientists and the wider culture to consider. He also offered his own solutions to such problems that accommodated and contributed to the reshaping of gender arrangements in modern times, while holding fast to the relations of power that structured the old sex/gender order.

HALL'S "BOYOLOGY": MALE DEVELOPMENT, MASCULINITY, AND MODERN ADOLESCENCE

When he offered his first description of adolescence as a distinct life stage in an article for the *Princeton Review* in 1882, G. Stanley Hall was on his way to becoming a leading figure in turn-of-the-century scientific and intellectual American life.[8] Having earned the first doctorate in psychology to be awarded in the United States from Harvard University in 1878, Hall went on to become a lecturer at Harvard; a professor at Johns Hopkins University; founder and president of Clark University; founder of the American Psychological Association, the *Pedagogical Seminary* (now the *Journal of Genetic Psychology*), and the *American Journal of Psychology*; and publisher of some 340 books, papers, and articles. Indeed, his influence in establishing the field of scientific psychology as an academic discipline at American universities was second only to William James's.[9] In addition to his academic accomplishments, Hall's renown also derived from his widespread popular appeal. Most notably, he earned the enthusiasm of numerous parents, educators, and reform groups as the preeminent spokesperson for the child-study movement during the 1890s. Despite the subsequent decline of child study, which reached its demise by 1911, Hall's influence as an academic pioneer and effective popularizer reverberated in a number of theoretical and practical efforts in the areas of parent education; child welfare; educational reform; and developmental, clinical, and educational psychology throughout the first decades of the twentieth century.[10]

Hall's training as a psychologist, which included studies in philosophy, theology, physiology, evolutionary biology, and the experimental psychology of German physiologist and philosopher Wilhelm Wundt, reflected his own diverse interests and inclinations, as well as the multidisciplinary origins of the field.[11] During his tenure at Johns Hopkins in the mid- to late 1880s, he succeeded as an experimental psychologist, where the attention of many of his colleagues in the new scientific psychology was finding its focus. However, by the early 1890s, his longstanding interest in the problem of development and his persistent efforts to reconcile natural science with deeply held philosophical and religious concerns, coupled with a series of professional disappointments and a personal crisis resulting from the tragic loss of his wife and daughter, led him to conclude that his energies as a psychologist would be best spent bringing the methods and theories of a scientific psychology to bear on the practical art of child rearing and on educational reform. Unable to formulate adequately a theory of child development and scientific pedagogy from laboratory psychology, Hall turned instead to the work of such thinkers as Jean-Baptiste Lamarck, Herbert Spencer, Charles Darwin, and German zoologist Ernst Haeckel to fashion a genetic psychology from the postulates of evolutionary biology.[12] His ardent commitment to understanding and explaining individual development and his accompanying prophetic call that the child be embraced as the key to humankind's past and the hope for its future put him out of step with mainstream psychology's adherence to rigorous scientific standards, which he had helped to pioneer. However, Hall was highly optimistic about his course. The study of childhood, he anticipated in 1896, "is likely to give us at last what we have longed for for a great while— something like a perspective, that will enable us to distinguish the deeper and older things of the soul from those that are of recent acquisition. I think the old adult psychology, that all of us teach even now, is to be, radically but gradually, transformed into a new genetic psychology."[13]

With his devotion to evolutionary science, Hall combined a romantic sensibility, derived from his own imaginative nature as a child, his education in romantic literature as an undergraduate at Williams College, his exposure to the German youth movement during three trips abroad in his twenties and thirties, and his affinity for the doctrines of European and American educational reformers of the mid- to late nineteenth century.[14] His ideas about child development were also influenced by his own boyhood upbringing in the 1840s and 1850s in the rural village of Ashfield, Massachusetts. Raised by pious Congregationalist parents, Hall was diligently schooled in the virtues of self-control and sexual restraint that defined manly character in mid-nineteenth-century America. As he recorded in his autobiography, *Life and Confessions of a Psychologist*, maintaining the self-discipline required to uphold the high standards for chastity, duty, and emotional containment set by his family and community was often excruciatingly difficult for him, but he imbibed these values nonetheless, and both they and his struggles against them figured prominently in his conceptions of developmental processes and outcomes.[15] The sometimes auspicious, but frequently uneasy, blending of evolutionary science, romanticism, and Victorian moralism produced both conservative and progressive strains in Hall's thought and helps to account for his broad cultural appeal. Described by Joseph F. Kett as "conceptually as well as chronologically poised between two centuries," Hall communicated mixed, ambivalent, and contradictory messages on a variety of themes related to children, education, gender, and sexuality. Such a posture resonated with a society in transition, in which many Americans of the white middle class, in particular, found themselves challenged to adjust to the social and cultural changes wrought by modernity, without entirely forsaking the "traditional" ways of life they expected to lend order, security, purpose, and definition to themselves as a group, and, through their leadership, to the nation as a whole.[16]

It was in the two-volume work, *Adolescence: Its Psychology and its Relations to Physiology, Anthropology, Sociology, Sex, Crime, Religion and Education,* that Hall provided his most thorough, if at times grandiloquent, explication of his genetic psychology and the adolescent stage of development. The book sold more than twenty-five thousand copies in the United States and in a revised and shorter version published in 1906 achieved marked popularity as a normal-school text.[17] For *Adolescence,* Hall reviewed vast bodies of literature and data from the fields enumerated in its subtitle, and he incorporated many of the findings garnered from responses to questionnaires investigating children's feelings, thoughts, and behaviors administered under the auspices of the child-study movement. In its most general tenets, Hall's developmental psychology asserted the determining influence of childhood and, especially, adolescence in the life of the individual; described the growing child as moving through a sequence of invariant stages, from the simple to the complex, with saltatory transitions between them; and emphasized the role of the instincts and emotions over the role of the intellect in the development of the mind. Hall's theory was descriptive and prescriptive, attempting to reconcile a warning that the natural progression of development must not be artificially manipulated, on the one hand, with a radically optimistic vision for the improvability of the human race, on the other.[18]

What most distinguished Hall's theory of human development was its enthusiastic embrace of the variant of evolutionary theory known as the *doctrine of recapitulation.* Uniting the biological disciplines of embryology, comparative anatomy, and paleontology, the idea that the development of the individual somehow repeated the evolution of the human race was introduced to Hall from the works of Haeckel, who formulated the biogenetic law asserting that ontogeny recapitulated phylogeny, Spencer, and Darwin. Hall quite literally interpreted the doctrine to mean that the progressive appearance of physical, psychological, and behavioral attributes in the individual child's development constituted a compressed process of reliving all the earlier stages of ancestral evolution. In this scheme, the embryo and young child first repeated all the stages of species development and then the child and the adolescent went on to relive the stages of human cultural evolution, from savagery to barbarism to civilization. According to the Lamarckian theory of acquired characteristics, to which Hall also subscribed, the most recent traits developed by the child's parents or grandparents were obtained by the child at the cusp of adulthood. With each generation, then, children had the potential to inherit more advanced traits and to foster and pass on higher traits of their own, thereby furthering evolutionary progress. Recapitulation and Lamarckianism began to fall out of favor among natural scientists around 1900, with the rediscovery of Gregor Mendel's genetic laws and the acceptance of August Weissman's theory that chromosomes, as opposed to ancestral behaviors, determined inheritance.[19] Nonetheless, in *Adolescence*, Hall continued to avow his conviction that the child and the race were "each keys to the other." Championing himself as a "Darwin of the mind," he argued that the study of one would illuminate important truths about the other and that both had to be comprehensively discerned by scientific inquiry if the developmental needs of the child and the promises of evolutionary progress were to be fathomed and fulfilled.[20]

At the heart of his genetic psychology, Hall positioned the adolescent years, the decade of life from ages 14 to 24. Ideally a time of "marvelous new birth" of body and mind, when the individual crossed over from a barbarous state to a civilized one, adolescence marked the nascency period, or emergence phase, of the "later and more precious" developmental acquisitions, including complex muscular organization, reasoning capability, social sensibility, consciousness of sexual feeling, high-level moral capacity, and religious sentiment.[21] In marrying recapitulation theory with Lamarckianism, Hall enthusiastically conceptualized adolescence as above all else a plastic stage of human development, during which the force of the instinctual endowment of the race expended its most advanced energies and then gave way to the potential for propitious environmental influence. If plasticity was the blessing of adolescence, however, it also portended its greatest danger. As incidences of juvenile crime and psychopathology revealed, the environment could just as easily lead adolescents astray as it could promote their growth toward higher ways of being. The task for those concerned with nurturing the development of the adolescent, then, lay in "providing the most favorable environment and eliminating every possible cause of arrest or reversion" when the higher capacities appeared. Hence, Hall was determined that adolescence be studied and monitored, with just the right mix of influence and noninterference by those who understood the peculiar contours of its physiology and psychology. Such supervision was essential, he repeatedly insisted, not only for the potential of individual development, but for the future of evolutionary progress, as well. As both the moment when heredity bestowed on the individual the most superior of the human traits and the moment when the individual was most susceptible to external conditioning, adolescence was the point of departure for the higher evolution of humanity, the very "bud of promise for the race." Alternatively, if such potential was unfulfilled, thwarted, or misdirected in some way, adolescence could also mark the beginning of racial retrogression, degeneration, and devolution.[22]

Hall's faith in the promise of hereditary endowments, his recommendations for appropriate environmental interventions, and his hopes for evolutionary progress, as well as his fears about how ontogenetic and phylogenetic development might go wrong, were all primarily focused on the development of the white middle-class American boy. His was at the forefront of a number of voices attempting to explain and to manage the civilized boy's social position and the definition of masculinity at a historical moment marked by significant social, economic, and cultural change.[23] During the decades surrounding the turn of the century in the United States, white middle-class male hegemony was challenged by claims, however forcibly repressed, of white middle-class women, African Americans, ethnic minorities, and the working-class to various forms of political, social, and economic power, as well as by the broader shift from a small-scale, entrepreneurial style of capitalism to a corporate, consumer-oriented economy. Arising out of and contributing to the major transformations of this period was a growing nationalism and an increasing commitment by political, business, military, and religious leaders to an ideology of expansionism and to policies and practices that would secure the position of the United States as a dominant power on the world stage.[24] The coeducation struggle was one of many cultural sites in which the meaning and direction of such changes were debated. Beginning in the 1880s, attention turned increasingly to the challenges faced by the civilized boy in negotiating these changes and to finding ways to enable him to meet them and to protect and perpetuate his claim to a position of superiority within the unfolding course of biological and cultural evolution.

Transformations in the lives of male youth that had begun in the late eighteenth century increased during the second half of the nineteenth century. Most notable was an ongoing differentiation in the opportunities available to middle-class and working-class boys. Under the conditions of advancing industrial capitalism, many jobs were opening up in the white-collar sector of the economy, including a demand for clerks, bookkeepers, accountants, engineers, lawyers, and business managers. Yet only those parents of sufficient means could afford to forgo the wages of their teenage sons and make the commitment to prolonged formal education these desirable new jobs required. Having once worked side by side on farms or in small machine shops throughout their youth, even while their adult destinations were unquestionably to be conditioned by their class status, middle-class and working-class boys now found themselves located in markedly different institutional settings during their teenage years. Middle-class boys attended public high schools, colleges, and professional schools in steadily increasing numbers, with these experiences typified by boys' economic dependency on their parents and increased supervision by those adults responsible for monitoring their route to success in their chosen profession. Negotiating economic opportunities around manufacturers' demands for unskilled cheap labor and craft and trade unions' fears of juvenile competition, working-class boys whose parents could not afford to support a lengthy education spent their teenage years in low-end blue-collar jobs. These boys found themselves in a semidependent relationship with the adult world. Their work life was carefully circumscribed by adult supervision, and their wages were expected to contribute to the family economy, even as some earnings were often used to support peer-oriented leisure activity, largely free from adult surveillance.[25]

As prolonged education increasingly became if not a clear determinant, then at least a necessary prerequisite of the boy's economic prospects in a bureaucratic, corporate economy, middle-class parents sought to inculcate in their sons a constellation of values that would guide them through their extended transition to manhood. Buttressed by a veritable industry in conduct-of-life literature and manifested in numerous organizations and institutions engaged in boy-work, middle-class family values of the late nineteenth century emphasized the manly virtues of rationality, self-control, persistence, and diligence that had abided in the antebellum period. Whereas earlier in the century some sanction had been given to the individual initiative and autonomy these values implied, they were now combined with a greater emphasis than ever before on the importance of youthful obedience, compliance, and conformity, essential virtues if the boy was to stay in school and the young man was to bide his requisite time on the bottom rungs of the corporate ladder. From this perspective, male youth was conceptualized as a lengthy moratorium phase during which the boy primarily learned the importance of suppressing rebellious impulses and deferring to benevolent authorities if he hoped to realize an adult status marked by career success.[26]

Paradoxically, however, ideal male identity now also incorporated a resistance to the weak body and passive character that were the feared result of undue manly self-restraint and compliance civilized life seemed to require. In his 1881 work *American Nervousness*, neurologist George M. Beard defined the problems associated with the excesses of manly virtues as a disease. The bodies of white middle-class men, he argued, were being dangerously depleted of their vital nerve force because of the overdeveloped intellect and the suppressed emotion that were necessary for achieving professional success in modern society.[27] Such problems, moreover, were perceived to have roots in current configurations of male childhood and youth. Boys receiving too much attention at home, especially from doting mothers, and boys spending too much time in school, primarily under the direction of genteel female teachers, were failing to acquire the physical and mental vigor that would equip them to compete in the struggle for survival and supremacy in the biological, economic, and social arenas later in their lives. As a response to what was derogatorily referred to as both the "overcivilization" and the "effeminization" of the middle-class white male's personality and body, male identity came to be characterized by a number of traits frequently associated with working-class and primitive masculinity, including physical prowess, aggressiveness, boldness, and spontaneity.[28]

The widespread popular enthusiasm in the late nineteenth century for male physical culture promoted and reflected the ambivalent prospects for Gilded Age masculinity by combining a keen antimodern sentiment with an enthusiastic receptivity toward the forms of economic success promised by an urban industrial society. For physical culture proponents, civilized life, with the changes in the nature of family, education, work, and leisure it proffered, threatened to weaken and corrupt the growing young male, especially, by fostering in him a desire for wealth, a love of luxury, a tendency toward bookishness, and an inclination toward dependency, passivity, and self-indulgence. Deliberate cultivation of the youthful male body, in the form of regular exercise, suppression of the sexual function, attention to diet, and plenty of exposure to rural settings, benefited the boy both by encouraging in him the discipline and control to resist the temptations and enervations of modern life and by befitting him with the strength and the drive to compete and to revel in his achievement of success in the same environment that was deemed to be so dangerous to his health and character.[29]

By the time Hall published his text in 1904, these multiple and contradictory expectations for male selfhood were manifest in a variety of middle-class cultural forms and social institutions, including advice manuals, literary texts, medical tracts, family relations, school curricula and organization, and formal leisure activities such as those sponsored by YMCA.[30] Hall's contribution was to locate the mandates for masculinity within the purview of the biological and psychological sciences and, especially, to frame the paradoxical expectations for male identity in developmental terms.[31] Relying on the logic of recapitulation, he declared it natural and normative for all boys to experience and to express the physical, emotional, and moral attributes of primitive masculinity during childhood and early adolescence. Comparisons between boys and savages had also been made in the antebellum period, mostly reluctantly, and the assertion of such parallels that set white middle-class boys apart from both their sisters and their fathers also competed with notions of an androgynous childhood and with expectations that all middle-class children acquire the habits and values of polite Victorian society from an early age.[32] With the predominating influence of evolutionary theory and the doctrine of recapitulation during the last quarter of the nineteenth century, such comparisons became more prevalent, drawn frequently by scientists endeavoring to measure and rank the peoples of the world by racial type, as well as by political and cultural authorities seeking to justify various forms of colonial intervention and domination.[33]

Hall was among the most outspoken in sanctioning savage sensibility and behavior in the individual boy's development. He also provided the most systematic explanation for how primitive boyhood was to be reconciled with ontogenetic and phylogenetic progress. Following from Spencer and his disciple, the American philosopher John Fiske, Hall held that the "length of the growing period is one of the most important factors in development" and asserted that "the higher the species the larger the proportion of its average life . . . spent in attaining maturity." According to this reasoning, primitive males finished recapitulating the more limited traits of their ancestors when they reached puberty. Civilized boys, in contrast, had a longer racial history to repeat and were able to use the abundant new energies that arrived with the onset of the sex instinct to continue to develop mentally and morally throughout their prolonged adolescence.[34] Boys in their late teens and early twenties were to devote this time of life to experiencing and integrating their protracted and increasingly complex racial heritages, to controlling the primitive passions and mastering the self, and to pursuing the higher learning that would promote the innovation of new racial traits. The final developmental outcome for Hall was a male self that was strong, self-directed, self-controlled, and socially adjusted, capable of preserving the evolutionary gains achieved by the white race and carrying those forth into even higher states of civilization. With a racial history marked by a violent struggle for survival in which males were the primary aggressors, the biology and psychology of the individual male body and mind, and the future of the white middle-class boy as its primary points of focus, Hall's developmental theory hereby conflated adolescence with the development of the civilized boy, leading one of his followers to refer to all of adolescent psychology as "boyology" and "boy analysis."[35]

Hall's discussion of play in *Adolescence* and other writings marked one such move to equate adolescent development with the maturation of the civilized boy, with both his race and gender accounting for his exclusive enjoyment of what Hall deemed to be the extraordinary possibilities of this life stage. Countering a contemporary notion among educators of play as "practice for future adult activities," Hall asserted that "true play never practices what is phyletically new." Rather, "it exercises many atavistic and rudimentary functions, a number of which will abort before maturity, but which live themselves out in play like the tadpoles [sic] tail, that must be both developed and used as a stimulus to the growth of legs which will otherwise never mature." Hall recognized that all human beings—primitive peoples young and old, past and present, and civilized girls as well as boys—engaged in some fashion the various "motor capacities" and "psycho-motive impulses" called "play."[36] Along these lines, he was particularly interested in challenging the perception of doll play as entirely a demonstration of "nascent parenthood" in girls with an understanding of it, at least in part, as an expression of "the wide spread animism, if not fetishism, of children and savages." He went on to contend that the "doll instinct," which reached its apex between the ages of 8 and 9, not only allowed children to express such fundamental religious impulses but also to experience a sense of superiority through relationship to an object smaller than themselves, to vent feelings of anger and fear, and to communicate their individual desires, uncertainties, interests, and talents to the adults around them. "Perhaps nothing so fully opens up the juvenile soul to the student of childhood as well developed doll play," Hall concluded.[37] For this reason, he wanted doll play to be encouraged by parents and educators, to be used as a tool both to gain insight into the child's development and to train the child's intellect, will, and heart in "wholesome direction[s]."[38] Declaring it "unfortunate" that doll play was seen as the exclusive province of girls, Hall sought to claim its benefits for boys as well. The passion for dolls was "naturally far less developed [in boys] than with girls," he readily admitted. Nonetheless, boys also showed a penchant for playing with dolls, particularly with "exceptional" ones such as "colored" dolls, animals, clowns, and fantasy figures, and ought to be given more opportunities in school and at home to do so.[39]

Hall also made it clear, however, that the growing boy's energies and capacities were not satisfied by mere doll play because his life "is naturally rougher and demands a wider range of activities" than the more staid and circumscribed domestic world of his sisters.[40] It was, he contended, only civilized boys who *fully* repeated the long, diverse, and difficult history of the human race in their necessarily more varied and vigorous play activities; only they, therefore, were able to reap the maximum developmental benefits to both "body and soul" that such play conferred. "[Play] gives a sense of superiority, dignity, endurance, courage, confidence, enterprise, power, personal validity, virility, and virtue," he declared. "To be active, agile, strong, is especially the glory of young men."[41] The playful activities in which Hall wished to see adolescent boys engaged—outdoor sports such as hunting, fishing, and swimming; combative practices such as fighting, boxing, and wrestling; team sports such as football and baseball; and all manner of military exercises—were important because they shored up boys' physical strength, enabled the release of the base instincts of anger and aggression, and cultivated the higher virtues of bravery, loyalty, cooperation, service, enthusiasm, decisiveness, and self-control. If these activities, instincts, and virtues were not given their due in the boy's development, then his achievement of an authentic, mature, male self was at risk. As Hall maintained in his explanation of the importance of play in enabling both the expression and the management of anger in the adolescent boy: "To be angry aright is a good part of moral education, and non resistance under all provocations is unmanly, craven, and cowardly. An able-bodied man, who can not fight physically, can hardly have a high and true sense of honor, and is generally a milk-sop, a lady-boy, or a sneak. He lacks virility, his masculinity does not ring true, his honesty can not be sound to the core. Hence, instead of eradicating this instinct, one of the great problems of physical and moral pedagogy is to rightly temper and direct it."[42]

Play served as a primary means of "perfecting" the body, mind, and character of the civilized adolescent boy not only by giving him the opportunity to repeat the history of the race but also by satisfying the distinct needs of the pubertal male body as well.[43] Hall argued that physical activity of all kinds at this age was driven by the boy's newly acquired sexual interest. His view of this interest, like so much else in his developmental thinking, was marked by ambivalence and by an effort to reconcile the moral absolutes of an older sexual order with the seemingly more flexible possibilities for responding to sexual desire in the modern age. On the one hand, his own Victorian upbringing held fast in his thinking about sex, and he continued to associate normal sexuality with heterosexual desire, marriage, and reproduction. On the other hand, the intense sexual anxiety he experienced during his youth (particularly his struggles with masturbation), his romanticism, and his interest in Freudian psychoanalysis and the new science of sexology led him to condemn "the old prudery and false reticence," to question the effectiveness of outright repression, and to conclude that sexuality was central to the development of the personality and essential for physical and psychological well being.[44] Thus did he determine adolescent male play to be, in many "subtle but potent" ways, a form of courtship that performed an important role in the process of sexual selection, with the boy engaging in play in large part to display his prowess to the "other sex." While Hall conceded that coeducation gave girls and boys some opportunity to participate in athletic games as partners, he much preferred that the girl, just as females in the animal kingdom and in savage societies, assume her "true role of sympathetic spectator" in adolescent sport. By passively observing male play and then exerting her domesticating influence upon it, the girl was to value the expression of physical power in the boy, while encouraging him, through her own promise of sexual and emotional response, to refine the virility, fierceness, and brutality of his savage nature into the more civilized qualities of tenderness, bravery, honor, magnanimity, and forbearance.[45] Play also met the needs of the pubertal male body by functioning as a more general form of sublimation. "Activity may exalt the spirit almost to the point of ecstasy," Hall declared, "and the physical pleasure of it diffuse, irradiate, and mitigate the sexual stress just at the age when its premature localization is most deleterious. Just enough at the proper time and rate contributes to permanent elasticity of mood and disposition, gives moral self-control, rouses a love of freedom with all that that great word means, and favors all higher human aspirations."[46] In acquainting civilized boys with the pleasures of the body and channeling those pleasures into nonsexual activities, play enabled them to experience the physical, emotional, and spiritual invigoration normally aroused by adolescent sexual awakening, while also training them in the equally important practice of self-mastery and enhancing their capacities for self- and social creation.

As Hall's conceptualizations of and prescriptions for play reveal, notions of catharsis and transformation were central to his thinking about the development of the civilized boy. Joining a longstanding fear of precocity among developmental thinkers with an equal measure of caution about undue repression, Hall warned that both the early (and hence excessive) development and also the underdevelopment of functions, capacities, emotions, and instincts in childhood and adolescence could result in the undesirable appearance, and even the criminal manifestation, of such behaviors and traits later in life. With this claim, he sought to legitimize the expression of a collection of behaviors for civilized boys that had been deemed unruly, annoying, or immoral by the nineteenth-century white middle class. There was, Hall declared, "something the matter with the boy in early teens who can be called 'a perfect gentleman.' That should come later, when the brute and animal element have had opportunity to work themselves off in a healthful way." More specifically, by this time in the boy's life, "he should have fought, whipped and been whipped, used language offensive to the prude and to the prim precisian, been in some scrapes, had something to do with bad . . . associates, and been exposed to . . . many forms of ethical mumps and measles."[47]

If, however, the biological dictates of childhood and early adolescence mandated that the white middle-class boy be "exposed to" a set of savage instincts and behaviors, the remainder of adolescence was to be devoted to the process of "recovering from" such exposure, with the earlier expression of the lower instincts now spurring the higher powers into existence, as was the case in the aforementioned transfiguring evolution of the tadpole's tail into the frog's legs. Now was the time when reason, judgment, and inhibition came to dominate the psychic life of the boy and when his mind became capable of "knitting together . . . all the new and old factors of personality" so as to maintain the proper tension between savage impulse and civilized self-control.[48] In discussions of mental illness and juvenile delinquency, Hall attested to the dangers wrought on normal psychological and social development if the regulative functions failed to emerge during this stage of life. Moving beyond a strictly hereditary view of deviance and its associations of "bad boy" behavior with ethnic and racial minorities and the working class, Hall claimed that all boys were at risk of going astray in this regard and all boys deserved assistance, ranging from sympathetic guidance to overt forms of surveillance and constraint, to prevent them from doing so. Even so, in his analysis, it was the civilized boy who was singled out as being most capable of "keeping alive and duly domesticating by culture the exuberant psychic faculties" of childhood and adolescence. In this way, the civilized boy would attain a "higher plateau" of individual existence and lead the way in humanity's ongoing evolution toward a race of the "super-man."[49]

HALL'S "BUDDING GIRL" AS THE
QUINTESSENTIAL AND PERPETUAL ADOLESCENT

Hall's concept of adolescence described and dictated the civilized boy's procurement of a male identity firmly established on the twin pillars of masculine virility and manly self-control. Per the requirements of the doctrine of recapitula-

tion, however, the route from savage boyhood to civilized manhood also necessarily entailed a movement through a life stage marked by a set of attributes of
which the civilized female, who occupied the evolutionary rung just below that
of the civilized male, was the chief exemplar. Thus, even as he volubly condemned the effeminization of modern life, reinvigorated earlier protests against
identical coeducation, and resolutely dismissed women's claims for full social
and political equality, Hall also positively ascribed to all adolescents such self-
described "feminine" traits as physical and mental volatility, emotionality, altruism, and religiosity.[50] Woman, he worshipfully professed, represented youth "in
the full meridian of its glory in all her dimensions and nature so that she is at the
top of the human curve from which the higher super-man is to evolve."[51] Biographer Dorothy Ross attributes Hall's efforts to reconcile femininity and adolescence to the influence of his individual psychological development and to his
longing to challenge strict expectations for masculine sensibility and behavior in
his own life. "Hall's development of a concept of adolescence," she writes, "was
part of his solution—an extremely creative solution—to a set of personal conflicts which had been reinforced by the specialization of occupational roles and
the rigidly differentiated sex roles of Victorian culture." Historian T. J. Jackson
Lears places Hall's construct of an "androgynous" adolescence within the broad
turn-of-the-century current of antimodernism, in which many artists and intellectuals sought release from the demands of a competitive, material, and incon-
stant world driven by "male ego ideals" by evoking an alternative vision of a
nurturing, spiritual, and eternal "feminine principle."[52] To these explanations, I
would add the significance of the history of ideas about the teenage girl and,
especially, the figurations of her development in late-nineteenth-century medical
and educational discourse. While Hall's concept of adolescence had roots both in
classical medicine and in romantic depictions of puberty and youth stretching
back to Rousseau, the contemporary debates about the teenage girl's body,
mind, and temperament also offered him a readily available and broadly recog-
nizable language and set of conceptual frameworks through which to think and
talk about adolescence in biological and psychological terms. Hall made ample
use of these frameworks, frequently holding the girl up as the model for a
particular adolescent attribute or behavior he was describing. He did so, however, while emphatically proclaiming the necessity of dichotomous sexual difference for evolutionary progress, as well as the primacy of the civilized male in
facilitating racial advancement. The result was a concept of adolescence that
broadened and then foreclosed certain possibilities in the boy's development,
while deeming that development normative and superior, and that valorized and

then marginalized both the girl and her feminine attributes, within the context of developmental thought and the society in which she came of age.

One way that the girl's "dimensions and nature" exemplified adolescent development generally was in the dynamics of her physical growth. Drawing on a range of growth studies conducted from the 1870s to the 1890s, Hall declared adolescence to be the "awkward age" in development during which physical growth normatively proceeded in irregular, disproportionate, and discordant fashion.[53] Contrary to Clarke's claim that boys' bodies grew steadily and equally from childhood to maturity, with girls deviating from this supposedly more auspicious pattern, Hall discerned disharmony and even difficulty in the growth of both boys and girls. After a rapid period of growth during infancy, he explained, boys' and girls' height and weight increased slowly and evenly up to age 8 or 9, at which point several years of "slightly diminished growth" was inaugurated. This was then followed, at age 12 in girls and at age 14 in boys, by a "veritable outburst of physical growth" and by the asymmetrical growth of the various body parts during the years of adolescence.[54] The data on growth patterns gathered by physicians and physical anthropologists was given meaning by Hall by way of the recapitulation theory. From ages 8 to 12, growth was retarded because, at this age, the child was reliving "some long stationary period during which life had been pretty fully unfolded and could be led indefinitely and with stability and security." Growth then speeded up and became imbalanced just before adolescence as the child began to repeat an epoch of racial history described by Hall as "some ancient period of storm and stress when old moorings were broken and a higher level attained." Changes in the individual body of the adolescent, then, were effect, sign, and cause of the larger evolutionary imperative toward racial progress. Rhetorically wondering whether this "period of rapid growth [was] advantageous" or whether it would "be better if the curve of height and weight were a straight line, as . . . it had been generally supposed to be," Hall characterized physical growth in such a way as to claim adolescence as *the* key period of both peril and possibility in human development. Growing "part by part" or by "nascent periods" was, indeed, fraught with danger and rendered the adolescent helpless and vulnerable, "far more in the need of protection, physical care, moral and intellectual guidance" than was currently recognized. However, uneven growth also prevented stagnation and served as the stimulus for more advanced development as "structure and function, body and mind" continually strove to restore their lost equilibrium with one another.[55]

Awkwardness in physical development explained a great deal about teenage boys, including their ungainliness, clumsiness, roughness, and even their poor manners.[56] Even so, the growth spurt was deemed by Hall to be more pronounced and more consequential in the girl's development. It was also both more intriguing and more vexing. Thus, while recapitulation theory explained why rapid growth generally occurred just before puberty, further explanation was needed about its earlier manifestation in the female. Whereas Spencer and Jacobi had offered competing explanations about the relationship between individuation and genesis in the individual girl's growing body, Hall delved deeper into the history of the race to speculate about likely causes. Very early in the development of the species, he conjectured, females were most likely prematurely impregnated because of the "hypertrophied sex passion" in the male. This was followed, in savage life, by the creation of institutions that tended to delay fertilization, with the cell development and vitality that had once supported reproduction now turning back to somatic growth. Alternatively, given that the instinct for modesty was "strongly developed" in the individual female before the onset of sexual desire, more vigorous girls would have been most able to resist impregnation, "even in an age of animal violence," thereby passing their size and strength on to their female offspring and "helping man onto a higher plane of greater maturity."[57]

Whatever the origin of the "interesting and challenging fact" of the earlier female growth spurt, though, girls experienced a greater measure of developmental disharmony because of it. "Hips, chest, and the reproductive system, and the instincts that go with them, constitute a far larger proportion of the girl's whole system in weight and function than is the case with boys," Hall declared. As it had for late-nineteenth-century opponents of coeducation, this meant that the girl's ability to develop capacities beyond her reproductive ones was comparatively "far more reduced" throughout the period of her adolescence, with the demands of the body reigning supreme during adult womanhood as well. Departing from Clarke, Hall warned that boys, too, had to be protected from overly taxing some "parts and functions" at the expense of others during the crucial teenage years. However, there was decidedly something wrong with the boy, in particular, in whom all developmental components were not ultimately brought into harmony and a higher "at-one-ment" was not finally achieved. As the better representatives of disharmony in physical development during adolescence, girls were thus also understood to be normatively pathological in their inability to overcome it. "The pubertal changes which take place in the male organs . . . have received far less attention than has been bestowed by morphologists, physiologists, and gynecologists upon those of the female," Hall readily admitted, "and the sympathetic reverberations of these changes upon the whole organism are far less known." Hall suggested he was going to offer a much-needed corrective to past ignorance about the dynamics and influences of puberty in male development, and indeed, drew on analogies with the girl in doing so. He nonetheless also reaffirmed the crucial distinction that while puberty explained much of what happened to the boy during his teenage years, for the girl, it was the exclusive and all-encompassing determinant of the rest of her life story.[58]

Storm and stress in the pubertal body mirrored and produced volatility in the adolescent mind and spirit, as well. "So, too, if the soul grows with every part of the body," Hall explained, "its development is not continuous, uniform, or proportionate, but with successive nodes, the earlier stages ever a little more strange and alien to the newer, like dimly remembered past lives to a transmigrationist."[59] With the flood of ancestral inheritance received at adolescence, there occurred a "loosening of the bonds between the manifold factors of our ego," as well as "a sudden and independent growth of single elements" in the personality that for a time were uncoordinated with and even engaged in a competitive struggle for survival against one another. Hall documented the recent research in psychiatry that revealed that for many young people the "all-sided mobilization" of old and new "psychic elements" at adolescence resulted in a wide range of psychoses and neuroses, including melancholia, mania, hysteria, anxiety, and moral perversities of various kinds. His larger goal, though, was to show that such morbid manifestations were "only exaggerated forms" of the psychological changes occurring during this period of life that was normally marked by "emotional strain" and even "repressed insanity."[60] According to Hall, all adolescents were engaged in an inner "strife of opposite moods" that rendered them alternatively hyperactive and lazy, euphoric and depressed, self-aggrandizing and full of self-doubt, selfish and altruistic, oriented toward solitude and sociability, sensitive and imperturbable, conservative and radical, wise and foolish. Adolescents were unstable, unpredictable, unreliable and largely incomprehensible, both to themselves and to the (nonexpert) adults around them.[61] Moreover, the universal difficulties of this life stage were compounded by a mixture of hereditary predispositions (individual and racial) and by the pressures and demands of modern life. The more civilized the child, then, the more stressful the adolescence, since civilized adolescents had more hereditary influences to integrate and also were faced with the myriad of temptations, corruptions, and distractions abiding in an urban, industrial society. More than previous developmental thinkers, Hall deemed the storms and stresses of adolescence to be both inevitable and desirable and called for an educational regime that tolerated and even encouraged the wide-ranging and often contradictory expression of adolescents' "multiplex personalities." For it was, he decreed, in the perturbations of the soul prevailing during this time that lay, also, the best possibility for a "new, larger, better consciousness" to emerge in the individual and to predominate in the race.[62]

While Hall described girls' and boys' experiences with such traits as enthusiasm, self-affirmation, and bashfulness in sex-specific ways, he associated the broader emotional intensity and moodiness of adolescence with a feminine sensibility. Thus, in both the more diseased psychological states and the general temperament of adolescence, the girl again served as a most apt representative, with her awkwardness of soul described by Hall as "more tumultuous and also more subterranean" than that of the boy. "In the transition from the grub to the butterfly state," he further declared, "the female is most liable to become psychologically upset, because her reproductive organs and functions are not only larger, but the changes are more rapid."[63]

The volatile changes in the girl's body and mind were also regularly recurring, with the menstrual cycle described by Hall as "one of nature's greatest rhythms," giving "an ebb and flow to all the tides of woman's inner life." Yet, in their vacillating moods and irregularity of activity (including the activity of the reproductive organs), boys, too, he noticed, recalled the ancient and female quality of being governed by the larger, rhythmic forces of nature, namely, those of the moon, the tides, and the seasons. "The spells of discomfort, distraction, irascibility, and depression in males thought to be of this [periodic] character are probably much more common than is generally supposed," Hall mused.[64]

As in the case of the girl's physical development, though, the key difference arose with the girl's inability to move beyond the adolescent tendency toward emotional fluctuation, with one abiding characteristic of woman being her reliable inconsistency. "Each day of the twenty-eight [day cycle] she is a different being," Hall affirmed, "and the wide range of circumnutation which explores the pleasures and pains of life, its darkness and light, its depressive and exalted states, its hopes and fears, its sense of absolute dependence and independence, . . . reveals her as a more generic creature than man, less consistent than he is if we compare days or hours, more so if we compare months as the units of her life." The boy, in contrast, while also "navigat[ing] a choppy sea" during this stage of life when his mind was in its "generalized form" and, not incidentally, benefiting from doing so by psychologically exploring "the maximum area possible of human experience," at some point near the close of the adolescent period also took "the helm of his own being" into his own hands. Now leaving the girl behind in his capacity to exert the very highest human powers of rationality, self-control, and self-determination, the boy alone succeeded in "striv[ing], fight[ing], and storm[ing]" all the way up the developmental trajectory to achieve the "mental unity" and "settled character" that were necessary if he were ultimately to break his way "into the kingdom of man."[65]

In addition to their more pronounced and lasting emotional volatility, girls also possessed a greater capacity for the deep feeling that constituted "the chief psychic ingredient" of the adolescent stage of life. According to Hall, civilization, with its overvaluation of the intellect and its insistence on emotional and so-cial refinement, was deadening the individual's capacity for any strong feeling, whether it was fear, anger, courage, joy, or love, leaving a "parched and bank-rupt" heart in its wake.[66] Along with this indictment, though, he expressed optimism that the new psychology would serve as a key tool in rectifying the problem, as psychologists were now "coming to understand that what con-stitutes life is the intensity and the variety and scope of what we feel." It was such feeling, Hall insisted, that made men, a process that necessarily entailed the young boy's cathartic expression of the primitive, masculine passions of anger and aggression. But it was men of heart, men such as Calvin, Luther, and, most exemplarily, Jesus Christ, more so than men of physical strength or even men of intellect, whom Hall deemed to be true geniuses and to be most capable of making the sort of history that would advance human evolution to a higher plane. Although such men were admittedly exceptional, however, he repeatedly noted that ordinary woman, nearer to the race and more governed by its in-stincts, most fully embodied the feeling heart on which the very soundness of civilization so greatly depended. Indeed, she, "who was once thought soulless, now comes nearer to having two souls than does man."[67]

In her greater capacity to feel, the girl also more fully exemplified what for Hall were the highest moral capacities to emerge during the adolescent stage—sympathy, altruism, religiosity, and an inclination to serve others. At adoles-cence, he decreed, "life is no longer ego-centric, but altro-centric."[68] In prescrib-ing this mandate for adolescent development, Hall joined with a number of other intellectuals and reformers of his day in condemning the materialism and self-interest that he saw as the driving forces in modern society and expressing a yearning for a social community in which selfish individuals realized their organic connection to all beings and subordinated their disparate desires to the needs of the common good. "It is just this [sympathy]," he decried at some length in his autobiography, "that the criminal, the egoist, the profiteer, the irreconcilable who cannot compromise or do teamwork, the undesirable citizen, the man always insisting on his rights and forgetting that every right must be created by a corresponding duty, the soulless corporation, the public-be-damned capitalist who regards labor as a commodity, the striker who feels no responsibility for the interests or comfort of the community—in a word the man who is dominated by selfish personal interests—lack."[69] While both Locke and Rousseau dictated that the boy's acquisition of virtue during his youth was essential for the stability and progress of civil society, political and scientific thought from the mid-eighteenth century onward had associated the capacity for feeling and selflessness almost exclusively with woman and her feminine nature. Hall both perpetuated and departed from this trend, singling out girls as models for emotionality, connectedness, compassion, and faith, while also emphasizing the imperative for such traits to be recognized and expressed in the boy's development as well. That boys affected "stoical and callous ways" and expressed an "instinctive shame" of feeling pity and compassion during their youth testified not to the "bad quality" of such feeling, as conventional wisdom had it, Hall argued, but rather to its very strength as a force in their psychic lives. Data from child-study questionnaires investigating what children hoped to be and do and whom they most admired confirmed that in both sexes "moral qualities rise highest and also fastest just before and near puberty and continue to increase later yet." Significantly, though, such studies also revealed that girls "far most" displayed an "increasing admiration of ethical and social qualities" throughout the teenage years.[70]

As Hall explained it, it was the onset of the sex instinct at puberty that prompted the individual's movement from the selfishness of childhood to the benevolence of adolescence. Here, he joined with such contemporary sexual romantics as Havelock Ellis and Swedish feminist and reformer Ellen Key in conceptualizing the sex instinct less in Freudian terms, as driven by the selfish desire for personal pleasure, than as motivated by love of (in order of ascending virtue) partner, offspring, community, the human race, nature, and the transcendent "absolute."[71] Thus, although expressing ample concern that pubertal sexuality might be sullied or led astray by pernicious environmental influences, Hall contended that in its natural state it was directed toward the higher moral purposes of reproduction, racial regeneration, and the organic connection of all being. In some of the most turgid passages in a tomb already laden with weighty prose, he so described the awakening of adolescent sexuality: "Life is now polarized, oriented, and potentialized. The soul is filled with a Titanism that would achieve a *vita nuova* upon a higher plateau, where the music of humanity is no longer sad but triumphant . . . Now the soul realizes the possibility of a new heaven and a new earth; that the highest dreams of human beatitude may be

real; that there is a *summum bonum* awaiting man on heights not yet scaled, and
that erethism and its calentures are prophecies of a higher human estate . . . The
flesh and the spirit are mated . . . Nature . . . is transcended in the soul's *natura
naturata,* and the extra and supernatural organ of faith comes into possession of
its kingdom."[72] Deeming the middle teens to be the age of conversion "the world
over, and at all periods of history and stages of civilization," Hall hereby linked
the appearance of the sex instinct and the religious impulse together at adolescence. "Religion," he averred, "has no other function than to make this change
[from "self-love" to "love of man"] complete, and the whole of morality may be
well defined as life in the interest of the race, for love of God and love of man are
one and inseparable." Because the maternal instinct preceded, and indeed helped
to give rise to, the paternal instinct in the evolution of the race and because
reproduction played a greater role in the life of the individual female, it was the
girl who most perfectly manifested both the "humanistic altruism" and the
religiosity that Hall prescribed as a normative developmental achievement of
adolescence. Just as with the expression of feeling, he dictated that the boy follow
her lead and come to heed that voice in his soul that called him to "find the joy of
sacrifice, get only to give, live for others, [and] subordinate the will to live to love,
or to offspring."[73]

Hall's Christian millennialist vision of an organic community in which "the
race, not the self" would reign supreme was, nonetheless, a decidedly racist,
elitist, and antidemocratic view, with his cues here taken from the social thought
of both Herbert Spencer and German philosopher Friedrich Nietzsche. Indeed,
in his writings on the "education of the heart," he deemed those most worthy of
pity, and of the educational incentives and social opportunities that ought to
follow from it, not the poor, sick, or criminal, but that small group of civilized
boys who were uniquely poised to "be of most aid in ushering in the kingdom of
the superman."[74] It was also ultimately a sexist vision, for however much Hall
waxed poetic about the ways in which the girl supremely modeled the adolescent sensibility, he also established clear limits on her growth. Modifying Rousseau's characterization of the female as perennial child, Hall's final assessment of
the girl's development was that she was both quintessential *and* perpetual adolescent. "Woman at her best never outgrows adolescence as man does," he asserted, "but lingers in, magnifies and glorifies this culminating stage of life with
its all-sided interests, its convertibility of emotions, its enthusiasm, and zest for all
that is good, beautiful, true, and heroic. This constitutes her freshness and
charm, even in age, and makes her by nature more humanistic than man, more
sympathetic and appreciative."[75] For this reason, Hall was comparatively less

ambivalent about the problem of maturity in relation to the development of the girl than he was in the case of the boy. Simply put, he did not have to regret her loss of his beloved adolescent sensibility because she never left it totally behind. As important, though, Hall also mourned the passage of the adolescent stage less in the girl because he was far less equivocal about dictating the singular and static outcome that her achievement of maturity, such as it was, was to entail—the manifestation of her sexual difference in the all-encompassing roles of wife and mother in the private domestic sphere.

Meanwhile, to assume his rightful place as a dynamic leader of evolutionary progress, the civilized boy had to be guided in growing through and then out of not only the savage proclivities of his childhood but also the feminine sensibilities of his adolescence. In this way, Hall endowed the white, western boy, specifically, with the best of all psychic worlds. In childhood, he was to relive the earliest stages of racial evolution, the virile attributes of savage masculinity, serving to inoculate him against the potentially enervating effects of civilization later in life. Then, with adolescence he assumed the more highly evolved traits of civilized femininity, which allowed him an even fuller range of expression of human emotion and morally equipped him to allay some of the worst excesses of modernity. While lamenting that growing up for the boy entailed a loss and a "forgetting" of his connection to the depth and breadth of both the primitive and feminine "collective soul," Hall recognized that for better and for worse such a process could never be complete. "The conscious adult person is not a monad reflecting the universe," he insisted, "but a fragment broken off and detached from the great world of soul, always maimed, defined by special limitations, like, yet different, from all others, with some incommensurability parting it off as something unique, well fitted to illustrate some aspects and hopelessly unable to exemplify or even know other regions in the cosmos of soul."[76]

Even so, whatever capacity Hall allowed for developing the fully mature, individuated, and self-actualizing ego, he required of and reserved for the civilized boy alone. For him, adolescence was not only the nascency period of the great bodily, emotional, and spiritual passions but also the time to assert an active, autonomous self into the world—the time to wander, to discover, and to experiment with new ways of knowing and being; to resist authority and to test his powers against others; to prove his individual strength and courage; to cultivate and exercise the intellect and the will in making wise, purposeful choices; to imagine a future particular to his unique talents and abilities; and to strive for greatness and recognition in his community.[77] The civilized adolescent boy emulated the girl in her awkwardness and vulnerability of body and soul, her intu-

itiveness and emotionality, and her compassion for others and her love of God.
He also finally surpassed her because he alone faced the challenges and reaped
the rewards of achieving what ultimately Hall and most twentieth-century psychologists deemed the supreme tasks of adolescent development—leaving home,
choosing a vocation, fostering a unique individual identity, and, forging an autonomous self capable of preserving social order and facilitating progressive
social change.

In a short story published in his 1920 collection of imaginative writings,
Recreations of a Psychologist, Hall returned to the theme of a feminized adolescence, exploring its possibilities and limits in relation to the boy's development
and, indirectly but no less significantly, its meaning for female development.[78]
"How Johnnie's Vision Came True" is the tale of the 14-year-old Johnnie Smith, a
"sturdy," "old-fashioned," and "already very useful" farm boy who is bored and
resentful of the "monotony and drudgery" of his rural life and in whom the seeds
of both ambition and love are beginning to germinate in his soul. Understood by
his mother, with whom he has "heart to heart" talks, Johnnie is also contemptuous of his father, whose predictable path in life he is determined not to
follow. One hot summer afternoon, he takes his new shotgun and makes his way,
in typical adolescent fashion, by "alternate sauntering, resting and steep climbing" up the nearby Mount Hatch, from which he surveys "all the world he knew
and far more" with "a vague unique sense of exaltation."[79] On the mountaintop,
he first calls on the sun, Hall's symbol for the adolescent boy's masculine intellect
and energy, to help him to understand life and to live it actively, deliberately, and
gloriously. "Shine into and through me," he begs. "I want to know all the world
as you do. You never saw a shadow. You could not. And all I know is shadow
darkening down into black ignorance. Don't set, but rise in my soul." In a fit of
ecstasy, Johnnie then "dedicate[s] himself to the sun as the loftiest, biggest, most
dynamic thing he knew." The sun, however, does set, and during the period of
sleep that ensues, Johnnie next finds himself kneeling before the moon, from
which radiates first a vision of his own mother and then of a more generic female
figure resembling the Virgin Mary. He calls out to the "queen of the night,"
telling her that he is homesick for her and feeling powerfully that she has a secret
to impart, not only to him, but through him, to the entire world. Gazing on her
with rapture and reverence, Johnnie "felt that she was wise in life and had known
its chief joys and sorrows, its high lights and shadows; that without the lore of
pedants or books she understood the world and, best of all, understood him."
Together, these female visions convey to him a deep feeling of peace and calm
that henceforth never leaves him, promise that all of the ideals now born into his

heart will one day be realized, and vow "sometime, somewhere, when the hour is ripe," to return to him once again.[80]

After having passed through this experience of "great crisis" and "conversion," Johnnie returns to his home, as a "man" with a "mission in the world." He does not know his precise purpose in life, although, with his gun left behind on the mountain, he is sure it is not to become the "great hunter and frontiersman" of his boyhood revelries. Following instead the humanistic inclinations aroused in him by his lunar visions, he pursues interests in music, novel writing, and poetry. He finally goes to college and, "by dint of hard, absorbing effort," becomes a respected professor of literature and a member of Congress, as well as a husband in a loveless marriage and a father of children who "grew up well, married and left him." Abruptly advancing the story through the details of Johnnie's adulthood to his status as a widower at age 65, Hall tells us that his life "had after all not been satisfactory, successful though it seemed to others." It is at this point in his old age that Johnnie encounters a young woman who is "in living flesh and blood the identical lady of his pubescent hilltop vision." Struggling between his "judgment" and his "desires," Johnnie decides to pursue her and comes to find out that she, too, "evolved a double and counterpart" in her youth that represented her "idea of manhood," of which Johnnie is the embodied version. Determining that their "doubles [had] been kept apart long enough," Johnnie and the woman marry and in so unifying "senescence and adolescence" commence to pioneer a new kind of marriage in which "love chronicles will not end but begin at the church door."[81]

Johnnie's passage through a stage of feminized adolescence clearly offers him possibilities for feeling and being in the world that a strictly masculine sensibility cannot afford. In opening his heart to the "unutterable love and yearning" imparted to him by the female visions of his youth and in later marrying the woman who embodies them, Johnnie achieves both a wholeness and a happiness that his father's example, worldly success, and the fulfillment of his manly duties to family and state cannot provide him.[82] And yet, however much Hall finds wanting in the achievements and responsibilities of Johnnie's adulthood, it is nonetheless significant that it is he, alone, who exercises a deliberate, individual effort in carrying them forth. Indeed, we learn even less of the adult activities of the woman whom he marries, who, despite her own admission of longings to express the masculine qualities of her self, stands as a generic representative of the feminine sensibility of adolescence, differentiated not even by a name as common as Johnnie's. The resolution of the story in marriage, too, conveys Hall's mixed messages about the meaning of a feminized adolescence for girls and boys. Metaphorically, the marriage of Johnnie and the woman boldly represents the union of the masculine and feminine parts of the self and suggests that the bisexual/bigendered nature of the soul must be honored for healthy psychological development in the individual and also for the ongoing progressive evolution of civilization. Literally, though, it is still a marriage, however modern, in which the masculine and the feminine complement, rather than fuse with and radically transform, one another through the side-by-side placement of self-contained opposites.

A "NEW AND BETTER WOMANHOOD":
HALL'S MAKING OF THE MODERN ADOLESCENT GIRL

Hall's depiction of the civilized girl as the archetypal feminized adolescent and his extension of that sensibility as a temporary developmental phase to be passed through by her brother relied on traditional Victorian conceptions of the feminine to offer up an expansive version of American manhood that seemed to Hall to be better suited to the modern age. His efforts to reconcile femininity and adolescence in other ways, however, also pointed the way toward a more modern set of concerns related to the girl's development, as well, even if his own resolution of these ultimately reiterated conventional imperatives for dichotomous sexual difference and separate social spheres. One way that Hall contributed to the construction of a more modern adolescent girl was in casting the crisis of female adolescence in at least partially new terms. Nineteenth-century physicians had deemed the teenage years to be dangerous and stressful ones for the girl because of the essentially hazardous nature of female puberty and warned of the dire consequences that would befall both girl and society if the onset of the reproductive function was not managed properly. Hall repeated similar claims and concerns about the physical difficulties inherent in becoming a woman, but he also dwelled more pointedly on the psychological struggles the civilized girl faced in assuming a feminine gender identity and conforming to the biological mandates of growing up female in a changing world.

Thus, from child-study questionnaires that polled girls and boys about their aspirations for their adult lives, Hall discovered that girls had more ideals than boys and, most significantly, that the majority of girls chose male figures as their adult role models, which he interpreted as signs of their greater "discontent at their lot." Adolescent girls were also, he noticed, susceptible to a kind of inward anger that was the result of the "thwarting of purpose and expectation, limitations of freedom, a sense of injustice, [and] invasions or repression of the self."

Such feelings of anger, he added, were often the cause of "vasomotor disturbances" that were especially likely to occur during and have deleterious effects on menstruation.[83] In addition, older adolescent girls were particularly inclined toward a "pathos of unrealized hopes and ideals," which attending college could postpone, but never allay altogether.[84] "In our environment," Hall averred, "there is a little danger that this age [of 18 or 20] once well past there will slowly arise a slight sense of aimlessness or lassitude, unrest, uneasiness, as if one were almost unconsciously feeling along the wall for a door to which the key was not at hand. Thus some lose their bloom and, yielding to the great danger of young womanhood, slowly lapse to an anxious state of expectancy, or they desire something not within their reach, and so the diathesis of restlessness slowly supervenes." Hall further ruminated about the causes and effects of female dissatisfaction in the account of his mother he provided in his autobiography. In reviewing a series of entries from his mother's religious diaries written as a young wife and as a mother, Hall was surprised to find almost no direct mention of her husband or children. Instead, she wrote of her struggles to develop her self and follow her own inclinations, which for her meant nourishing her relationship with God, in the face of her all encompassing domestic duties on the family's mid-nineteenth-century Massachusetts farm. "I sometimes feel burdened with care," Abigail Hall admitted, "and think that with less, and consequently more time for self-communing, for the cultivation of the heart and intellect, I could serve God better and enjoy his service more." Hereby did Hall recognize that his mother was motivated by "an impulse to keep on growing" and a "longing for a larger, fuller life" that might indeed have been thwarted by the "excessive domestic duties" she was required to perform.[85]

For Hall, as for previous developmental thinkers, puberty marked the onset of the sexual difference. One of the most important tasks of the teenage years was securing biologically determined, dichotomous gender identities, roles, responsibilities, and relationships that would govern the man and woman throughout adult life. More than others, however, he also conceived of adolescence as a plastic moment of personality formation during which gender identity was not fully organized or integrated, when the boy or girl might inadvertently fail to achieve or might deliberately resist and reject the destiny set for him or her by nature.[86] "Not only in the body, but in the psyche of childhood," he explained, "there are well-marked stages in which male and female traits, sensations, and instincts struggle for prepotency. Here, too, the instincts peculiar to the opposite sex may not vanish as they normally should so that we have bisexual souls." Hall's positing of a feminized adolescence for the boy heightened the danger that "instincts in the male predominantly feminine" might fail to be "relegated to the background" of the psyche as they properly should, therefore threatening the boy's degeneration into inversion and homosexuality.[87] However, it also worked to mitigate this threat by allowing the boy a designated period of life during which he might cathartically explore, express, and duly outgrow the feminine aspects of his nature. As was the case with so many other attributes of adolescence, gender identity confusion was more prevalent and more prominent in the case of the girl, who, Hall acknowledged, was far more likely to assimilate boys' ways during her youth than vice versa. Although recognizing the psychological strain caused by living in a society in which, "after ages of seclusion, woman has suddenly emerged into a larger life than her heredity prepared her for," however, Hall nonetheless expressed very little tolerance for the girl's experimentation with and integration of masculine sensibility and behavior in the development of her self.[88] Thus, in completing his analysis of his mother's discontent, he concluded that to her credit, she eventually settled into "more complacency with herself" and accepted that the best way to satisfy her desire for self-fulfillment through service to God was to embrace the role of the true woman and do her duty to others in her immediate family. In contrast, for too many modern girls, he noticed disapprovingly, "the divorce between the life preferred and that demanded by the interests of the race is often absolute." This widespread dissatisfaction was giving rise to such aimlessness, restlessness, and emptiness among young women, he lamented, that "it seems that the female character [is being] threatened with disintegration."[89]

Hall declared that the adolescent girl's discontent with the requirements of womanhood was sad and even unnatural, although admittedly very common in the modern age.[90] Freud would go further than this to deem the struggle to secure gender identity—and heterosexual object choice—to be universal, normative, and motivated by the young child's discovery of anatomical sexual difference in his theory of the Oedipus complex, which he fully elaborated in his major psychoanalytic writings of the 1920s.[91] For both, though, the solution to this psychological crisis was entirely the same. During adolescence, the girl was to overcome her envy of male prerogatives and complete her development according to nature's intention, by coming into full knowledge and acceptance of her distinct domestic destiny. More than Freud, who consigned the achievement of mature femininity to the girl's recognition of her lack of a penis and her realization of her essential inferiority, Hall celebrated this acceptance as the key to her own personal happiness and to the unique contribution she was to make to the advancement of civilization. Identical coeducation was such a problem for him, then, not only because it depleted the physical reserves the girl needed to fulfill her destiny but also because it obscured the source of her true power and worth to her self. Indeed, one of the effects of modern life was that it allowed what were for Hall self-evident truths about the girl's nature to remain deep in the recesses of her unconscious, which required effort by the psychologist to get her to uncover, admit to, and embrace. "The more we know of the contents of the young woman's mind the more clearly we see that every thing conscious and unconscious in it points to [maternity] as the true goal of the way of life," he asserted. "Even if she does not realize it, her whole nature demands first of all children to love . . . and perhaps a little less, a man whom she heartily respects and trusts to strengthen and perhaps protect her in discharging this function . . . All ripe, healthful, and womanly woman desire this, and if they attain true self-knowledge confess it to themselves, however loath they may be to do so to others . . . Nothing can ever quite take its place, without it they are never completely happy, and every other satisfaction is a little vicarious. To see this is simple common sense and to admit it only common honesty."⁹² Early-twentieth-century feminist psychologists would disagree. Some rejected the premise that girls were inherently dissatisfied with their femininity, while others insisted that the problem arose not from within the female body or psyche but from an outmoded set of social expectations for women that could and should be changed. Such debates over whether, why, and with what consequences the achievement of femininity was fraught with difficulty for the girl constituted an initial, not the final, word on the matter and would become a mainstay in the psychological literature on female development of the twentieth century.⁹³

Hall also helped to formulate a more modern vision of the adolescent girl by raising the possibility that along with the maturation of the reproductive organs and the corresponding salience of sexual difference at puberty there arose in her normal feelings of sexual desire.⁹⁴ As in the case of the boy, Hall's conceptualization of female sexuality also stood suspended between two centuries. Such straddling of the old and the new was not unique to Hall but rather was emblematic of the earliest intellectual current that rode the wave away from what Freud would term the "civilized morality" of the nineteenth century toward the sexual "liberalism" that would characterize the twentieth century. The Victorian sexual ethos encompassed by Freud's phrase functioned as an integrated system of economic, social, and religious norms that were to be instilled opportunely during childhood by clergymen, parents, teachers, and physicians. It described models of masculinity and femininity rooted in the ideology of separate spheres and prescribed a regimen of sexual hygiene that required reticence, purity of thought, conservation of energy, and self-control. More specifically, men and pubescent boys were understood in this scheme to be driven by aggressive sexual urges that could and must be regulated by the internal mechanisms of the will and the conscience. Women and pubescent girls were conceived of as passionless, as naturally pure in mind and body and thoroughly passive during the consummation of the sexual act. Prepubescent children of both sexes were thought to be paragons of sexual ignorance, moral innocence, and spiritual goodness. As one facet of the nineteenth-century white middle-class struggle to define itself conceptually and to secure and maintain material prosperity, civilized morality pointed the way for men toward individual achievement and economic success in a competitive marketplace; for women toward moral superiority and a measure of control over sexual relations within the domestic sphere; and for both sexes, adult and child, toward physical health and spiritual salvation.[95]

During the late nineteenth and early twentieth centuries, however, the social context that had given rise to civilized morality was being rapidly transformed. The same economic, cultural, and social changes that were reshaping the meanings and experiences of gender roles and identities were, concomitantly, reworking Victorian sexual norms, as well. Changes such as the growth of a consumer economy that depended on the buyer's desire for immediate gratification and personal pleasure, the massive influx of immigrants from southern and eastern Europe, and the migration of African Americans from the rural South to northern cities threatened the Victorian sexual order by moving commercialized sex from its previous underground location into the worlds of mainstream advertising and entertainment and by making alternative forms of family patterns and sexual mores far more visible than in the past.[96] The challenge to female sexual purity, in particular, initially came from working-class girls experimenting with new forms of heterosocial relations on the job and in their leisure time and from middle-class Bohemian women seeking to emancipate women in all areas of life. During World War I and into the 1920s, middle-class teenage girls began to self-consciously fashion their own version of a sexualized female self, attending to and displaying their bodies in new ways, talking more openly about sex, and engaging in such formerly taboo practices as petting and necking.[97]

At this transitional moment, Hall presented a concept of adolescence that legitimized some new sexual manners and mores for white middle-class girls as well as boys, while also managing to preserve many of the key tenets of the older sexual order. He drew on and helped to garner acceptance for Freud's emphasis on the fundamental role of sexuality in human life, as well as his attention to the maintenance of the proper balance between the expression of the sexual impulse and its sublimation and control. Even more than Freud, Havelock Ellis paralleled the tone and content of Hall's depictions of the sexuality of the "budding girl."[98] Recognized by historians as a "central figure" in the emergence of a modern sexual ethos, Ellis has been credited with leading the way in articulating both the transformative and the conservative aspects of that ethos's attitudes about female sexuality.[99] As part of that ethos, Ellis helped to pioneer a modern standard for female adolescent sexuality, specifically. He acknowledged the teenage girl's sexual appeal and even recognized the possibility for her experience of sexual pleasure, while also posing a new set of dilemmas about what this implied for the manifestation of sexual autonomy, responsibility, and danger during the period of her development into adult womanhood.

Ellis's most important work, *Studies in the Psychology of Sex,* consisted of seven volumes, six of which were published between 1890 and 1910.[100] Making the case for the scientific study of "normal" sexuality, this work also reset the bounds of normal sexual experience to encompass enjoyment, variety, expressiveness, and reasonable restraint, as well as the gradual development of the sex instinct during childhood. Ellis both drew on the hierarchical assumptions of nineteenth-century evolutionary thought and anticipated the more egalitarian inclinations of early-twentieth-century cultural anthropology. Thus, the multitudinous comparisons of sexual attitudes and behaviors across cultures referred to throughout the *Studies* reasserted the primacy of western civilization, although now with a marked reversal from earlier sexual ideology. In this series of works intended to valorize the influence of the sexual impulse in individual human development and collective social life, the sexuality of savages was now explained to be *less* developed, in both quantity and quality, than among civilized peoples. However, Ellis also used cross-cultural comparison, along with the case from animal behavior, to further the notion of an essential, "universal" sexuality, hereby promulgating the idea of a singular and common "human" nature. This reasoning about race made possible the bolder assertions about gender in the *Studies.* The notion that the sex impulse had actually increased with civilization allowed him to extend its prerogatives to white middle-class girls and women by countering prevailing anxieties that female sexual expression of any kind was a threat to evolutionary progress. At the same time, Ellis's acknowledgement of a shared human nature across racial types also enabled him to champion unequivocally the biological, intellectual, and social equivalence of the two sexes, even as he continued to explore as a central focus in his analysis the meanings and implications of sexual difference.[101]

As did the Victorian physicians and moralists who came before them, turn-of-the-century sexologists conceived of sexuality as a central component of gender identity. That is, what it meant to be a man or a woman was reflected in and constituted by male and female sexual nature, respectively.[102] Ellis helped to break new ground in this regard by discerning the sexual impulse in man and woman to be both quantitatively similar and qualitatively different. According to Ellis, a vibrant, abundant female sexuality found its source in woman's unique capacity for maternal love and creation, which he exalted as essential for the emergence and ongoing progressive evolution of civilization. Also modern in his analysis was his entertainment of the possibility that sexuality was a gradually developmental phenomenon and his concomitant attention to the relational unfolding of sexual desire and gender identity over the course of the life cycle. Quoting the gynecologist Braxton Hicks on the appearance of the sex instinct during infancy and childhood, Ellis gestured toward what would come to constitute one of the more contentious claims of twentieth-century sexology. Nonetheless, he still maintained puberty as the more meaningful originating moment, if not the absolute cause, of both sexual desire and gender difference: "I venture to think . . . that those who have much attended to children will agree with me in saying that, almost from the cradle, a difference can be seen in manner, habits of mind, and in illness [of boys and girls], requiring variations in their treatment. The change is certainly hastened and intensified at the time of puberty; but there is, even to an average observer, a clear difference between the sexes from early infancy, gradually becoming more marked up to puberty. That sexual feelings exist [it would be better to say "may exist"] from earliest infancy is well known, and therefore this function does not depend upon puberty, though intensified by it . . . The changes of puberty are all of them dependent on the primordial force which, gradually gathering in power, culminates in the perfection both of form and of the sexual system, primary and secondary."[103] Such an analysis marked an important difference from Freud, who, in his *Three Essays on the Theory of Sexuality*, shockingly contended that the girl and the boy were more alike than different in early childhood, not because of their shared sexual innocence, as nineteenth-century developmental thinkers had it, but because they both possessed an aggressive, selfish, pleasure-seeking, "masculine" sexual impulse, which found expression through the male and female child's manipulation of the "homologous" organs of the penis and the clitoris, respectively. Suggested by Freud here, and fully developed in his subsequent writings on penis envy and the Oedipus complex, was the contention that the link between the sexual impulse and the reproductive imperative in the girl was only established during adolescence, the result of a complex psychological struggle that ultimately entailed the renunciation of her original sexual/gendered nature.[104] For Ellis, rather, the presence of a "primordial force" of gender-specific sexual sensation in girls and boys unfolding from infancy onward proclaimed the equivalence of capacity for sexual feeling in both sexes, while also positing the permanence of a distinctive feminine essence that was timelessly rooted in female nature.

Ellis's consideration of the potential existence of the sexual instinct in young children notwithstanding, it was the manifestation of sexual desire in the pubertal girl that he offered up as primary evidence for his claims for the existence of the normal sexual impulse in adult women. "How natural the sexual impulse is in women, whatever difficulties may arise in regard to its complete gratification," he asserted, "is clearly seen when we come to consider the frequency with which in young women we witness its more or less instinctive manifestations." Moreover, these manifestations "not only occur with most frequency in young girls, but, contrary to the common belief, they seem to occur chiefly in innocent and unperverted girls." Ellis did recognize that the "common belief" previously held about girls' natural purity had not gone entirely unchallenged before his own investigations into the matter, and he pointed to earlier medical sources that mentioned the appearance of sexual desire in girls as one of the normal signs of puberty.[105] Whatever his judiciousness in this regard, though, he largely fashioned his own rendering of the normal adolescent girl as a sexual being as marking a break from the older sexual order, which, in many important ways, it did. In fact, what was most novel about Ellis's depiction of the adolescent girl was not simply his acknowledgment of her sexual desire but rather his attempt to join together this recognition with the more widely held assumption about her essential purity, so as to constitute a set of uneasily corresponding, as opposed to mutually exclusive, attributes of female adolescent psychosexual development.

In his discussion of "the evolution of modesty," Ellis made the case for the concomitance of innocence and sexual abandonment during female adolescence. Modesty, he explained, was "the chief secondary sexual character of women on the psychical side," which originated from an ancient organic tension in the female psyche between a fear of male sexual violence and a need to rouse male sexual attraction. A double-edged trait, it encompassed both sexual awareness and repulsion. Understanding modesty as an ontogenetic, as well as a phylogenetic, phenomenon, Ellis maintained that during childhood girls naturally lacked this trait because they also were largely lacking in consciousness of sexual desire. Civilized adult women were responsible for putting modesty in the service of refining the rituals and emotions of romantic love, but, unlike primitive women, they also ideally abandoned the more rigid elements of the trait that undermined rationality, artistic expression, and personal pleasure. It was, then, during gig-gling, blushing, virginal adolescent girlhood when modesty reached its fullest flowering, with its greatest psychic implications. Thus did Ellis's modern adolescent girl fluctuate between a "shrinking reticence" that underscored her sexual innocence and vulnerability and a sexual precocity that at moments even surpassed the boldness and intensity of the supremely aggressive teenage boy.[106] The girl's psychic vacillation matched Ellis's own, and that of a culture beginning to witness profound changes in the sexual behaviors of girls from across the social spectrum. In the same breath, then, Ellis acknowledged that "[t]he girl at puberty is usually less keenly and definitely conscious of her sexual nature than the boy . . . [Yet,] [e]ven in the matter of conscious sexual impulse the girl is often not so widely different from her brother."[107] Legitimizing and promulgating the mixed messages the mass media and consumer culture were beginning to send to teenage girls, which impelled them to foster their sexual desirability (if not their desire), on the one hand, while continuing to function as the arbiters of sexual morality, on the other, Ellis decreed the period of female adolescence to be fundamentally shaped by the emotional instability produced by the concurrent influence of these two potent forces in the body and the psyche.

Ironically, what Ellis presented as one of the more "liberating" of his contributions—that the girl's essential sexual passivity and naiveté were complemented by normal feelings of desire—turned out to be an approach that tended to reduce the girl almost entirely to her sexually charged innocence. Throughout the *Studies*, Ellis waxed romantic about the psychic, spiritual, and even somatic potential of sexual energy to foster development in the individual and promote progress in the culture. His optimistic interpretation that sexuality was a powerful force that could be transformed into other forms of positive energy was, however, distinctly gendered. Thus, he understood the generative process of sublimation to be a uniquely male phenomenon. The containment of female sexual energy, it seemed, was limited to girls' natural inclination toward an internal reserve and because of an array of negative external inducements, toward repression as well. Although coeducation opponents worried about the possibility that the adolescent girl's application of her newly acquired sexual energy to purposes other than the development of her reproductive organs could undermine future maternal function, Ellis was more interested in establishing that such energy was diffused throughout both her body and mind. "A number of converging facts," he pointed out, "tend to indicate that the sexual sphere is larger, and more potent in its influence on the organism, in women than in men."[108] Less likely than the boy to express her sexual energy in spontaneous sexual behavior and also less likely to transform it into autonomous, nonsexual, creative action, the girl was nonetheless more likely to be thoroughly influenced by her sexual nature, though this was often largely unbeknown to herself and to those adults who ended up utterly bewildered by the unpredictable moods and behaviors marking her adolescence.

Perhaps the most illuminating illustration of the contradictory implications this view of sexuality held for the girl can be found in Ellis's discussion of adolescent dreams. The phenomenon of nocturnal emissions had long been viewed as one of the few visible signs of the onset of male puberty. Ellis examined the recent scientific studies on the topic to conclude that from a medical and psychological point of view, this form of autoeroticism was both common and normal in boys and men and was reflective of the spontaneous and focused nature of the male sexual impulse. The erotic dreams that usually accompanied nocturnal emissions were, Ellis noted, comparatively rare among adolescent girls, and when they did occur, they were much more "irregular, varied, and diffused." Moreover, unlike boys, whose waking life was largely unaffected by erotic dreams, girls rather profoundly felt the influences of such dreams on their conscious actions and emotions, and the effects sometimes confused their perception of reality.[109] Much more widespread than these erotic night dreams among girls, though, was their greater tendency to engage in reverie or daydreaming during their waking life. Ellis admitted that this "very common and important form of auto-eroticism" had thus far attracted little attention among scientists, and he himself did much to legitimize girls' conscious erotic fantasies as a topic worthy of study. In addition, his interpretation of this phenomenon suggested that in daydreaming, girls became the agents in their own erotic development and the authors and subjects of their own sexual life stories. Nonetheless, Ellis presented his daydreaming adolescent girl as a predominantly passive figure, who whiled away otherwise productive hours consumed by the imaginative pleasures of love and romance.[110] With his sexual energies more efficiently focused, compartmentalized, and channeled, the boy awakened from his erotic night dreams refreshed and ready to engage in autonomous creative activity. The girl, in contrast, more constantly coped with a diffuse sexual energy that impinged on her thought and conduct in a vague, but all-encompassing, fashion. Now sexual as well as innocent, the adolescent girl Ellis described was thus freer to claim a passionate nature, indeed, in such a manner and to such a degree that it was this facet of her self that trumped all other aspects of her personality.

Ellis was not unaware of the ambiguous consequences his interpretation of female adolescent sexuality held, if less for modern girls themselves than for the society in which they came of age. One possibility, he acknowledged, was that a general acceptance of youthful female sexuality could lead some abnormal minds to overeroticize barely pubertal girls, to the detriment of the distinct sort of "sexual charm" girls at this age were meant to contribute to the culture.[111] Other implications of his analysis posed additional dilemmas for the treatment of the adolescent girl in modern times. Ellis made the case that once it was recognized that the sexual impulse held such profound sway over boys and girls alike, adolescents of both sexes had the right not only to an information-based sex education but also an erotic education as well. "Even in the great revival of sexual enlightenment now taking place around us," Ellis lamented, "there is rarely even the faintest recognition that in sexual enlightenment the one thing essentially necessary is a knowledge of the art of love." If informed and skilled, he further argued, girls as well as boys had the right and the responsibility to make autonomous decisions about their sexuality. Age-of-consent legislation, especially those laws that set the age at which a girl could legally agree to a sexual relationship in the upper teenage years, therefore visited an injustice on the adolescent girl. "To foster in a young woman who has long passed the epoch of puberty the notion that she has no responsibility in the guardianship of her own body and soul," Ellis decried," is out of harmony with modern feeling, as well as unfavorable to the training of women for the world."[112] Also "out of harmony with modern feeling" (though how "unfavorable" for women and girls would remain open for debate) was the habit of sympathizing with girls who reported cases of violent sexual assault. Once girls were recognized as sexual beings, Ellis suggested, their role in fabricating such accounts or initiating sexual encounters also always had to be considered.[113] Was the adolescent girl to be perceived primarily as a sexual actor or a sexual victim? As responsible for her own sexual self-management and self-control or as needing the external protection of others? As entitled to express feelings of sexual desire or merely her sexual desirability to others? Ellis's conceptualizations of the modern girl's psychosexual development raised a set of new and vexing questions with which the culture would be wrestling for some time to come.

Hall's appraisal of the sexuality of the budding girl kindled similar complications. Ellis and Hall were avid devotees of one another. Ellis peppered his writings with references to and quotations from Hall's works, and he recognized Hall as an important authority on the science of child development.[114] Hall, in turn, was an enthusiast of Ellis's, with *Adolescence* conveying a similar spirit of sexual idealism as that radiating from the *Studies*. Like Ellis, Hall emphasized the connection between the maturation of the reproductive organs at puberty, the establishment of sexual difference, and the awakening of heterosexual desire. "There is," he contended, "great reason to look to sex for the key to far more of phenomena of body and soul at this [stage of adolescence] as at other periods of life, than we had hitherto dreamed of in our philosophy." Also like Ellis, he contended that because the reproductive organs were "both more inward and relatively larger in size and function," girls were, perhaps, even more susceptible to the "sentiments of strange, nameless yearning, aimless unrest, moments of rapture and fullness of life and joy abounding" that characterized the emergence of the sex instinct at adolescence. Thus, in his discussion of periodicity, he suggested that it was normal for "the desire and the flow [to] coincide in time," but that such a possibility had been perverted early in the evolutionary process by primitive aversions to and superstitions about menstruation, which had led to undue repressions of the female sexual instinct.[115] He also noted that adolescent girls were, for a time, more sexually forward than boys, deeming this behavior "a rudiment of the age when woman was the active agent in domesticating man," responsible for holding him "by her own attractions" to his paternal responsibilities "in the long ages that preceded marriage which clenched these obligations." "On this view," he concluded, "woman must once have had courtship proclivities for a prolonged period after as well as before motherhood."[116]

In making such claims, Hall joined Ellis in both formulating the modern expectation that adolescent girls were sexual beings and in begging new questions about the limits of youthful female sexual energy and expression. For his part, Hall assured that the girl's sexual feelings were kept in check by a "deep-seated" female tendency toward "anatomical, physiological, and psychological modesty," which enabled her to fulfill her role as domesticator of male sexuality in the evolutionary process of sexual selection. "All the backfisch does may be directly calculated to provoke proposals," he allowed, "but it is at the same time so unconscious that if anything approaching a tender declaration came, she would draw back frightened lest she had betrayed her heart; and rather than do this she would prefer to die on the spot." Even when she seemed to be taking an active role as sexual pursuer of man, it was always to the end of family formation and to the fulfillment of her self-defining maternal function.[117] An advocate of sex education in the public high school, Hall nonetheless fell far short of Ellis's calls for training the young in the "art of love." He did, however, obliquely encourage another kind of female sexual agency by contending that girls needed to be taught how to protect themselves by being made aware "of the commonest wiles and arguments used for their betrayal." He also hoped they might learn to weigh the "risks, dangers and degrees of permissible liberty" they assumed as both they and society came into a recognition of their sexual nature. Even so, Hall also directly insisted that what girls most needed to gain from sex education was "hygienic instruction concerning their monthly regimen" and, supremely, insight into the meaning of that function and the feelings of "yearning" and "rapture" that accompanied its onset so that they would be sure to recognize and to realize the ultimate purpose toward which their future life was aimed.[118]

With an even greater measure of comfort and confidence than in his treatment of female adolescent sexuality (although related to his endorsement of the adolescent girl's desirability), Hall also sanctioned other psychological traits and social behaviors that nineteenth-century architects of girlhood of all persuasions had deemed to be incompatible with the achievement of their idealized versions of Victorian womanhood. Hall expected the civilized girl to assume the guise of the true woman early on in her adolescence by acquiring and modeling such attributes as piety, selflessness, maternal feeling, and compassion. However, whereas such diverse thinkers as Blackwell, Clarke, Duffey, and Jacobi had decried the girl's inclination toward "precocious" sensibilities and behaviors, including her attraction to fashion, her capricious appetite, her longing for social intercourse and excitement, and her indulgence in sentimental reveries and states of "morbid sensitivity," as detrimental to her physical, psychological, and moral growth (albeit for very different reasons), Hall now rendered these orientations to be largely benign and thoroughly charming attributes of adolescent girlhood. Indeed, he deemed their manifestation to be essential for the girl's development into the sort of woman required by the modern age. Initially, in "The Budding Girl," and even more emphatically in his subsequent "Flapper Americana Novissima," published in the *Atlantic Monthly* in 1922, Hall self-consciously lent scientific legitimacy to manners and mores that middle-class teenage girls had been experimenting with for some time, largely in the context of their school experiences, and that the mass media and commercial advertisers were concurrently engaged in representing and selling to girls and to the culture at large.[119] His "objective and concrete" voice of science also, however, offered ample assurance that the modern adolescent girl's ways of being did not portend any fundamental change in woman's domestic identity, role, or status and, indeed, served to contain the possibilities for female adolescent development within a set of parameters that were decidedly new and remarkably familiar.[120]

"What greater joy has life to offer," Hall marveled in prescribing the daily pursuits of the modern adolescent girl, than "just going to school and coming home again on a fine day, loitering along for an hour or two, strolling by roundabout ways to take in new sights and with other girls, looking at shop windows, enjoying all the sights of the street and of companionship and of freedom, hearing and telling all the news." Such girls, he further elaborated, normatively possessed a "clothes consciousness" that was rooted in the necessities of sexual selection and was borne out by statistics garnered from child-study questionnaires that revealed that many teenage girls for a time planned to become dressmakers or milliners. While her taste tended to be "rather loud" and completely lacking in "a sense of fitness of place, occasion, time of day, or season, or weather," it was, Hall declared, entirely developmentally appropriate that she should seek to express it. Likewise with her appetite, which was "full of whims, freaks, and niftiness" that were, he advised, better recognized as essential to her nature and at least partially indulged at home, rather than dismissed or opposed outright. A highly imitative creature, the adolescent girl was also "vulnerable to scores of fads" adopted by her peers, who dominated her supremely social existence and exerted the most significant influence on her "little store of habits, tastes, viewlets, on life, character, conduct, [and] morals."[121] As much as the adolescent girl was sensitive to her girlfriends' opinions and devoted to cultivating their affections, however, Hall also cautioned that she was to be expected on occasion to belie the more noble aspects of her feminine nature in her heightened capacity for selfishness, irresponsibility, and even cruelty toward those around her.[122] Drawing on his own conversations with an admittedly unrepresentative sample of teenage girls, he further asserted that with fashion, friends, and boys fittingly occupying so much of the adolescent girl's attention, she had little interest remaining for academic pursuits. "What does the backfisch care in her heart of hearts about the shop-worn school studies for their own sake?" he questioned. "She accepts them, with a more or less equanimity somewhat as a necessary evil, but if she is normal she does not put her whole soul into them. 'When I get mad and want to swear,' said one, 'I say "Decimal Fractions!" for that is the dreadfulest thing I know.' "[123] Here shrewdly refashioning the protest against coeducation in more modern terms, Hall posited that the practice was untenable simply because teenage girls had far less intellectual ambition than either its earlier opponents or advocates had been able or willing to recognize.

As anticipated as they were to be, however, Hall also warned that attributes such as the girl's preoccupation with her appearance, her erratic appetite, her gregariousness, her self-centeredness, and (albeit less so) even her anti-intellectualism were not without potential pitfalls and, as with the mandates of male development, needed to be rightly understood and properly managed by girl-workers, parents, and educators, all of whom were to be informed by insight from psychological science. One consequence of the teenage girl's nature receiving its proper due, which Hall noted without much apprehension, was that she was becoming a distinct market for sellers. They sought to appeal to her unique and pronounced tastes in sweets, jewelry, perfume, and clothing.[124] Somewhat problematic was that the girl's expression of such traits was inevitably to engender "perpetual strife with her mother" throughout the teenage years. While denying that the girl possessed the sort of sexual jealousy toward her mother identified by Freudians, Hall nonetheless deemed the mother-daughter relationship as unavoidably difficult and "infinitely complex." Hall argued that this relationship must assume a central place in assessing a range of behaviors in the developing girl, from misbehavior and rebellion to undue expressions of filial devotion and responsibility.[125] Of greatest concern to Hall, though, was that in relating coyly with boys, devoting herself to her friends, and pursuing her many-sided interests, tastes, and enthusiasms, all of which were normative during this stage of her life, the modern girl was nonetheless at risk of becoming lost to her essential self and to the fundamental purpose in life that was necessary for her own happiness and for the ongoing advancement of civilization. "In her physiological and affectional life," Hall contended, the teenage girl had always had "much to conceal," and therefore was, by her very nature, "not what she seem[ed]" to be. Under the conditions of modern life, however, which both gave rise to and indulged the plasticity of adolescence, her innate "passion to deceive" and her "passion for secrets" had threatened to become pathological. "Thus, girls, who have peculiar need of self-knowledge of their sex and themselves, at just the nascent period which nature has provided for the acquisition of that knowledge, now escape it," Hall volubly sighed. "And hence it comes that there was never such disassociation and disintegration of soul before possible as that to which young women are now exposed."[126]

As dire as such a state seemed, however, Hall accepted the "dualization of soul" as particularly emblematic of the youthful female self in modern times. His goal was to make adults aware of the existence of the "upper" and "lower" selves of the adolescent girl and to impress on them the necessity of their accommodation to her "conscious self," all the while assisting her in the uncovering of her "deeper ego" by the time she reached her early twenties. If this seemed too tall an order, though, he also, paradoxically, urged adults not to worry very much about what were, merely, the "surface phenomena" of adolescent girlhood, for if read properly, these indicated that the girl was well on her way up "the steepening, old, well-worn but flowery highway" that was leading her "straight into the paradise of ripe womanhood, never so glorious as now and in this country."[127]

The adolescent girl, Hall promised, had not, after all, changed all that much, or if she had, it was only for the better. Not nearly as independent, self-confident, or defiant as she seemed, the flapper was a mere fledgling whose undeveloped pinfeathers prevented her from straying too far from the nest. In addition, beneath her "bundle of inconsistencies," she still possessed a "fundamental unity," rendering her (to shift metaphors) "simply like a climbing vine in the stage of circumnutation, before it has found the support by which it can raise itself toward the sun." Moreover, in all of the adolescent girl's frenetic flapping and peripatetic sprouting, Hall also discerned the potential for continued racial progress, which was to be rooted in the heightening, if sometimes in surprisingly new ways, of dichotomous sexual difference. "Underneath the mannish ways which [the adolescent girl] sometimes affects," he insisted, "she really vaunts her femininity, and her exuberance gives it a new charm. The new liberties she takes with life are contagious, and make us wonder anew whether we have not all been servile to precedent, and slaves to institutions that need to be refitted to human nature, and whether the flapper may not, after all, be the bud of a new and better womanhood."[128]

Again and again in his writings on girlhood, Hall declared there to be "in all the wide domain of psychology perhaps no such *terra incognita* as the heart of the adolescent girl."[129] Presaging Freud's famous closing remark in his 1933 essay, "Femininity," in which he resigned that what he had not so far been able to explain about the topic using the tools of psychoanalysis would have to be inquired about from women's own experience or the poets, Hall frequently appealed to "common sense" and "common honesty" in describing and prescribing the development of the adolescent girl's body, mind, and soul. In rendering the adolescent girl such an enigma, he also issued a call to his fellow scientists to take her up as a topic of particular concern. Writing at a time of tumultuous social change, Hall applied his own scientific authority and expertise to the invention of a concept of adolescence that looked to the attributes of Victorian girlhood to proffer some critique of civilized masculinity and also allowed for some new possibilities for female development, all the while shoring up the power relations of an older order. In his efforts to resolve the multiple contradictions inherent in his developmental thinking, he ended up reinforcing an already firm foundation for modern psychology's embracing the boy as the normative adolescent. Against such a model, the girl, when she was to be found at all in the psychological literature of the twentieth century, would most often be determined to be lacking. However, by valorizing the girl's traditional feminine qualities, seeking to account for her seeming discontent, raising the specter of her sexuality, and sanctioning some new ways for her to express and assert the self, Hall's concept of adolescence also portended that the challenge of reconciling the experiences and expectations of adolescence and femininity would be a "problem" to be reckoned with—by his fellow psychologists, the culture at large, and girls (and boys) themselves—for some time to come.

CHAPTER FOUR

"New Girls for Old"

Psychology Constructs the Normal Adolescent Girl

Writing in 1916 on the subject of "girlhood and character" for a series of religious education manuals, Mary E. Moxcey registered the wry musings of one father at a southern Sunday school convention on the possibilities for a *female* adolescence. "I don't know much about this 'adolescence,'" the man opined, "and I don't know whether gals is supposed to catch it too; but I do know that the gals is just as ornery as the boys, or a leetle more so."[1] Moxcey's mildly amused determination of the aptness of this man's commentary reveals something about the construction of the adolescent girl that followed in the wake of G. Stanley Hall's *Adolescence*. For Moxcey and others writing on the subject of female child development in the mid-1910s and 1920s, the "girl problem," which was the common euphemism for female juvenile delinquency during the Progressive era, no longer referred only to protecting, controlling, and reforming those "deviant" immigrant and working-class girls who failed to comply with the ideals of Victorian femininity. Rather, such commentators took notice of the rapid diffusion of "ornery" behaviors among female youths from across the social spectrum during the early years of the twentieth century. They also identified the "girl problem" with the challenge of understanding, explaining, and responding to the development of the "normal" modern adolescent girl.[2]

Trained in psychology and active in "club, settlement, Young Woman's Christian Association and Church school work," Moxcey promised to make available to a popular audience of parents, teachers, and girl-workers like herself existing scientific knowledge about girlhood garnered from physiology, psychology, pedagogy, and social work. She hereby continued Hall's link between an interdisciplinary science of child development and the practical application of scientific knowledge, making claims as to both the social necessity and the exclusive authority of that knowledge. As Moxcey would have explained it to the concerned father at the Sunday school convention, adolescence, with its accompanying dangers and possibilities, was something every young person was indeed "supposed to catch," boy or girl, rural or urban, working or middle class. Her aim, then, was to identify those "great underlying uniformities" in the aspirations and problems of all adolescent girls, along with the *"general laws or principles"* that would help in guiding and solving them.[3] In doing so, she joined other developmental thinkers of the period in problematizing the normal, psychologizing the deviant, and re-encoding certain cultural expectations for gender, race, and class difference and hierarchy.

Previous work by historians on early-twentieth-century girlhood has overwhelmingly focused on the cultural constructions and social experiences of the female juvenile delinquent. Historians have also focused on the popular formulation and activities of the delinquent girl's more privileged counterpart, the flapper.[4] Missing from the historical record, however, is a consideration of the psychological construct of the normal female adolescent, a concept that during the 1920s bridged the delinquent and the flapper with a uniform set of explanations regarding the mandates of female development. Locating all girls on a psychological continuum from the normal to the abnormal, the psychological sciences of the 1920s pursued what historian Elizabeth Lunbeck calls the "metric mode of thinking" in their approach to the study of the young female. The metric paradigm preferred age and gender characteristics and differences at the expense of class analysis, identifying the universals of female psychological development as the primary motivators of all forms of female adolescent social behavior.[5]

In the normal female adolescent, American psychologists of the 1920s produced a complex and often contradictory figure that both embraced and tempered some of the more socially threatening behaviors of the delinquent and the flapper. Rooted in gender ideologies of the past, yet also priding themselves on their distinctly modern sensibility, these experts placed the developing girl at the center of intellectual, social, and cultural problems concerning female subjectivity and sexuality. Individual and female, independent and selfless, sexually confident and sexually vulnerable, the adolescent girl that emerged from the psychological literature of the decade was generally meant to be a sympathetic figure, whose behaviors were to be explained, tolerated, and duly adjusted, rather than condemned or punished. Such sympathy was not always forthcoming, however. Nor did it serve all girls equally well. Whereas the new psychology of female adolescence did offer explanations of the growing girl that encouraged broader expectations for female development and greater compassion for deviance than in the past, such explanations came to rest on a range of interpretations that reestablished certain limits on what girls could become, while also laying exclusive claim to the sort of knowledge that would best enable the wider culture to understand them.

Initial scientific conceptions of female adolescence emerged from nineteenth-century medicine within the context of concern for changes in the lives of the white middle class. Rendered traditionally vulnerable and newly vibrant, Hall's turn-of-the-century budding girl also embodied qualities that reasserted and redefined the parameters for white middle-class femininity. Even so, for much of the Progressive era, it was the working-class and immigrant teenage girl who took on newly heightened significance in the scientific literature, as the changing circumstances of *her* work and leisure unfolded within a climate of anxious and idealistic social reform. Some of the ensuing discourse about the female juvenile delinquent encouraged new understandings of the development of the girl. We begin, then, by considering some of the key contributions of the Progressive period. Next, we turn to the years following World War I, when changes in the experiences of girls of the white middle class, again, became an important factor in prompting the literature on the psychology of female adolescence to proliferate in earnest.

CASTING THE "GIRL PROBLEM": FEMALE JUVENILE DELINQUENCY IN THE PROGRESSIVE ERA

In the wake of the heated debates by physicians such as Edward H. Clarke and Mary Putnam Jacobi over the effects of coeducation on the civilized girl's development, white middle-class Protestant reformers named their own version of the "girl problem" as a menace deserving widespread attention and collective response. As young, urban, immigrant working girls sought to forge autonomous and sexualized identities for themselves in the increasingly heterosocial worlds of work and commercialized leisure, reformers of the late nineteenth century warned against what they perceived to be socially disintegrating threats to the Victorian moral code. Drawing on the legacy of antiprostitution work from the antebellum period, wherein fallen women and girls were perceived as either victims of unbridled male lust or dire socioeconomic circumstances outside of their control and from contemporary medical discourse that depicted the teenage girl's body as inherently vulnerable, reformers' efforts to raise the age of consent, combat venereal disease, eliminate prostitution, and regulate commercialized leisure took on a distinctly protective nature. Solutions to the plight of working-class and immigrant girls focused on the creation of institutions to shelter them from the perilous urban environment, including boardinghouses, employment agencies, and working-girls' clubs offering education and wholesome recreation, as well as passing laws to punish evil men for seducing the innocent.[6]

The language of female victimization continued to be a powerful tool for reformers well into the twentieth century. The new generation of reformers of the Progressive era, however, also adopted several other interpretations of the "girl problem." Like their immediate predecessors, Progressives drew their ranks primarily from the white, Protestant middle class and expressed much anxiety over the changes in American society due to the forces of industrialization, urbanization, and immigration. Pursuing a broad array of economic, social, and political reforms from many perspectives, Progressives rallied around problems associated with working-class children and youths. Infant health, child labor, and juvenile delinquency all garnered significant attention within Progressive reform. Progressives' diagnosis of and response to social problems set them apart from previous generations of middle-class reformers. Some, influenced by the ideology of Social Darwinism, adhered to a biological interpretive framework and blamed the innate moral depravity of individuals and groups for the many social ills they saw around them. Others were more likely to identify factors in the environment as the cause for human suffering and social disorder. However they approached or amalgamated these perspectives, Progressives called on the power of the state to reform and regulate both individuals and the social and economic conditions in which they lived. In addition, to identify and understand the social problems that needed attention and to craft and implement policy solutions, they also relied heavily on the research techniques, findings, and authoritative expertise of the burgeoning biological and social sciences.[7]

In responding to the problem of female juvenile delinquency, Progressives both drew on existing assumptions and proffered some new insights into the development of the adolescent girl. One explanation some Progressives offered to account for the worrisome sexual assertiveness of working-class and immigrant girls was "feeble-mindedness." Feeble-mindedness was redefined within the context of the early-twentieth-century eugenics movement. Like the progressivism in which it flourished, eugenics encompassed a range of meanings, which helps to account for its broad-based appeal. Rooted in evolutionary theory, eugenics gained legitimacy following the turn of the century with the rediscovery of Mendel's laws of segregation and independent assortment, which became the basis for the science of genetics. Eugenicists claimed that the physical, moral, and intellectual qualities of human beings were determined by heredity and believed that racial/social progress could be brought about through the encouragement of reproduction by the biologically "fit" and the limitation of reproduction by the "unfit." Given the focus on reproduction as the cause of and solution for the problems of modern life, women (as proponents of eugenics ideas and practices and as recipients of eugenics treatment) and systems of thought about gender and sexuality were of central importance in the early-twentieth-century eugenics movement.[8]

Eugenicist psychologist Herbert H. Goddard was at the forefront of defining the condition of feeble-mindedness and associating it with the dangerous sexuality of the adolescent girl. A former student of G. Stanley Hall and director of research at the Training School for Backward and Feeble-minded Children in Vineland, New Jersey, Goddard played an important role in introducing and popularizing the use of intelligence testing in the United States and argued for a link between inherited mental deficiency and criminality. "We are fast approaching the day when we shall realize that disease and defect, mental and physical, are conditions favorable to the commission of offenses against the public," he wrote in a treatise on juvenile delinquency. "We shall accordingly ascertain the mental and physical conditions of all people and recognize the fact that the persons suffering from abnormal conditions of body or mind are particularly liable to commit a crime."[9] Goddard was responsible for coining the term "moron" to refer to those individuals who never developed beyond a mental age of 8 to 12 years. This group of mental defectives was especially threatening, he averred, because they so often seemed to be "normal" to those around them. For Goddard, one of the primary signs and effects of feeble-mindedness in female morons was sexual depravity. By this, he meant that those girls measured to be of low mental ability on intelligence tests always threatened to commit acts of sexual deviance because they were incapable of developing the capacity for self-control that emerged in late adolescence. In addition, any untoward sexual behavior, including prostitution, unwed motherhood, or trading sexual favors for the "treat" of participating in commercial amusements, called forth a diagnosis of feeble-mindedness as its cause. Joined by others in the eugenics movement, Goddard accused moronic girls of indiscriminately seducing hapless boys and endangering the future of the human race by passing on their debauched physical, mental, and moral traits to the next generation of offspring. Since "negative" eugenicists such as Goddard perceived the cause of female deviance to be the product of inherited traits that could not be changed, the solutions they offered to the girl problem were permanent incarceration and sterilization. By 1923, forty-three states had established institutions to contain the menace of the feeble-minded, and by 1931, thirty states had passed laws authorizing the sterilization of mentally defective persons.[10]

Progressive era psychiatrists cast their own version of the young, female sexual predator in the form of the hypersexual female. Many American psychiatrists were enamored with eugenics, but when girls engaged in sexually deviant behavior tested normal or above on intelligence tests, they reinterpreted the girl problem through the lens of the newly formulated psychiatric category of psychopathy. Psychopathy referred to a broader range of behaviors and emotional expressions than intellectual capacity and was a less sharply defined or concretely measurable condition than feeble-mindedness. It was also, therefore, more flexible and potentially more far-reaching in its explanatory power. Psychopathy thus contained intimations of the "metric mode of thinking" and the turn to the problems of the normal personality in modern psychiatry, which became the basis for its cultural authority in the twentieth century. Even so, for many Progressive era psychiatrists the hypersexual female, which encompassed not only sexually aggressive teenage girls but adult women who claimed any measure of sexual and social autonomy, was clearly defective and diseased and posed an aberrant danger to society that required psychiatric treatment to control and contain it.[11]

Those who made use of the feeble-mindedness and psychopathic explanations for female juvenile delinquency interpreted the sexual misconduct of the adolescent girl as a product not of conscious sexual desire and autonomous sexual decision making but of inherent pathological conditions that could be managed only through the implementation of external social controls. Other experts and reformers of the period, however, drew on, furthered, and redirected some of the assumptions about female adolescence in Hall's work to see girls as willfully engaged in deviant activities for the purposes of personal pleasure and self-determination. This group of Progressives acknowledged the emergence of a normal female sexual instinct during adolescence, recognized its potential for sublimation under proper environmental conditions, and argued for remedies that emphasized education, opportunities for wholesome recreation, improved economic circumstances, sound family dynamics, and individual psychological guidance over punishment and repression.

Jane Addams set the tone for this approach in her 1909 text, *The Spirit of Youth and the City Streets*. Born in 1860, Addams was raised in an upper-middle-class, socially conscious family and educated at the pious and academically respectable Rockford Female Seminary in Illinois. During her early years, her thinking about herself and the world were most influenced by nineteenth-century domestic ideology, with its valorization of civilized women's superior capacity for moral insight, feeling, and agency, as well as romanticism, which emphasized both individuality and the organic connection of all beings. What historian Dorothy Ross calls this "composite domestic/romantic epistemology" propelled Addams to become a leader among the first generations of turn-of-the-century white, Protestant, middle-class new women in seeking to expand the application of domestic values from the home into the public sphere and to align the goal of female self-realization with the fight for social justice.[12] In 1889, she helped to found the Hull House settlement in Chicago, which became a model for social reform efforts in the Progressive era. Through her work at Hull House, Addams continued to derive her principles of democratic social ethics and put these into action by providing direct services to the poor, advocating on their behalf, facilitating their political organization, and encouraging solidarity across lines of class and culture. In all of these endeavors, she also engaged in the practice of what Ross calls "interpretive sociology." In contrast to contemporary academic sociology's emphasis on social knowledge that was derived from abstract reflection and the purportedly objective observation of specialized problems, Addams undertook the creation of knowledge about social groups and relations that was "interpretive, socially situated, relational, warranted by personal experience, and gendered."[13]

The knowledge about youth conveyed in *The Spirit of Youth and the City Streets* falls under this rubric of "interpretive sociology." Here, Addams sought to understand working-class immigrant boys and girls on their own terms and then to interpret their experiences to her own class of reformers to elicit sympathetic response and collective social action. She derived her insight into adolescence from the records of the juvenile court but also, more importantly, from her own intimate relationships and experiences with the young people she had encountered at Hull House. This experience with the young authorized her to tell their stories and required her to consider the multiple perspectives on youthful (mis)behavior held by working-class youths themselves, their parents, and the respectable readers of her book. Her conception of "socially situated" knowledge notwithstanding, Addams nonetheless also established the starting point for understanding the adolescent that all of these perspectives first had to concede: the recognition of the powerful "emotional force" that "seizes" all girls and boys with the onset of the sex instinct at puberty. Without attending to the upheavals in the history of the race that occasioned and were reflected by this force, Addams decidedly joined Hall in seeing the emergence of the sex impulse as the fundamental source of vital energy, compassion, and idealism in the development of the individual, as well as of a distinctive individuality that reached its fullest potential in the service of humankind. In comparison with Hall, she asserted that problems during adolescence arose less because of the inherent volatility of the sex instinct or because civilization inevitably had to repress it, than because modern society had not taken enough initiative to provide alternatives to the crass, commercial outlets ubiquitously available for its expression. In Addams's analysis, working-class parents, whom she had sympathetically observed making untold sacrifices for their children according to their own understandings of their children's best interests, were far less to blame for the behaviors of their children than was the failure of the collective will to provide the young with opportunities for "more adequate public recreation" and meaningful, well-paid forms of industrial labor.[14]

Addams's construction of particular knowledge about the adolescent girl also differed from Hall's in important ways. For Hall, as for Spencer and Clarke, the onset of puberty occasioned some measure of both immediate weakness and sustained deficiency in the girl's development, which justified limiting her educational opportunity during the teenage years and prevented her from making an innovative contribution to racial and social progress as an adult. Hall certainly lauded the girl's exemplary capacity for feeling and compassion, but he also deemed that her altruism was to be normatively manifested in her roles as wife and mother in the private home. Departing from nineteenth-century critics of precocity in the girl's behavior, Hall looked bemusedly on the modern girl's various assertions of independent selfhood, but he also distinctly divorced these from her powers of empathy and worried that they might thwart her realization of her true self and, consequently, compromise the maternal endowment she was to bequeath to the ongoing evolution of civilization.

Addams, in contrast, found nothing debilitating about the onset of female puberty and discovered that the same "quest for adventure" was the primary force driving male and female adolescent development alike. To be sure, she echoed Hall in noting that the roots of such a quest most likely lay in "the unrecognized and primitive spirit of adventure corresponding to the old [male] activity of the hunt, of warfare, and of discovery," and she noted that the universal adolescent search for excitement was most often misdirected in sex-specific ways. However, she also decisively deemed the adolescent prerogative of "yearning towards the world" to be as much a province of the girl as her brother.

Addams then went on to set the girl apart again by aligning her longing for individuality and her capacity for altruism in such a way that she would not only privately exemplify the compassion necessary for the creation of a truly democratic social life, but simultaneously achieve individual purpose and social change by publicly acting on it.[15] Beginning her description of the modern girls she observed on the city streets with the same sort of benevolent posture assumed by Hall, Addams thus went further in seeing in these assertions of youthful female independence the mutually reinforcing potential for self-realization and social progress:

> As these overworked girls stream along the street, the rest of us see only the self-conscious walk, the giggling speech, the preposterous clothing. And yet through the huge hat, with its wilderness of bedraggled feathers, the girl announces to the world that she is here. She demands attention to the fact of her existence, she states that she is ready to live, to take her place in the world. The most precious moment in human development is the young creature's assertion that he [sic] is unlike any other human being, and has an individual contribution to make to the world. The variation from the established type is at the root of all change, the only possible basis for progress, all that keeps life from growing unprofitably stale and repetitious.[16]

The advocates for coeducation in the 1870s had made a similar link between civilized adolescent girls' potential to "rule themselves" and to participate in the "regeneration of the world." Addams's more radically democratic vision now counted immigrant and working-class girls as among those who experienced "youth's most obvious needs" for independence and self-expression and were driven by the "old desire . . . to bring about juster social conditions" for all. Like G. Stanley Hall, she was more willing than Clarke's initial critics had been to accept the frivolous giggling and outrageous fashions as part of adolescent girlhood. She, however, neither defused nor heightened anxiety over such enthusiasms by perceiving them as mere "surface phenomena" that also threatened to conceal from the girl her essential domestic destiny. Rather, for Addams, girls' desire "to appear finer and better and altogether more lovely than they really are" was a longing that was "pregnant with meaning," which had to be fully mined for its deep social significance and also "properly utilized" to realize the Progressive ends of "bring[ing] charm and beauty to the prosaic city and connect[ing] it . . . with the vigor and renewed life of the future."[17]

Numerous works of the period affirmed that the working-class and immigrant adolescent girl possessed a normal sexual instinct that was too often misdirected into inappropriate behavior because of the economic and social conditions in which she lived. Sophonisba Breckinridge and Edith Abbott's *The Delinquent Child and the Home* and Ruth True's *The Neglected Girl* employed more systematic research methods than Addams, but like her, they also made a firm connection between the formulation of social knowledge and its application to projects for broad-based social reform. Members of the Hull House female network, Breckinridge and Abbott received their doctoral degrees in the social sciences from the University of Chicago and were employed as researchers for the Chicago School of Civics and Philanthropy. Their study was based on an analysis of ten years of court and reformatory records from the Cook County Juvenile Court and on interviews with court officials and families of delinquents. Likewise, True, a social investigator for the Bureau of Social Research of the New York School of Philanthropy, drew conclusions about the causes of female juvenile delinquency from the extensive interviews she conducted with sixty-five problem girls. As the titles of their works suggest, Breckenridge, Abbott, and True placed particular emphasis on the role of family dynamics in shaping adolescent misbehavior. According to True, girls whose "impatient desire for action and experience" was thwarted by undue parental control or was allowed to languish under parental neglect were at particular risk of "social, moral and physical deterioration." "The girls' instinctive idealism, a wild thing here, unnurtured, is as elusive and fleeting as it is beautiful," she declared. "It is foredoomed to fade swiftly in the midst of unfriendly reality." Like many other maternalist experts and reformers of the period, Breckinridge, Abbott, and True saw the solution to the girl problem as residing at once in working-class mothers' adoption of the habits, morality, and child-rearing sensibilities of the private white-middle-class family *and* in the structural reform of the conditions of industrial capitalism that would make it possible for individual mothers to achieve those ends.[18]

While Addams and her fellow female expert reformers made significant, if also selective and strategic, use of Hall's developmental psychology in their renditions of the girl problem, it was under the influence of physician William Healy that Progressive era approaches to juvenile delinquency took their most pronounced turn toward psychology, with important consequences for the treatment of juvenile delinquency and normal adolescence alike. Healy was born in 1869 to a poor tenant farming family in Buckinghamshire, England, and immigrated to the United States when he was 9 years old. In a remarkable story marked by hardship, personal initiative, and the benevolence of others, he eventually made his way to Harvard University, Harvard Medical School, and Chicago's Medical College, where he received his medical degree in 1900. After doing postgraduate research in Europe, Healy returned to Chicago to join a private neurology practice. In 1909, a committee organized by Judge Merritt W. Pickney of Chicago's juvenile court appointed him to direct the newly founded Juvenile Psychopathic Institute (JPI), whose mission it was to use the medical and psychological sciences to investigate and treat the dismaying problem of repeat juvenile offenders. Such a goal was impelled by the initial assumption, held by Ethel Sturges Dummer, the child saver, wealthy philanthropist, and leading patron of the JPI, and others on the committee, that recidivists were mentally abnormal and that psychological expertise was necessary to understand and to address the "root" cause of these children's recurring misbehavior.[19]

The most important direct influence on Healy's work in carrying out the purpose of the JPI was the eminent psychiatrist and leader in the concurrently burgeoning mental hygiene movement, Adolf Meyer. In his capacity as director of the Henry Phipps Psychiatric Clinic at the Johns Hopkins School of Medicine, Meyer pioneered in turning psychiatry away from a somatic approach to mental illness to a "psychobiological" framework. Mental illness, personal unhappiness, and socially deviant behavior, he maintained, were all caused by a maladjustment between the individual and the environment, the result of a broad array of intersecting biological, psychological, familial, and social factors. Along with other reformers and psychiatrists associated with the National Committee for Mental Hygiene, Meyer saw childhood and adolescence as crucial moments when such maladjustment appeared and therefore could be most effectively remedied. Healy brought Meyer's mental hygiene framework to bear on the recidivists to Chicago's juvenile court. He played a leading role in founding a child guidance movement that took as its primary subject the whole child who was best understood and treated as an individual "mind at risk," rather than the social and economic context in which juvenile misconduct occurred.[20]

In 1915, Healy published his first book advocating for a professional child guidance approach to the problem of juvenile delinquency. *The Individual Delinquent: A Text-book of Diagnosis and Prognosis for All Concerned in Understanding Offenders* was a massive tome reporting on his analysis of 1,000 case studies of repeat offenders, the average age of whom was between 15 and 16 years old.[21] At the outset, Healy informed his readers of his fundamental finding that delinquency could not be reduced to a diagnosis of mental abnormality and declared that his primary interest was in "the complexity of human nature in relation to complex environmental conditions." He also emphasized the importance of a developmental approach to comprehending and responding to this multifaceted dynamic. "Just because the delinquent's character is the result of a long-continued process of growth," he asserted, "one needs to regard him as the product of forces, as well as the sum of his present constituent parts; one must study him dynamically as well as statistically, genetically as well as a finished result." Many of the "causative factors" Healy found to be associated with delinquency—including inherited biological defects, physical ailments, stimulants and narcotics, incompetent parenting, poverty, bad companions, and commercial amusements—were commonplace in the Progressive era discourse of juvenile misconduct. Healy's main contribution was his focus on the interactions among these factors and also the primacy he gave in diagnosing and treating the "mental life" and personality of the individual coping with them—what he succinctly summarized as "the complexity of causation, determinable through study of the individual case."[22]

Like Hall, Healy made much of adolescence as a delinquency-prone stage of life and singled out puberty as a fundamental cause of the serious "developmental physical abnormalities" and "mental aberrations," as well as the milder "character changes and peculiarities" that often precipitated juvenile crime at this age. He explained that teenage boys were particularly prone to misbehavior if they suffered from "poor general physical development" or delayed puberty. Teenage girls, however, were apt to act out in sexually inappropriate ways if they experienced overdevelopment or premature development of the sex characteristics. In the case of girls, Healy was surprised to find that "[i]n the great majority of instances the instabilities of adolescence are not at all centered about the menstrual period." He nonetheless concluded that even in normally developed girls, the "tendency towards restlessness and perhaps misconduct," and especially the "inclination towards sex misdemeanors," was frequently exacerbated during the regular recurrence of menstruation. In reporting these latter findings, Healy highlighted a problem he believed plagued those who endeavored to understand either adolescence or juvenile delinquency. "The line of demarcation between the normal and the aberrational during the adolescent period," he cautioned, "is very difficult to maintain." Healy asserted that while it was important to distinguish between the "usual storm and stress" and the pathological during this time of life, it was by no means easy to do so. With this admonition, he suggested that at least some elements of his approach to the problem of the relationship between mental life and (mis)conduct in the "individual case" might have broader applicability beyond working-class and immigrant youths who found themselves under scrutiny by the juvenile justice system.[23]

In juvenile courts and female reformatories around the nation during the Progressive era, these strands of thought regarding the female adolescent's subjectivity and sexuality competed and intertwined to produce a complex mélange of advocacy, paternalism, and social control. They also yielded varied and mixed results in the lives of individual girls and their families.[24] What each of the Progressive era interpretations of the "girl problem" had in common, though, was a continued endorsement of the ideals of nineteenth-century civilized morality, particularly its valorization of female chastity, and a widespread focus on the immigrant and working classes. Progressive experts and reformers emphasized a qualitative difference between the compliant and the deviant, between those who conformed to the desirable norms of social behavior and those who, for whatever of a number of possible reasons, did not.

Yet, as Hall's, Addams's, and Healy's work suggested, the distinction between "good" and "delinquent" girls was not always as evident as some reformers would have liked. The blurring of this line was in part realized by changes in the social activities of the "good" girls themselves. The movement of the white middle-class girl away from the Victorian ideals of girlhood began in the late nineteenth century with the threat she posed to the male privilege of higher education. It continued into the early years of the twentieth century as more girls of the middle class found their way into higher education and the labor market, as they mimicked the sexualized and autonomous behaviors of their working-class sisters, as they cultivated modern manners unique to their particular social environments, and most notably as they succumbed to "khaki fever" and sought out romantic encounters with servicemen during the First World War.[25] The result of this unsettling conduct on the part of the middle-class girl was that by 1920, she too was being steadily drawn into the gaze of the social commentator and the social scientist. Far less likely to receive public punishment or be subjected to state intervention than her working-class or immigrant sister because of her privileged status, she instead became the catalyst for psychological conceptualizations of a normal modern female adolescence.

To acknowledge the white middle-class girl's capacity for "deviant" behavior, however, meant that deviance itself needed to be reexamined, with the resulting new interpretive frameworks circled back onto the figure of the female juvenile delinquent. Historian John R. Gillis describes this relationship between ideas about normalcy and deviance as "dialectically inseparable in their origins and development."[26] In the case of the girl, it was the psychological formulation of a female adolescence that effected the dialectical negotiation between the resolutely working-class image of the juvenile delinquent of the Progressive era and the baldly middle-class figure of the flapper that emerged in full force during the 1920s. The female juvenile delinquent remained an important figure in the social scientific discourse of this period, only now she appeared amid and in relation to a proliferation of knowledge production about the development of the normal adolescent girl.[27]

RECASTING THE "GIRL PROBLEM":
NORMAL FEMALE ADOLESCENCE IN THE 1920S

Psychologists' recasting of the "girl problem" in the 1920s took on meaning in relation to large-scale transformations in American society and, more specifically, to changes in young people's lives, in intellectual and professional developments in the science of psychology, and in meanings and politics of gender. The introduction of new industrial technologies in the 1920s made mass production more efficient and fueled the widely touted economic prosperity of the decade, although this was unevenly experienced by different groups of Americans. Most of the economic growth was in the consumer goods industries, including the manufacturing of automobiles and household appliances such as refrigerators, stoves, and washing machines. Increasingly, American business was controlled by large corporations, which created new kinds of jobs that standardized the labor of both working-class and middle-class employees. Corporate success depended on the efforts of marketers and advertisers, who made use of new research techniques and retail strategies to sell not only consumer goods themselves, but also a consumer ethos that emphasized the values of personal pleasure, leisure, and individual status and success over older ideals of productivity, collective responsibility, and self-restraint. Americans imbibed the values of the consumer culture, as well, in their engagement with the new forms of media communication that burgeoned during the decade, most notably the radio, the movies, and the mass-market press. In part because of the migration of African Americans and the immigration of eastern and southern Europeans to urban areas, the majority of U.S. residents lived in cities for the first time in the 1920s. The mass marketing of the automobile promoted the growth of suburbs, as well. Many Americans embraced the modern values and lifestyle fostered by the mass consumer economy, seeing these as signs of and routes toward ongoing social progress. Others, however, expressed ambivalence, anxiety, and outright opposition and sought to stem the tide of change through the forces of political conservatism, nativism, and antiradicalism that also took firm hold during the course of the decade.[28]

These developments provided the larger context for the significant changes in schooling, work, and leisure that took shape for girls and boys from all social milieus in the 1920s. The combination of compulsory school attendance laws and technological innovations that diminished the need for unskilled laborers meant that more children and adolescents from diverse backgrounds were attending public primary and secondary school than ever before. As historian Joseph M. Hawes explains, this led to the emergence of "an expanded, centrally important public-school system . . . that would socialize all its pupils into the modern metropolitan culture of postwar America." Attendance at colleges and universities among white middle-class youths was on the rise, as well. At the same time, the expanding urban industrial economy that supported the growth of a mass culture emphasizing entertainment and consumption was aimed in large measure specifically at young people. Girls and boys from across the social spectrum now eagerly frequented the commercial dance halls, amusement parks, and movie theaters and adopted the risqué modes of fashion and style formerly associated with the working class and ethnic and racial minorities. Together, these transformations worked to both regulate and standardize the lives of the young to an unprecedented degree *and* to foster the creation of distinct and autonomous peer cultures through which the young rebelled against a range of social and cultural conformities. Certainly, the youth culture of the high school and college was never homogeneous. Nevertheless, more than ever before, it seemed that young people had little in common and, indeed, were fundamentally at odds with adults. They also had more in common with each other, irrespective of their class, ethnic, racial, or gender differences.[29] It was, then, as white middle-class girls increasingly appropriated the manners and mores not only of their middle-class brothers but of their working-class, ethnic, and African American female peers as well, that they became renewed objects of interest for scientific experts, who now sought to articulate a set of expectations for a normal psychology of female adolescence that could be relied on to interpret and direct the entire universe of teenage girls' behaviors.

Such scientific conceptions of the normal adolescent girl emerged within the broader purview of the increasingly professional and popular field of psychology in general and developmental psychology in particular. Beginning in the 1920s, the scientific study of the child became formalized, located in particular institutions, associated with certain credentials and professional identities, and tied to specific funding sources. Hall had done much to inaugurate a formal science of child development. However, even before the publication of *Adolescence*, he faced criticism by his colleagues in psychology for his lack of scientific rigor in relying on mothers and teachers to gather data for his child-study questionnaires, as well as for his speculative reliance on recapitulation theory. In addition, progressive

educators attacked his support of corporal punishment, his opposition to coeducation, his sanctioning of savage behavior in boys, his antidemocratic sentiments, and his perceived prurient attention to sexuality.[30] With the waning of the
child-study movement, a new impetus to study the child arose in 1909 when
former mental patient Clifford Beers founded the mental hygiene movement,
spurring a national public interest in mental health. Mental hygienists incorporated insights from Freudian psychoanalysis, the behaviorism of John B. Watson,
and the holistic psychiatric approach of Adolf Meyer. Placing particular emphasis
on the environmental (as opposed to neurological or inherited) causes of mental
health problems and on the possibilities for prevention, they focused on the role
of early childhood experiences and the influences of family dynamics and parental responsibility in promoting or inhibiting sound mental health and appropriate
social behavior.[31]

From 1910 to 1919, mental hygienists contributed to the development of
new approaches to the treatment of juvenile delinquency. Viewing delinquent
behaviors as the product not of inherent mental defects but of personality disorders, mental hygienists such as William Healy contended that psychological
treatment ought to replace the incarceration and punishment of juvenile offenders. This approach led to the foundation of child guidance clinics associated
with the juvenile courts in cities across the country.[32] The psychiatrists, psychologists, and social workers who staffed the child guidance clinics faced
persistent difficulties in solving the problem of juvenile delinquency, however.
As a result, they increasingly turned to what they imagined to be the more treat-
able minor behavioral and emotional problems of children from modern middle-
class families as their primary object of concern. Their motives, explains historian Kathleen W. Jones, were mixed. Child guidance practitioners of the 1920s
were driven by a need "to legitimate their professional authority and thereby
expand the clientele for their services," as well as by their humanitarian hopes
"to engineer a more perfect society by placing their faith in the promise of
science."[33]

Child guidance experts depicted every child as essentially normal, with the
capacity for experiencing some degree of maladjustment during development.
What individuals needed to ensure their personal happiness and well being, and
what society needed to guarantee both the stability and progressive dynamism
of its citizenry, was for parents, teachers, and girl and boy workers of all kinds to
have access to expert knowledge about the expectations for normal development
and the means by which it was apt to go wrong. As Phyllis Blanchard, psychologist for the Philadelphia Child Guidance Clinic, attested:

Gradually, the layman is becoming educated to the real facts in regard to these questions that normality is largely an abstract concept, that abnormality is not a separate entity but differs only qualitatively and by degrees, that each one of us has his own individual differences from the hypothetical normal type, and that, in a sense, we are all abnormal—some more, others less; some in one way, some in another. With this better understanding as to the meaning of abnormal trends, we shall be more prone to seek advice for ourselves and our children, and the psychiatric clinic will thus have, in time, a far-reaching influence upon individual and social welfare.[34]

In Blanchard's framework, every child was an essentially *ab*normal child, with abnormal referring not to inherent defectiveness, but to the high degree of vulnerability to the many minor, but nonetheless consequential, pitfalls of normal development.

To child guidance practitioners like Blanchard who centered their attention on the abnormalities of the normal child, "normal" remained a highly "abstract concept," with multiple and often contradictory meanings. Referring to the usual, the desirable, and the necessarily elusive, the idea of the normal posed particular problems for those devoted to thinking about the adolescent stage of development. "Normal adolescence," wrote psychologist Jessie Taft, "is a combination of terms that may perhaps be considered contradictory. If by normal one means average, and at the same time implies painless adolescence or adolescence without conflict, then certainly there is a contradiction. For the adolescence that occurs without stress and strain is too unusual to be called normal, and if it were the usual thing, it would have no mental-hygiene problems to be discussed."[35] For Taft, Blanchard, and other psychologists of the 1920s, one of their primary tasks became the enumeration and explication of those "mental-hygiene problems" of "normal adolescence," along with the developmental characteristics, needs, and mandates that underlay them.

Scientists of the 1920s ascertained developmental norms and identified mental hygiene problems by employing rigorous empirical research methods at universities and child welfare stations nationwide.[36] Their findings were disseminated through a vigorous parent education movement that sought to enlighten mothers of their role in both causing and ameliorating the conflicts and difficulties in their child's development. Women received such information from a variety of popular venues, including child-rearing manuals, nursery schools, women's clubs, college courses, newspaper columns, and the newly founded *Parents* magazine.[37] The work in child guidance, child development research, and parent education in this period was supported by state and federal tax monies and by two private philanthropic foundations, the Commonwealth Fund and the Laura Spellman Rockefeller Memorial. As with previous endeavors by scientific experts to understand and explain the child, much of the impetus for this work was decidedly inclined toward a concern for child welfare and hope for social reform and progress. Such goals also existed in concert and tension with experts' interest in enhancing their professional status and influence, their anxiety about preserving social order and stability, and their increasing focus on the therapeutic treatment of individuals and families, as opposed to a broad vision for social change.[38]

Finally, the incursion of the "new woman" into public and intellectual life and the emergence of modern feminism also played a role in shaping new scientific conceptions of female adolescence.[39] White men dominated the newly codifying and professionalizing social sciences of the early twentieth century. College-educated white women, however, also participated in shaping some of the questions, methods, and knowledge claims of the disciplines of sociology, psychology, and anthropology in these formative years.[40] Jane Addams challenged the purportedly objective standards of academic sociology with an "interpretive" approach to the formulation of social knowledge that allowed for the influence of personal experience and gendered perspective. In contrast, many other women, especially by the 1920s, were drawn to professional social science precisely because they saw in its "supposedly neutral and meritocratic ideology" their best hope for escaping the sex stereotyping that limited their claims to intellectual and social authority and power. Unfortunately, the same hallmarks of meritocracy and objectivity that women saw as windows of opportunity were used to regulate access to and advancement within the professions as well, with standards of training, credentializing, and licensing often working to further male privilege.[41]

Ironically, it was because of the sexism women faced in seeking jobs in mainstream academic departments, as well as the low prestige associated with working with children, that some were able to find a home in the field of child development studies.[42] Indeed, such was their influence during the 1920s that the editor of the first edition of A Handbook of Child Psychology felt the need to defend against it. "[M]any experimental psychologists continue to look upon the field of child psychology as a proper field of research for women and for men whose experimental masculinity is not of the maximum," Carl Murchison lamented. "This attitude of patronage is based almost entirely upon a blissful ignorance of what is going on in the tremendously virile filed of child behavior. The time is not far distant, if it is not already here, when nearly all competent psychologists will recognize that one-half of the whole field of psychology is involved in the problem of how the infant becomes an adult psychologically."[43] Murchison's statement reveals the degree to which many men persisted in coding the scientist as male and the scientific enterprise as a masculine endeavor. It also exposes the anxiety experienced by some men in the psychological sciences as their disciplines became increasingly focused on mediating the problems of everyday life, including those related to child rearing and family relations. This marked an appropriation of domestic knowledge from women to the expert that became an important source of male psychologists' professional and cultural authority. Despite women's struggles by the late 1920s to hold onto gains made earlier in the century in the fields of professional social science, however, men's achievement of that authority did not occur without some concessions to the contributions of women scientists.[44]

For all the experts, men and women, who heeded Hall's call to map the *terra incognita* of the psychology of the adolescent girl, it was almost impossible not to position themselves in some relation to the various perspectives on feminism and women's rights that had culminated in women winning the vote in 1920. Indeed, the struggle for suffrage, the subsequent passage of the Nineteenth Amendment, and the ensuing battle over the Equal Rights Amendment inflamed a political, intellectual, and popular debate over equality versus difference in gender relations that came to inform virtually all conceptualizations of female development in one way or another. If women were no longer commonly excluded from public life by virtue of their shared disfranchisement, both proponents and opponents of feminism in the postsuffrage era wondered, were they now to be regarded as different from or the same as men? If different, did that difference connote deficiency, exceptionality, or vulnerability? If the same, did that mean women were entitled to equal opportunity, treatment, or outcomes in private and public life?[45] In pondering these questions, many psychologists rejected feminism outright and used the tools of the discipline to dismiss feminists' claims to social power and their critique of male domination. Some women in psychology found it hard to sustain a feminist critical stance in the face of the commitment to objectivity within social science ideology; they also sometimes deemed it undesirable to claim a feminist identity lest it jeopardize their authority as disinterested professionals.[46] Even so, some female psychologists who took up the problem of the girl's development managed to sustain both a devotion to empirical science and some measure of a critical stance against gender inequality. Notable among them were Phyllis Blanchard, Leta Hollingworth, and Lorine Pruette, whose descriptions of and prescriptions for adolescent girlhood are discussed next. Representing a range of positions on the equality/difference paradox that defined modern feminism, Blanchard, Hollingworth, and Pruette drew on varying currents in feminist thought to name and normalize a category of modern female adolescence. They also constituted the "girl problem" as a particular venue on which the meanings and merits of feminism were to be discerned and debated, at a time when the organized political activity on behalf of women's rights crested and then declined.

In 1928, Leta Stetter Hollingworth published a textbook entitled *The Psychology of the Adolescent* that constituted a notable synthesis of the psychological literature on normal adolescent development that had accumulated over the course of the decade. Hollingworth was born in 1886 to a doting mother who expressed much interest in her daughter's early development. In the baby book Margaret Danley Stetter kept before her death when Leta was just 3 years old, she presaged two of the themes that would shape her daughter's career as a psychologist when she wrote about the sense of injustice her 8-day-old baby girl felt when her father expressed his wish that she had been born a boy and recorded young Leta's intellectual giftedness and precocity. Following her mother's death, Leta experienced a painful childhood under the sway of a difficult stepmother and sought refuge in academic pursuits, at which she excelled. After graduating from high school at age 15, she went on to study literature and writing at the coeducational University of Nebraska, where she met her future husband, Harry L. Hollingworth, who would become a leader in the field of applied psychology during the early twentieth century. Following a brief attempt at a writing career and stint as a teacher in Nebraska, she moved to New York City, where Harry was working as an assistant to the Columbia University psychologist James McKeen Cattell. Discouraged by the lack of opportunity available to women to pursue an advanced degree in literature or even, because she was married, to teach in the public schools, she turned in 1911 to graduate studies in psychology at Teachers College, Columbia University. She and her fellow female students experienced considerable hostility. She managed to endure it in part because of her husband's financial support and intellectual encouragement.[47]

Leta Hollingworth devoted her graduate career to countering reigning assumptions about women's limited mental abilities. Some of the foundation for her work was laid by her advisor, the educational psychologist Edward L. Thorndike, who was a firm believer in women's intellectual inferiority and was therefore not terribly enthusiastic about training women graduate students. Nonetheless, Thorndike, along with such psychologists as Robert Woodworth and Clark

Wissler, sought to replace the evolutionary suppositions of G. Stanley Hall with a rigorous experimental psychology whose findings were beginning to challenge numerous notions about the determining influence of the body over the mind.[48] Hollingworth went further than her male colleagues in applying the methods of experimental psychology to investigating the question of sex differences in mental capacity. Her first contributions to the field were a study that disproved the deleterious effects of menstruation on women's mental abilities and research that challenged the notion from evolutionary theory that males were more intellectually variable than females.[49] By the time she received her doctorate in 1916, Hollingworth had already made several foundational contributions to research on the psychology of women. Her first academic job, as instructor of educational psychology at Teachers College, however, took her in another direction, and as a result, she came to devote most of her attention for the rest of her career on the social and psychological problems of mentally defective and intellectually gifted children.[50] Her consideration of the normal adolescent in *The Psychology of the Adolescent* thus represented somewhat of a departure from her focus on deficient and exceptional children. Hollingworth did not, however, depart from her allegiance to or faith in the methods of empirical social science. The current "lore" of adolescence, she criticized, rested "upon the mere opinions of professional observers" rather than quantitative research, "which would give observation the status of scientific fact." While acknowledging that much work still needed to be done in this regard, Hollingworth's text nonetheless quickly became a standard in the field, offering an authoritative paradigm of the mandates of normal adolescence that thoroughly emblemized the current mode of thinking on the topic.[51]

Hollingworth began her book with the problem of providing a definition of adolescence that would mark the distance psychology had traveled over the twenty-five years since the publication of Hall's foundational text. Like most psychologists of child development by the late 1920s, she gently dismissed Hall's "voluminous pioneer works on adolescence" as being "of historic value primarily, rather than of scientific or practical value today."[52] Indeed, in the most comprehensive analysis of individual human development published since *Adolescence*, Harry Hollingworth decisively rejected recapitulation theory as an explanation for any aspect of postnatal development. Instead, he endorsed a widely held view among social scientists that understood development to be the product of a complex interaction between the forces of nature and nurture. The study of individual development, he maintained, consisted of a multifaceted consideration of "the general facts of evolution and embryology, modified . . . by the determination of education and training, and under constant influence of social heritage and established institutions."[53] That said, pressing questions remained about how and in what proportions nature and nurture interacted in the child's development. As in the past, these questions carried both scientific and ideological import, as experts approached them from a variety of perspectives on the problems of individual freedom and social equality.

The orientation of mental hygiene certainly marked a change in focus to the influences of the child's immediate environment on development, with many of its adherents professing egalitarian attitudes toward social difference.[54] At the same time, scientists continued to pay attention to the biological influences shaping child development under several guises during the 1920s. Arnold Gesell, Hall's former student and director of the Yale Psycho-Clinic, provided the most pronounced continuity with the organic developmental paradigm with his maturational model of development, which charted the biologically propelled, normative stages in the child's physical, mental, and social development. Gesell's approach created universal standards and expectations for development against which all children were to be ranked and measured, thereby supporting an ethos of conformity in an era of mass consumption and political conservatism. However, he also profoundly criticized the dangerously antidemocratic tendencies of behaviorism, which he accused of subjecting children to indiscriminate parental manipulation and power at the expense of fostering their inherent individuality.[55] Other liberal social scientists rejected the group classifications based on race and gender difference that predominated in late-nineteenth-century science with appeals to a common human nature or to the individual inheritance of mental capacity and temperament. Still others made gestures to the new science of endocrinology and its effects on the growth of the child. By the late 1920s, a mixture of environmental and biological factors even made its way into the mainline eugenics movement. Faced with criticism from geneticists for their simplistic understanding of the workings of heredity, as well as for their conservative political agenda, and cognizant of the environmentalist orientations of much of social science, mainline eugenicists ensured the survival of the movement and secured the expansion of its sterilization policies by shifting their focus to mothers' responsibility in child *rearing* as playing the vital role in impeding or facilitating racial progress.[56]

Within this context, Leta Hollingworth's *The Psychology of the Adolescent* took up the thorny task of disentangling biological puberty from social adolescence, with the effect of challenging and reaffirming the expectations for gender, race, and class difference and hierarchy inscribed by earlier architects of the concept of adolescence. Thus, one point of Hall's with which Hollingworth did partially agree was that adolescence was a stage of life that was unique to modern society. According to Hall's evolutionary biological paradigm, adolescence emerged via ontogeny's recapitulation of phylogeny as a product of and facilitator to the acme of modern civilization. In his analysis, it was the civilized boy who experienced the longest period of auspicious moratorium between the achievement of biological maturity and the realization of adult social status, thereby accruing the greatest benefits conferred by this stage of life. Likewise, Hollingworth turned to descriptions of puberty rites in primitive cultures to argue that while bodily changes during the teenage years were universally experienced (differing most markedly by sex), "[t]he years of growth and change which follow [puberty], and which civilized peoples call *adolescence,* were and are usually disregarded in the practices of savage tribes." Hollingworth nonetheless also departed from Hall in emphasizing that it was the developmental difficulties afforded and posed by western civilization for the child to "build up habits of self-determination and self support," rather than sudden physiological change, that accounted for the storm and stress that was preeminently experienced by youths in modern society.[57] Hollingworth also differed with her former advisor Edward Thorndike's views on the matter. Thorndike was a rather lone voice during the early twentieth century in questioning Hall's proposition that adolescence was necessarily, and beneficially, a period of emotional upheaval for any group of young people— civilized or primitive. Many more social scientists aligned with Hollingworth in attributing adolescent stress to the social environment rather than biology, but continuing to accept it nonetheless, and, indeed, valorizing it as a particular virtue of modernity.[58]

Even as they attributed the stresses of adolescence primarily to the demanding pressures of advanced civilization, however, experts also located the normal developmental mandates of the adolescent stage, irrespective of their environmentally induced stress factor, as largely beyond or before the influence of social forces. Thus, the central contribution of Hollingworth's *The Psychology of the Adolescent* was its identification and explication of four "life-problems" or adjustments that she declared to be essential to the adolescent stage of development. Claiming to have found rudimentary evidence of these adjustments in the puberty rights of primitive cultures, Hollingworth identified them as "psychological weaning," the struggle to achieve independence from the family; "mating," the realization of heterosexual desire and its manifestation in marriage and parenthood; "seeking self-support," the choice of vocation and the achievement of economic independence; and "achieving a point of view," the "finding" of the self and the rational determination of its place in society.[59]

These mandates constituted an organizational starting point for many considerations of adolescence during this period.[60] In elaborating on them, experts expressed less interest in the role of biological puberty in shaping adolescent behavior than in the influences of early childhood experiences, family dynamics, peer pressures, the climate of schooling, and the demands of living in a fast-paced, heterogeneous, modern society. Reluctant to replace biological determinism with environmental mechanism, however, optimistic and enlightened psychologists also identified adolescence as the moment in the life cycle when the developing individual would take an active role in negotiating external influences and forge an independent self who would make an innovative contribution to social progress. They hereby deemed autonomy, choice, and self-determination to be the essential hallmarks of the adolescent stage of development. As Phyllis Blanchard and her colleague sociologist Ernest R. Groves explained, adolescence was above all else a period of "personality readjustment," during which developing individuals were at last able "to acquire insight into their own [emotional] reactions" to parental and social influences and pressures and "remold themselves in harmony with newly acquired ideals."[61] While Hall, too, expected boys in late adolescence to exercise these capacities for individual selfhood, he also extravagantly enthused about the "feminized" qualities of adolescence that were rooted in the dynamics of the pubertal body—(inter)dependence, emotionality, religious enthusiasm, and altruism—of which he depicted the girl as the chief exemplar. As their interest in biological stress waned, psychologists also turned their attention away from these feminine characteristics to focus more fully on the masculine mandates for the adolescent to wrest out emotional, economic, sexual, and intellectual independence in a complex modern world. In deeming these mandates universal to the adolescent life stage, experts only incompletely attended to the ideologies of gender, race, class, and culture that significantly informed them. One result was that the conceptual adolescent girl was freed from an all-encompassing biological determinism. At the same time, her development was increasingly assessed according to her capacity to negotiate the opportunities and problems arising out of freedom of choice and identity formation, mandates derived from the struggles of the modern boy to realize his society's highest achievements of rationality and individualism.

As we will see in the examination of each of the four mandates for adolescent development below, some architects of the "new girl" saw such expectations as filled with possibility and honed in on the positive changes in the modern female adolescent's psychology and social experiences. For them, modern girls were now finally poised to assume the masculine privileges that had long been associated with the adolescent stage of life. Others responded with anxiety and sought to offer reassurances about the enduring "feminized" qualities of the female psyche. A few attended to the girl's experience of development as a way to challenge the dominant expectations of the adolescent stage itself. Even as no clear consensus reigned, psychologists of the 1920s tried hard to engender an authoritative set of requirements for normal female development that was meant to prescribe the surest route for every girl to personal happiness and social adjustment in the modern age.

THE FOUR "LIFE-PROBLEMS" OF NORMAL ADOLESCENCE
The "Captain of Her Own Soul": The Problem of Female Adolescent Autonomy

Scientific conceptualizations of child development have historically served as one important site for formulating and contesting the ideology of American individualism. The sociocultural context of the early twentieth century, into which a formal science of child development emerged to claim more authority than ever before, was particularly overwhelmed by rapidly changing ways of experiencing and defining the individual self and its relationship to the social order. Among such changes that Americans witnessed in these years were the growing anonymity of large cities; the massive movements of migrants and immigrants across internal and national boarders; the increased standardization of labor under mass production; the unsettling challenges to traditional religious beliefs by the certainties of science; the rising divorce rate; the abundant promises of mass culture to satisfy desires for personal pleasure; and the lamentation over the condition of human alienation within cultural modernism.[62]

Experiences of these changes differed, of course, along the lines of gender, race, class, and age. By the 1920s, opportunities for autonomy for many men of both the middle and working classes had notably declined as work in both white- and blue-collar jobs allowed fewer occasions for self-management and individual discretion. Middle- and working-class women, too, faced this increased standardization of labor in the workplace. Yet they also assumed both a real measure and the appearance of greater autonomy as they embraced expanding opportunities for consumption, leisure, work, and political participation.[63] Likewise, working- and middle-class boys and girls coming of age in this period escaped some older forms of dependency as they cultivated all-important peer relationships, fostered autonomous youth cultures in dynamic urban environments, and assumed active roles as workers and consumers in an advanced industrial economy. Many adolescents, however, were also subjected to greater adult surveillance than ever before as they came into conflict with new sorts of limits imposed on individual freedom by such institutions as the high school, the juvenile justice system, and the mass market. They also faced new pressures to conform to certain kinds of behavior by their own agemates.[64] The socioeconomic forces that contributed to a decrease in personal autonomy for some and an increase for others gave rise to cultural anxiety that both worried about the personal implications of individuals having too little freedom *and* the social consequences of individuals having too much.

The newly codifying social sciences of the early twentieth century offered up many interpretations of these social changes and took on the problem of the relationship between the individual and the social as fundamental to the purpose of their inquiry. Sociologist William I. Thomas conceived of the relationship between individualization and socialization as a dialectic, each to be pursued for the sake of the other. He explained:

> The problem of the desirable relation of individual wishes to social values is twofold, containing (1) the problem of the dependence of the individual upon social organization and culture, and (2) the problem of the dependence of social organization and culture upon the individual. In practice the first problem means: What social values and how presented will produce the desirable mental attitudes in the members of the social group? And the second problem means: what schematizations of the wishes of the individual members of the group will produce the desirable social values, promote the organization of culture and society?[65]

Psychologists involved in the applied areas of child guidance and clinical psychology overall claimed to be optimistic about the compatibility of the dual mental hygiene imperatives for personal fulfillment and social adjustment. Yet, as Thomas's questioning alludes, anxiety about the potential irreconcilability of these developmental goals often percolated beneath the surface of their confidence. Radical social theorists V. F. Calverton and S. D. Schmalhausen quoted the playwright Ibsen as the man who perhaps best understood this "sick paradox of the age": "Suppress individuality and you have no life; assert it and you have war and chaos."[66]

When such anxiety was registered by social scientists, questions about the nature and role of woman were never far behind. Indeed, the paradoxes of postsuffrage feminism prompted such questions because woman's collective and unique responsibility to domestic and social welfare now competed with women's diverse demands to exercise individual freedom and pursue personal fulfillment. In the context of these concerns, the figure of the adolescent girl posed her own set of problems peculiar to her age and her gender. Thus, in his introduction to Phyllis Blanchard and Carlyn Manasses' New Girls for Old, Calverton belied the generally enthusiastic tone of the work to chastise the twentieth-century new girl for her seemingly overwhelming capacity for selfish, amoral individualism:

> The new girl of today . . . seldom sees herself in terms of civilization. Instead she sees herself in terms of herself, without concern for values that are other than purely, or impurely, personal . . . Scornful of the moral ideas and social philosophy of the old generation, she has no anxiety to create new moral ideals or a new social philosophy. She does not realize that the rejection of an old set of moral values necessitates the erection of a new set. She does not see that her conduct must be determined by more than a pleasure-pain principle; that as a social being living in a social world she must think of herself in terms of social rather than individualistic behavior.[67]

Together with Schmalhausen, Calverton served as editor of the Modern Quarterly, the successor to the radical journals The Masses and The Liberator, and of three major interdisciplinary anthologies published between 1929 and 1931 that attended to problems of the relationship between the individual and the social, changes in gender and sexual norms, and intergenerational conflict in modern times. In his writings, he issued a strong critique of patriarchy and its damaging masculine sensibility. Despite his harsh indictment of the modern girl, then, he held that she was justified in proclaiming her freedom from an old set of manners and morals that were "unquestionably masculine in their emphasis." The challenge of the contemporary adolescent girl, Calverton asserted, was to foster a new morality that would at once "exalt herself" and expose the pretenses and sufferings of a decadent patriarchal and capitalist culture. In this way, she would help to open avenues for expanded possibilities for "communal fellowship" and the greater "humanization" of civilization.[68]

If some of the more radical implications of Calverton's conceptualization of individual freedom and social progress were beyond the reach of most psychologists of the 1920s, the basic problem of the modern adolescent girl's claims for autonomous selfhood was not. Indeed, it was a small group of psychologists who lent a calming voice to what could sometimes be an unforgiving characterization of the flapper as she emerged in the popular parlance of the decade. Now looked on as an authority emeritus on these matters, G. Stanley Hall set the tone by recasting his budding girl into the "flapper Americana novissima" in his 1922 *Atlantic Monthly* piece. If the girl in her teens "seems to know, or pretends to know, all that she needs, to become captain of her own soul," he proclaimed, "these are really only the gestures of shaking off old fetters."[69] Hall's rendering of the flapper charted the breadth of the terrain from defense to dismissal that psychologists would travel over the course of the decade in their accounting for the possibilities of female adolescent autonomy. Explaining the figure of the flapper by way of the mandates of female development, psychologists both attested to the imperative of the expression of the girl's individuality and offered assurances of the containment of the independent female self by the essential feminine qualities of vulnerability, sociability, and selflessness.

The first American psychologist to address in-depth the question of female adolescent autonomy was Phyllis Blanchard, whose 1920 publication of *The Adolescent Girl: A Study from the Psychoanalytic Viewpoint* was to be widely cited throughout the decade. Blanchard was born in 1895 in rural New Hampshire, where by her own account she experienced a "lonely and isolated" childhood. She credited her mother, who was deeply unhappy with her marriage and with the "dull routine" of farm life, with spurring her on in the pursuit of higher education and a professional career. She graduated from the University of New Hampshire in 1917 and earned her PhD in psychology from Clark University in 1919, where she studied with Hall and read widely in the works of Freud, Alfred Adler, Carl Jung, Havelock Ellis, and Ellen Key. Following a brief stint at the New York Reformatory System and Bellevue Hospital, she joined the faculty at the University of Pennsylvania, where her husband, Walter Lucasse, was a professor of chemistry, and began what would be an illustrious career at the Philadelphia Child Guidance Clinic.[70]

In her first publication, Blanchard was eager to acknowledge her debt to Hall, "without whose unerring insight . . . this book would never have been written."[71] In fact, Blanchard's several writings on female adolescence served as a bridge between Hall's depiction of an adolescent girl almost wholly circumscribed by the Victorian mandate for a selfless femininity and emerging views that would try to reconcile female self-effacement with a universal adolescent mandate for personal independence and individual achievement. Along with other feminist psychologists of her generation, Blanchard's interest in investigating and en-abling the possibilities for such a reconciliation in the developing female psyche was rooted in personal experience. As a contributor to *The Nation*'s series on "These Modern Women" (1926–1927), in which she was invited to reflect on

the autobiographical origins of her feminism, she admitted, "The long struggle between my own two greatest needs—the need for love and the need for
independence—probably had its effect upon my final choice of a profession." In
elaborating on this point, she explained that her parents' difficult marriage,
which was interpreted to her exclusively through her mother's perspective, had
fostered in her a deep "antagonism toward men" during her youth and young
adulthood. As a result, she came to live under the "guiding fiction" that only one
aim was worth pursuing—"the attainment of distinction by my own efforts
without the need to love, honour, or obey any man." It was the study of psychology that led her to recognize that women were capable of sexual desire and that
"women needed men even as men needed women." This insight led her to seek
both love and freedom in her own life, a goal at which she ultimately counted
herself to have been successful.[72]

In her first go-round at squaring femininity with adolescence in *The Adolescent
Girl*, Blanchard attempted to synthesize the various schools of psychoanalytic
thought to formulate her own unique thesis about female development. Following Hall, she conceded that the view of adolescence as a period of "organic
instability" was "correct as far as it goes." She also picked up where Hall left off in
emphasizing the psychological nature of the crisis the girl faced at adolescence.
To do so, she drew on the works of Freud, Adler, and Jung to posit the existence
of an adolescent conflict between "love and ambition" that was peculiarly female
in nature. The psychological dangers that threatened the adolescent were not,
Blanchard argued, the product of a contest between the sexual instincts and
outer repressive forces, as Freud proclaimed, nor of the struggle of the will for
power, as Adler contended. Rather, the real basis for psychological conflict at
adolescence lay, as Jung had explained, in the vying *between* these two motives for
simultaneous realization in the life of the individual. Although individuals were
always susceptible to the neuroses that could result from the imbalance of these
two motives, the adolescent was particularly vulnerable to their competing
influences because it was as this age when the exclusively "egoistic" proclivities
of childhood were challenged by the "altruistic" impulses that first made their
appearance with the onset of the sexual instinct. Furthermore, while the adolescent boy experienced the vying between self-realization and service to others
largely as a stage to be passed through, for the girl the competition initiated a
psychological stress that held serious consequences for her development. "The
struggle is more profound in the girl than in the boy," Blanchard asserted,
"because she must learn to achieve complete subordination of the egocentric
tendencies which have been the sole guide of her conduct up to this time. The

boy, on the other hand, after a brief period in which his emotional life is flushed with this new impulse to serve his fellows, returns to a great extent to his old condition of egoism and self-aggrandisement."[73]

Of course, Hall had also made a claim for a feminized stage of development during which boys and girls alike were to awaken to racial imperatives that were to move them beyond their own selfish needs and desires. Civilized boys then went on to assume a leadership role in the advancement of biological and cultural evolution, while also cultivating the maximum possibilities for individual self-realization. Girls, however, ideally remained entirely devoted to the selfless sexual imperatives of marriage and motherhood. Blanchard's analysis emphatically reinscribed the gendered bifurcation of these developmental mandates, now displacing the altruistic impulses of adolescence almost entirely on the girl. Indeed, she asserted that only a modicum of selflessness accompanied the onset of the sex instinct in the boy, for his role in procreation, which involved no sacrifice comparable to that of the woman in childbirth or child rearing, was "intimately connected with his own pleasure" and therefore remained "in entire harmony with his egoistic tendencies." Blanchard then went on to follow through on Hall's suppositions about female adolescent discontent. She asserted that because of the female "inferiority complex," which, according to Adler, was induced by the organic changes that occurred at puberty and gave rise to a protest against femininity, and because of the unprecedented opportunities for modern girls to exercise freedom and independence in their daily lives, girls too often ignored, denied, resisted, or thwarted their essential feelings of compassion. The result was the emergence of a "bitter conflict" in the female psyche between the individualistic and social energies of the libido and an adolescent girl who was, inevitably, in a state of war with herself.[74]

Unlike Freud and Adler, who resigned girls to an acceptance of their inferiority, or Hall, who consigned the expression of their altruism to the private nuclear family, however, Blanchard's accounting of the adolescent girl entailed a critique of male standards of development. She also issued yet another rallying cry for girls to exercise their capacity for individual freedom by extending their compassion into the public sphere. Given the jeopardy posed to social progress by the ravages of the recent world war, Blanchard was especially eager to proclaim the vital contribution that an unselfish female nature was to make to social healing and advancement at this moment in modern civilization. She thus dismissed the girl's current "antagonism to the established order of things" as a misguided attempt "to follow the man-made path instead of blazing the trail for herself." Indeed, according to Blanchard, it was because the "emphasis upon the egocentric ideals of life borrowed from a man-made set of values" were "so utterly foreign" to female nature that feminism had never received universal support among young women. A "true feminism" would not focus on the rights of women for economic independence or political equality, she argued, but rather would seek to make possible the fullest expression in the family and society of the uniquely female capacity for unselfish love and devoted service. With this reaffirmation of the importance of female difference to social progress, Blanchard identified the crisis of female adolescence as a defining moment for both the individual girl and the future welfare of humankind. " [I]f [woman] is to be successful in making the love ideal an integral part of the world mind," she maintained, "she must first be certain of its supremacy in her own nature."[75]

Blanchard's influential first book thus renewed older claims about the psychological and social benefits of female selflessness, now filtering them through the modern lens of psychoanalytic psychology. Whether and in what fashion the girl might become the "captain of her own soul" also found respondents among psychologists with expectations for the possibilities for female autonomy that differed from Blanchard's. Such varying propositions for female adolescent selfhood made their appearance in two psychological discussions of the decade: the first regarding the tenor and boundaries of the parent-child relationship and the second concerning the role of individual personality and capacity, as opposed to group classification, in determining the educational and welfare needs of the child.

The question of female adolescent autonomy arose first out of a larger psychological discussion about the nuclear family. The current state and future fate of the family as a social institution was a prevalent topic of social scientific investigation and popular discussion during the 1920s.[76] As the forces of economic, cultural, and sexual modernism seemed sure to continue to alter its traditional forms and functions, psychologists of various intellectual persuasions stepped forward to reaffirm the family's primary influence on all aspects of the child's development. While vestiges of expectations of the family as an economic unit with a hierarchical structure remained, increasingly family life came to be conceived and experienced as a site in which egalitarian relationships fostered self-expression and emotional intimacy among husbands and wives and parents and children. Potential conflict between the sexes and the generations also came to characterize the modern family, and as such, psychologists offered up their authority as the best source of insight and negotiation in the pursuit of personal happiness, family strength, and social welfare. In this spirit, they overwhelmingly located the cause of social problems associated with rebellious, dependent, or neglected children to be rooted in domineering, indulgent, or unaffectionate parenting. To some degree, psychologists were willing to exonerate what they saw as well-meaning middle-class parents from blame, by determining parental inadequacies to be rooted in unresolved mental conflicts from the parents' own childhoods. Most important, they were eager to assure that all parents could make use of psychological expertise and learn better ways of caring for their children to nurture the mental health of each individual child, while at the same time promoting family cohesion and social stability.[77]

One of the most important developmental mandates psychologists of the 1920s expected parents to enable was "psychological weaning," the dependent child's achievement of emotional separation from the nuclear family during adolescence. A core prescription in all schools of developmental psychological thought of the decade, behaviorist, psychoanalytic, and mental hygienic alike, psychological weaning was described in terminology afforded a biological imperative. Thus, in her explanation of this task, Leta Hollingworth both likened it to the process of physical weaning and, with a certain measure of humor, decreed it an essential attribute of growing up in all times and places. "In primitive life," she asserted, "there was no question of the mother's apron strings, not only because apron strings had not at that time been invented, but because release from the family situation was then accomplished by formal public action." In the absence of such formalities in modern society, this dictate was to be facilitated by individual parents in the private home, who were expected to be neither too indulgent or protective nor too distant or domineering, so as to further in their children a capacity for autonomy that directed them toward both self-realization and social responsibility.[78] The ideal of psychological weaning established expectations for self-reliance and individual freedom in a modern world that appeared to be dangerously eroding both individuals' obligations for self-control and their opportunities for self-determination. While such real and perceived erosion of the possibilities for the cultivation and expression of individual selfhood were rooted in the complex constellation of socioeconomic factors discussed earlier, however, psychologists primarily attended to this problem by focusing on the dynamic relationships within the nuclear family. Seen as an essential function of private family relations, the individual subjectivity achieved there was variously touted by psychologists as having the potential to compensate for, adjust to, and resist against those forces threatening to undermine personal autonomy in broader social relations.[79]

Only a very few social theorists outside of psychology addressed the problem of psychological weaning to attempt a structural critique of the nuclear family. Thus, for the radical Schmalhausen, who described himself as a "socialistic logician, believing profoundly in the communization of many vital aspects of life," psychological weaning was less the route to autonomous individualism than to the cultivation of humanistic love in social relationships broader than those of the pathological nuclear family structure. According to Schmalhausen, the nuclear family was the sinister product of Christian capitalism, with the mother/child relationship a site of emotional oppression marked by domination, narcissism, and particularism from which children emerged incapable of fostering a sense of true social solidarity with their fellow human beings. "There is but one real problem in our lives: to seek liberation from neurotic bondage," he proclaimed. "To be free—to be free—from those who would enslave and crucify us—with their love! . . . But what weapons are subtle enough against the insidious power of love? Especially mother love." For Schmalhausen, the only lasting remedy for such danger, as well as the surest path to universal personal happiness and social evolution, lay not in any individualistic therapeutic remedies, but in the conversion of the entire social system to socialism. "Sanity," he succinctly counseled, "is more profoundly a social than an individual problem."[80]

As Schmalhausen's revilement of the mother/child relationship blatantly revealed, the concept of psychological weaning was never a gender-neutral proposition. Traversing quite a distance from Blanchard's valorization of female selflessness, which she rooted in women's capacity to mother, it could be used as a tool to disparage both developing girls and loving mothers for their failures either to achieve independent selfhood in themselves or enable it in others. "This problem of throwing off infantile dependency is especially hard, it seems," asserted Leta Hollingworth, "with only children, with youngest children, with physically delicate children, and with girls. Also, the difficulties seem to arise quite largely from the possessive attitudes taken by the *mother*."[81] The misogyny in many such explications of the concept of psychological weaning was palpable. Behaviorist John B. Watson accused emotionally needy mothers of coddling their sons and daughters for their own selfish reasons, such that as adults, they would be unable to achieve economic independence, marital happiness, or the capacity to parent their own children successfully.[82] At the same time, Freud's psychoanalytic account of the Oedipus complex deemed adolescent girls to be essentially lacking in the psychological wherewithal to overcome infantile fixations, hereby relegating them to perpetual emotional and social childishness.[83] For many psychologists of the period, the first problem was to be remedied by women seeking greater intimacy with their husbands, even as they continued to attend, with appropriate reserve, to their children's emotional needs. In addition, the second was to be solved by girls transferring their emotional dependence from their parents to their husband in marriage.[84]

For Leta Hollingworth, though, who devoted much of her early professional career to disproving accepted notions of inherent female psychological deficiencies, and for a few other psychologists eager to expand the possibilities for female autonomy, the problem of female individualization was recognized largely as a product of social factors that were subject to change. Questioning the universality and uniformity of the maternal instinct among women, Hollingworth argued that mothers held on so tightly to their children simply because society allowed them few other alternatives for meaningful social activity. "While affirming the essential nature of woman to be satisfied with maternal duties only," she proclaimed, "society has always taken every precaution to close the avenues to ways of escape therefrom." She also categorically rejected a Freudian interpretation that conceived individualization to be the product of intrafamilial psychosexual dynamics. Some degree of "homesickness" was biologically natural in all maturing children, she matter-of-factly averred, as they mourned the loss of physical comfort and solicitude that were theirs in infancy and early childhood.[85] Psychologist Lorine Pruette kept with this line of analysis and critique and attributed the greater prevalence of "homesickness" among adolescent girls not to psychopathology or unresolved Oedipal conflicts but to the conventional idealization of the qualities of duty, sacrifice, and loyalty inherent in the image of the good daughter. To free the adult woman from the "dogma of the child," and hence to promote less oppressive mothering, and to relieve the adolescent girl from her greater susceptibly to the emotional pain of psychological homesickness, both Hollingworth and Pruette advocated that the growing girl be encouraged to cultivate her own unique capacities and emotional satisfactions exclusive of her role either as daughter or potential mother.[86]

In *New Girls for Old*, Blanchard and Manasses also acknowledged the imperative for girls to separate from their families of origin, albeit with some important qualifications on the mandate. Here, they stepped forth in taking the bite out of some of the more virulent attacks on mothers that predominated in the psychological literature of the period. "Because of [mothers'] love and anxiety," they benignly asserted, "it never quite seems to them that their daughters have grown old enough or wise enough to be the sole arbiters of their fate." They also questioned how absolute the requirement of psychological weaning ought to be for the developing child, citing their findings of reports from perfectly well-adjusted girls who nonetheless preferred to live with their parents until they married, and carefully qualifying the ideal of the self as an isolated individual. "To be sure, since we are erotic and social beings, we can not be altogether self-sufficient," Blanchard and Manasses pointedly maintained, "but it is vitally necessary that we become so to a point where we are not entirely dependent upon the reactions of another individual for our sense of security and well-being."[87] Blanchard thus joined Hollingworth and Pruette in appropriating the concept of psychological weaning away from its association with female failure and deficiency to create a space in psychological theory for the possibility and value of female self-realization. Whereas Hollingworth and Pruette criticized the social arrangements that made it impossible for the girl to achieve the essential goal of an independent female self, Blanchard continued to hope that girls and women would at least partially reject the "egocentric ideals of life borrowed from a man-made set of values" and instead embrace the "love ideal" that was at the essence of their nature. It was a difference of opinion that would resonate in feminist psychologies of girls and women for a long time to come.

Along with the imperative for psychological weaning, the effort by psychologists of the 1920s to consider the whole child as an individual case—to approach every child as an "individual in the making"—also raised questions about the prospects for female adolescent autonomy. Following from William Healy, psychologists attested that whether in the family, the school, or the juvenile court, the assessment of the particularities of the individual child's physical, mental, and emotional potentialities and limitations was the surest route to healthy development, effective education, or beneficial treatment. "The true picture [of modern youth]," diagnosed Blanchard and Manasses, "is not one of an insurrectionist younger generation, all following the same line of conduct, as modern journalism would have us believe; it is rather that of distinct individuals struggling with very human problems and desirous of making some endurable adjustments to the demands of living." For the growing girl, such a perspective intimated that the group classification of sex difference might not always serve to define wholly her capacity or determine her destiny. "The matter of first concern, in a consideration of the education for the adolescent girl," stated psychologist Winifred Richmond, "is the girl herself."[88] Psychologist Willystine Goodsell emphatically agreed. In her 1923 treatise on girls' education she reviewed for her readers the range of empirical studies on female intelligence and health conducted over the past fifty years that soundly disproved the claims of opponents of coeducation that bodily difference determined the girl's abilities or necessarily circumscribed her education. Admittedly, some questions remained about whether girls differed from boys in their "fundamental taste[s] and interest[s]." For Goodsell, though, this was not a sign of essential sexual difference but simply an indication that more attention needed to be paid to fostering the girl's individuality from an early age. "Until women choose their field of work and have the same opportunity for self-development and self-expression within it as their brothers have," Goodsell declared, "it seems idle to estimate temperamental differences, which may in generous measure, if not wholly, be explained by the diverse life experiences and training of the two sexes from babyhood."[89]

Most psychologists of the period were compelled to recognize that the diversity and pace of modern life had considerably broadened the scope of the "feminine milieu," such that all women no longer conformed to a single life pattern. During the years of the girl's adolescence, therefore, it was imperative that she receive the proper guidance by parents, teachers, and child guidance professionals to help her to negotiate her array of educational, vocational, recreational, and mating choices. Objective scientific tests to measure intelligence, skills, interests, and temperament would direct both the average girl away from expectations beyond her capacity and the exceptional girl toward challenges appropriate to her talents. The mandate of individualization thus proposed that sex was no longer the only gauge on which to chart the girl's life course. To offset the risk that all of this would devolve into female selfishness, however, psychologists emphasized that the mandate of individualization encompassed a dual aim: that every girl find self-satisfaction and individual achievement in venues that promoted her own personal happiness, while at the same time optimizing her chances for contributing to the efficiency, stability, and progress of the wider social order.[90] "To say that the culture of personality, the liberation of individual powers, is the supreme end of education is to state only a half truth," insisted Goodsell. "[I]ts complement is the fact that this cultivation cannot reach full fruition in isolation—partial or complete—from the living currents of social life around us." Unlike Blanchard, Goodsell did not deem a "nurturing sentiment" to be inherent in the girl's nature. Nonetheless, she did argue that it was vital that this sentiment be encouraged through the modern girl's education, as indeed it had been "through many centuries of evolution." For it was in the adolescent girl's "feeling for human life" that lay the world's best hope for "progress toward more healthful, happy, and beautiful living."[91]

The cultural work performed by this body of 1920s psychological thought that reified both the universality of developmental mandates and the individuality of the human subject was varied and contradictory. In one regard, the focus by psychologists on universally shared developmental norms, on the one hand, and the uniqueness of the individual, on the other, reinforced one another and contributed to a larger effort by social scientists during the early decades of the twentieth century to reject group classifications as insidious and detrimental to the realization of human equality. Indeed, many social scientists worked hard in these years to replace longstanding assumptions about essential racial and gender difference with the notion of a common human nature marked by endless individual variations.[92] In other ways, though, developmental psychology's dual focus on the universal and the individual also managed to effect an obfuscation of the categories of social difference that made it difficult to probe the ideological underpinnings of developmental thinking. Thus, to deem psychological weaning a universal mandate of adolescent development, and not to examine conditions in the family or society that made its achievement more difficult for girls, left girls especially vulnerable to diagnoses of psychological failure. Likewise, to proclaim that girls had the same individual opportunities as boys and only had to be psychologically tested to determine their unique capacities failed to address the problems of gender bias in intelligence and personality testing, the various kinds of conditioning that shaped girls' "native" abilities going into these tests, and the array of social obstacles facing those girls who attempted to pursue the possibilities that the tools of psychological assessment charted for them. Some feminist psychologists took it as their task to scrutinize such proclamations. However, they, too, contributed their own sorts of unexamined assumptions about social categories to the mix. Thus, when Blanchard registered her forceful critique of the patriarchal sensibility that informed the universal value of individualism, she only went so far as to make girls the sole repositories of an alternative value system built on the ideals of altruism and service. Such an analysis continued to leave the influence of patriarchal values on ideas about *male* development largely immune to critique.

Moreover, despite the recognition of the importance of individualization for the girl's growth into maturity, most psychologists in these years also continued to reify gender differences of one kind or another as the primary distinguishing quality between human subjects.[93] Thus, women and girls were still understood to be different from men and boys, but now they were also recognized as different, as individuals, from one another. While freeing girls and women from all-encompassing deterministic assumptions about femininity, the latter conceptualization also allowed for continued exploitation of notions of female mystery and incomprehensibility, as well as theoretically denied women a basis for social solidarity based on gender. The difference/difference dilemma had the potential to catch girls and women in a double bind perhaps worse than ever before. The adolescent girl was now expected to be both selfless, by virtue of her gender, and self-realized, by virtue of the mandates of her life stage, although psychologists were not always clear about how she was to negotiate these two developmental demands. Indeed, the best that some psychologists did in these years was to perceive these motives in conflict, with this particular burden coming to define in large part what it meant to be an adolescent girl. The ramifications of the double bind were marked by deep ambivalences. Those who, like Blanchard, championed the expression of female difference as the key to the girl's development and to social salvation also unwittingly offered justification for the limitation of female opportunities for psychological growth and social participation. Conversely, those who, like Hollingworth, argued for female attainment of a masculine-style individualism, however critical of the social relations that worked to block such a goal, also opened the girl up to ridicule at her inability to achieve selfhood, to accusations of selfishness and frivolity, and to denial of a previously acclaimed unique female role in the world and of the kinds of personal and social power such a role afforded. Such ambivalences were particularly palpable in discussions about the developing girl's sexuality, and it is there that we now turn.

"Mating": Female Adolescent Sexuality in the "Age of Compulsory Expression"

In the 1920s, psychologists began to speak collectively of a normal female adolescent sexuality. The shift in the psychological discourse was in part descriptive, for a science of the normal was compelled to take into account the increasingly sexualized behaviors of adolescent girls and women in a variety of venues across the social spectrum. Indeed, while women's and girls' adoption of the practices and ideologies of sexual modernism was a gradual phenomenon that began before the turn of the century, there was no denying that by the 1920s, times had changed. Under the influences of an expanding market economy, changing patterns of work, schooling, and leisure, and the contributions of feminism, young women, especially, now asserted the same rights to sexual expression and pleasure that their brothers had long claimed. As sexologists' analysis of this cohort revealed, adolescent girls and young women born after 1900 were far more likely than previous generations to take part in premarital petting and intercourse. Among college educated women, the changes in behavior were the greatest, with those born after 1900 for the first time being more likely than their less educated agemates to engage in premarital sex.[94] Moreover, even greater numbers of girls in this period pursued sexual freedom in their manner of dress and use of cosmetics, their enthusiasm for the suggestive dances of the Jazz Age, and their eager consumption of new forms of entertainment and advertising, where the ideal of heterosexual experimentation and expression was sold in countless subtle and not-so-subtle ways.[95]

In addition to registering its observations of these behaviors, the psychological discourse about female sexuality in the 1920s was also normative and served as a site of struggle over the organization and regulation of the power dynamics of sex/gender relations amid current forces of social change.[96] Thus, as women and girls of all classes increasingly made appeals to traditional male prerogatives of higher education, wage labor, political participation, and sexual autonomy, some psychologists sought to reestablish the grounding of a conventional domestic femininity in large part on the terrain of sexuality. For Hall, who drew on Freud and Ellis to make some significant gestures toward the possibilities of a normal female adolescent sexuality, the emancipated and immodest flapper constituted a relief rather than a threat, with her newfound sexuality serving as one of the surest signs of her enduring femininity. "[T]rue progress," he asserted, "demands that sex-distinctions be pushed to the uttermost, and that women become more feminine and men more virile. This need modern feminism has failed to recognize; but it is just this which flapperdom is now asserting. These girls not only accept, but glory in, their sex as such, and are giving free course to its native impulses." In Hall's account, as in Ellis's and Freud's, the natural sexual impulses the adolescent girl was so eager to express, and which served as the foundation for her sex/gender difference and identity in adulthood, were marked by volatility and passivity, aimed at making herself desirable to adolescent boys and older men, and oriented toward marriage and reproduction.[97] Feminist psychologists of the 1920s also expected that the "native impulses" of youthful femininity now included, and even mandated, some semblance of heterosexual desire. They also, however, raised as many questions as they answered about what composed woman's original (hetero)sexual nature and how it developed over the life course. As a direct counterpoint to Hall, Leta Hollingworth noted that the female sexual impulse was so influenced by social conditions that it was hard to determine what constituted its essence; it was, therefore, inadvisable to interpret it as a sign or cause of other aspects of essential sexual difference.[98]

Uniting this range of perspectives, though, was a general agreement among child guidance experts that underlying the social phenomenon of adolescence was biological puberty and the various problems of sexual adjustment this physiological event posed for the teenage boy and girl. No longer deemed the primary cause of the modern adolescent's volatility, biological puberty nonetheless was recognized as fundamental in properly organizing the erotic sensibilities of the developing child. "These four or five years [following the onset of puberty] hold the only chance the average boy and girl will have to establish their hetero-sexuality," mental hygiene leader Frankwood E. Williams ominously warned. "Once prevented, it can never come naturally and normally again."[99] Williams's attention to puberty in this regard depended on two other suppositions about sexuality made by psychologists in this period, both of which were derived from psychoanalytic explanations of the psychosexual development of the child. The first, widely endorsed premise was that the sex instinct made its appearance during infancy and affected the developing child long before the onset of puberty. For many child guidance experts, this childhood sex instinct manifested itself not as feelings of desire, but as curiosity about the body, sexual difference, and reproduction. Its imperatives were, therefore, best satisfied by an education that provided the child with knowledge about these aspects of sexuality in develop-mentally appropriate ways. The second premise was that girls in particular passed through a "homosexual stage" in early adolescence, during which they were prone to "crushes" on their female peers and older women. Compelled to account for the homosocial orientation of much of girl culture, many develop-mental psychologists deemed the emotional and even physical attraction be-tween girls and women to be normal. However, they also warned of the danger to the adolescent girl's psychological health if she failed to move on and achieve the developmental goal of heterosexual adjustment. Both suppositions, while containing the potential to conceptualize the sexuality of the developing fe-male child in new ways, ultimately worked to heighten awareness of puberty as the moment when the maturing sex instinct focused on desire for the "op-posite" sex.[100]

The role of puberty in establishing heterosexuality, along with the older concomitant view of puberty as the fundamental determinant of dichotomous sexual difference, was also reinforced by findings in the emerging science of sex endocrinology. As historian Nelly Oudshoorn explains, in the first two decades of the twentieth century, gynecologists and biologists in Europe and the United States developed "the concept of sex hormones as chemical messengers of mas-culinity and femininity." Sex endocrinologists thought of the sex hormones in dualistic terms, with the gonads of each sex secreting singular, "opposite, an-tagonistic" substances. These scientists thus made important new discoveries about how the body worked, while also remaining "very close to common-sense opinions about masculinity and femininity."[101] Prominent American physician William P. Graves incorporated the findings and interpretative framework of sex endocrinology in the 1929 edition of his gynecology textbook. Emphasizing that an understanding of the hormones was "essential in the study of every branch of medical science," he also admitted that "much of the knowledge gleaned in this new line of research still remained "contradictory and confusing." The high degree of speculation regarding the functioning of the endocrine system did not, however, stop this physician from asserting the primacy of the influence of the "glands of internal secretion" over the entire functioning of the female organism. Substituting the idea of the metabolic chemical reaction for what was previously perceived as interorgan relationships structured by the workings of the female nervous system, Graves embraced the use of the term "female sex hormone" to refer to the secretions produced by the ovaries, "since it implies the wide distribution of the substance, and at the same time begs the question as to the place or places in which it is elaborated."[102] The term also begged the question as to the *timing* of the elaboration of the sex hormones in individual human development. Gynecologists and physiologists recognized the important, if still poorly understood, role of the endocrine system in governing growth and sexual differentiation from the earliest weeks of embryonic life. At the same time, they reestablished puberty as the crucial period for the formation of the "sex complex," the "metabolic synthesis" that produced the traits of biological and psychological masculinity and femininity, as well as gave way to the possibilities for heterosexual eroticism in normal adolescent boys and girls.[103]

The new science of endocrinology thus provided some continuity with earlier conceptualizations of adolescence as a biologically driven phenomenon. Even so, references by psychologists regarding the influence of the hormones over female adolescent development remained markedly tentative during the decade less because of the lack of existing scientific knowledge on the subject than the commitment by social scientists to discerning the interaction between the forces of nature and nurture in human development. Given this focus, psychologists accepted the imperative of heterosexual expression as a biological fact and avidly pursued what they saw as more compelling questions regarding the socioeconomic causes of sexual taboos, the social and psychological consequences of sexual repression, and the proper ways to educate children about the developing body and its normal sexual urges.[104]

Psychologists' contention that the sexual impulse existed as a normal biological urge for the girl as well as the boy was, nonetheless, a significant one, as was the corollary that it held similar potential to cause difficulties in adjustment during her adolescent years. The boldest claims about what this meant for adolescent behavior came from feminist psychologists who advocated that the adolescent girl ought to be able to satisfy her "avidity for new sensations" through physical encounters with boys. Thus, Blanchard endorsed various kinds of sex play, short of intercourse, that would give the girl "a direct physical relief to the fundamental impulses of sex." She and Manasses also enthusiastically recommended petting as perhaps the most effective way for the psychologically healthy girl to expend her sexual energy, while still remaining within the bounds of existing social expectations for female behavior. The maintenance of a balance between expression and control was, in fact, the key to successful adolescent adjustment and, not insignificantly, to her future domestic bliss:

> Thus the girl who makes use of the new opportunities for sex freedom, when not driven by other emotional maladjustments to carry her behavior to too great lengths, . . . is likely to find her experiences have been wholesome. Contrary to a section of popular opinion, she may be better prepared for marriage by her playful activities than if she had clung to a passive role of waiting for marriage before giving any expression to her sex impulses.[105]

Still, to manage the contradictions inherent in such recommendations and to negotiate the labyrinth of dangers and possibilities born from both the biological expression and the cultural repression of sexuality, many psychologists opted to revisit an interpretive approach with a long history in ideas about female sexuality. Thus, the psychological consensus in these years continued to hold that female adolescent sexuality should be manifested as something other than autonomous desire. Indeed, before she endorsed petting as the best way for the girl to express and contain her sexual impulses, Blanchard drew from psychoanalysis and endocrinology to assert that the adolescent girl was metabolically better equipped than the boy to divert her newly awakened sexual ardor into other emotional outlets. This was because she experienced "no direct source of constant [sexual] stimulation such as that furnished by the accumulation of spermatic fluid in the male." Keeping with Ellis's fascination with the "diffusion" of the sexual impulse throughout the girl's temper, Blanchard's early work described the girl's newfound sexual energy as a "general change of feeling-tone," conditioned by the functioning of all the glands of internal secretion, that was more likely to be felt as love, compassion, or religious ecstasy, and less so as anger or fear, than autonomous sexual desire.[106]

Others who subscribed to a psychoanalytic approach focused on the adolescent girl's capacity for sublimation, on her ability to convert her sexual energy into either high creative pursuits or more mundane practical employments. In her foreword to William Thomas's *The Unadjusted Girl*, one of the works of the decade that located the cause of female juvenile delinquency in the misplaced expression of universal psychological wishes, Ethel Dummer lauded such possibilities:

> Those who, in Freud's teaching of the danger of sex repression to mental health, find merely sanction for license, miss the point of his wonderful message. This theory that life force, libido, creative energy, follows the Law of Conservation true of Physical force—that as motion may become heat, light or electricity, so this inner power may be transmuted from procreative effort to creative work of hand and brain—would seem to explain much of the modern success in the rehabilitation of the young prostitute.[107]

Moreover, if the girl's sexual impulse was more often experienced and expressed as something other than physical pleasure, her sexual (mis)behavior, when it did occur, was primarily motivated by underlying emotional conflicts that had little to do with the willful satisfaction of bodily desire. In numerous case studies of girls from a range of social backgrounds, child guidance experts marshaled the evidence for such a claim. Hungry for love or good times, girls were thought to pursue sexual encounters to compensate for inadequate parental affection or to satisfy the more benign youthful cravings for freedom, adventure, and recreation. Likewise, other behaviors that seemed to bespeak a female adolescent sexual confidence, such as smoking, drinking, dancing, wearing cosmetics, or dressing suggestively, almost always could be traced, with the help of a trained expert, to extrasexual emotional motivations that were to be approached not with condemnation and punishment but with compassion and clinical treatment.[108]

These sorts of analyses partly echoed past formulations that inextricably tied female passion (or, rather, passionlessness) to romantic love, the desire to serve others, and the yearning for motherhood. What was different in the psychological discourse of the 1920s, though, was that the proposed substitutions for the girl's sexual desire were increasingly seen as the distinct product of an adolescent, as opposed to a mature, psyche. Thus, even as Freud's theory of female psychosexual development focused on the transference of the girl's aggressive sexual impulse into the desire for a baby, some feminist psychologists were beginning to speculate on whether the maternal instinct was present during girlhood at all. Keen as she was on valorizing the "love ideal" as an essential aspect of female nature that was rooted in the capacity to mother, Blanchard recognized the manifestation of maternal feeling in the girl in her care for her younger siblings, her love for strange children, and her interest in baby animals. However, she also acknowledged that there were strong social pressures on the girl to feel such solicitousness and suggested that whatever maternal instinct existed in the adult woman arose out of the biological experience of mothering. "That there should be even these suggestions of maternal instinct during adolescence," Blanchard wondered," is remarkable when we consider that at best it can only be faintly prophetic of the powerful impulse to come, since it lacks the complete physiological background which only motherhood itself can give." "It is possible," Lorine Pruette mused in a similar vein, "that a fuller development of the girl's emotional life is necessary before she can feel so strong an interest in children as in sweethearts." Even more boldly, Hollingworth wondered whether the sex instinct, if left uneducated, would become the reproductive instinct at all, in girls or boys. "As thinkers about youth's problems," she suggestively contended, "it is important for us to bear in mind that each human being is born without knowledge of the causal connection between sexual response and reproduction; that each has to learn it, either by instruction from others, or as primitive man had to learn, by sheer experience of cause and effect." For these psychologists, it was not the longing for motherhood that compelled the girl into sexual exploits but rather emotional deprivations that had their roots in the needs of childhood and eminently immature cravings for excitement and fun.[109]

Within the wider culture, such a view allowed for at least two possible interpretations. For some, such as the feminist and birth control advocate Margaret Sanger, who published several influential sex education manuals in these years, the distinction made between the adolescent and mature female psyche provided justification for both the sex education and the protection of girls during childhood and adolescence and the right of adult women to make autonomous decisions regarding their sexuality.[110] Many others preferred not to distinguish the adolescent girl from the adult woman but to conflate the two with the flapper serving as the emblem of female sexuality. Depicted in many popular venues as possessing a layer of sexual bravado barely disguising her essential ignorance, dependence, and vulnerability, the flapper was embraced for her daydreamy, giggling eagerness to please and to respond to male sexual desires. With her sexual impulse understood as the product of juvenile yearnings for independence and enjoyment, at best, and childish longings for emotional solicitude, at worst, the normal adolescent girl set a standard for female sexuality that in many ways served to reinforce and create anew links between femininity and the powerlessness of immaturity.[111]

Feminist psychologists' renderings of female adolescent sexuality made some contributions to this view. However, they also distinguished themselves in the apprehension they expressed about the psychological and social implications for the growing girl of what Lorine Pruette termed the "age of compulsory expression."[112] "What kind of girl, then," Blanchard asked, "is she who is victimized by this new freedom of modern youth, who finds that her trust in her boy companions has been misplaced and that they have taken advantage of her condition?" She found an example of such a potential victim in one 14-year-old girl who revealed her discomfort at the heightened expectations for female eroticism: "All the other girls let the boys kiss them," the girl remarked, "and if I don't they will think I am a poor sport . . . But I don't really like it very much, especially with some of the boys."[113] In one respect, Blanchard's musing and description of this clinical case resonated with those depictions of the adolescent girl that attempted to render her vulnerable and to diffuse her claims to social and sexual power. Her concern also sought to expose the power relations structuring the supposedly natural and beneficent imperative for female sexual expression during adolescence. Indeed, along with their enthusiastic commendation of female sexual expression, Blanchard and Masasses also acknowledged that girls were pursuing sexual freedom in a social world still strongly influenced by a double standard for sexual conduct. These psychologists regretfully affirmed that girls suffered far greater consequences than boys for their inability to maintain the appropriate balance between the pursuit of erotic pleasure and socially sanctioned forms of sublimation and self-control.[114]

Lorine Pruette further suggested that those promises to girls and women for greater sexual and economic freedom made by both feminists and the seductions of a modern capitalist culture had fallen far short of their claims to improve the status of women in society. Pruette, who was born in 1896 in rural Tennessee, described herself as a "lonely, bitter" child who by her teens had developed the tenets of what she referred to as "a sort of perverse feminism"—"a well-developed dogma on the world's injustice toward women because they could not have everything they wanted, on nature's injustice toward women because they have to bear the children, and a lack of interest in God because he was a man." She studied at Clarke, with Hall, and at Columbia, where she received her PhD in 1924. In addition to her work as a college teacher, she also served as a psychological consultant at R. H. Macy and Company and to various New York hospitals. She identified herself as a writer as well as a psychologist and was well known in the 1920s for her publications on the modern woman in professional journals, as well as in such venues as the *New York Times* and *The Saturday Evening Post*.[115] Like Hollingworth, Pruette was critical of the social conditions that hampered women's opportunities for individual achievement. She struck a much more disillusioned tone than either Hollingworth or Blanchard in her several writings on the adolescent girl. In a world in which, for good or for ill, the adolescent girl viewed feminism as hopelessly anachronistic, in which she was expected to labor in a workplace rife with gender discrimination, and in which her femininity was assessed on the basis of her sex appeal, Pruette warned, she found herself dangerously adrift in a sea of compulsory freedoms, with no larger vision of self-development or social progress to guide her. "She is a very small and untried knight, riding off to the jousting," she wryly observed of the flapper. "She is uncertain of her weapons . . . She is uncertain of her antagonist. Is it Man, or men, or the world? She does not even know the prize, but she thinks it must be something wonderful if it is to be worthy of her."[116]

Highly ambivalent about the role of the modern adolescent girl in achieving progress for the female sex, Pruette posed as one possible solution to the problems she raised a developmental view of femininity. "We have been having an interlude of women as perennial adolescents," she ruefully noted, " . . . but there are signs that the interlude is about over." The modern girl, whose femininity was conceived of largely as a function of youthful heterosexuality, represented but one necessary stage in the development of the human female. According to Pruette, and to feminist writer and lecturer Beatrice Forbes-Robertson Hale, it was this generation of girls who would finally enable woman to "grow up"—to become both more feminine, by drawing on the traditional strengths of the female sex as wives, mothers, and social caretakers; and less so, by deflecting currently exaggerated sexual energies into socially meaningful intellectual and political work. Resisting the current tendency to equate the flapper with mature womanhood, but also recognizing that the modern adolescent girl was most likely an enduring social fixture, Hale turned from a focus on the evolution of the female of the species to a consideration of individual human development. Female "exhibitionism," "narcissism," and "sexual free-booting," she suggested, were perhaps episodic behaviors in the woman's life, appropriate to the adolescent stage of development, but by no means emblematic of adult womanhood.[117]

Ironically, while formulations such as Pruette's and Hale's did little for the image of the adolescent girl, they were potentially useful in redeeming the character of the modern woman. However begrudgingly, Pruette and Hale accepted the dominant psychological characterization of female sexuality in these years, a sexuality exuberant and vulnerable, as well as highly susceptible to influence by extrasexual emotional motivation. Where they parted company with other psychologists and the culture at large was in their refusal to accept the youthful flapper as the womanly ideal for modern times. Indeed, they relied on the logic of developmental psychology to offer up the adolescent girl herself as a buffer to this ideal. Thus, as modern psychology made it increasingly clear that adolescents were not adults, new possibilities opened up for the identification of qualitative differences between girls and women. Recognition of these qualitative differences became a way for Pruette and Hale to return to the adult woman some measure of social power that they perceived to be lost in the grounding of femininity on youthful eroticism. Puberty, they understood, did give rise to a female sexual impulse that sought expression in the various behaviors the flapper was so prone to exhibit. Nonetheless, the maturing adolescent could be expected to outgrow the excesses of her youthful sexuality and to embrace a form of adult femininity marked by the more exalting characteristics of conjugal love, maternal devotion, and social service.

The normalization of female adolescent sexuality thus became a way for thinkers from a variety of perspectives to express and at least partially resolve ambivalences about modern gender roles and sexual mores for the larger society. Across the board, psychologists recognized puberty, with its newly discovered flourishing of hormonal activity, to be the initiating moment for normal heterosexual eroticism, in boys and girls. Some celebrated the ingratiating, puerile sexuality of the flapper as an emblem of modern femininity. Others objected to this conflation and instead contributed to the discussion about female development to claim for the adult woman a fully autonomous sexual subjectivity or to reclaim for her the ennobling powers derived from the more traditional female virtues. In any case, the very manifestation of the sexual impulse during the girl's adolescence was at last undisputed by most psychologists.

Another point of some psychological consensus in these years was the view that sexual control did not, in fact, pose the biggest problem for the girl during her adolescence. Indeed, even in the "age of compulsory expression," many psychologists continued to deny the normalcy of manifestations of youthful female sexuality that ultimately belied the "natural" female tendency toward restraint or its yearning toward marriage and motherhood. Thus, as Blanchard first suggested in 1920, and as other psychologists reaffirmed over the course of the decade, the defining conflict of normal female adolescence was not between sexual license and sexual control but rather between "love and ambition." Psychologists' assertion of the centrality of this psychic tension in the life of the girl could be relied on to deny the girl the capacity for full sexual agency *and* to rescue her from becoming utterly reduced to a sexual persona. It also became a way for such experts to foreground the individual psychological, as opposed to the collective social, dimensions of the relationship between the worlds of public work and private family relations that increasing numbers of girls and women found themselves compelled to negotiate in modern times.

"Seeking Self-Support": Navigating the "Multiple Choice Situation" of Career and Home

Experts' treatment of the distinctly female conflict between love and ambition took on its most concrete form in discussions over the girl's capacity to manage the relationship between home and work. That this was now a "multiple choice situation," as Blanchard and Manasses put it, was widely touted as marking an important break with the past. Thus, in 1916 Mary Moxcey echoed the predominant cultural sensibility of the Progressive era in declaring that the choices girls and women faced between home and career were mutually exclusive. "The disconcerting part of it all," Moxcey regretted, "is that the choices are so few, and yet so complicated in their results—and that nothing will wait." During the 1920s, psychologists posited that the conflict between the girl's allegiance to individual aspirations and to the needs of her future family was not absolute, but they also acknowledged that the "freer environment" only made her difficulties of adjustment that much greater.[118]

Navigating the multiple choices related to home and work was articulated by social scientists and experienced by girls and women at a time when the marriage rate was rising, the age of marriage was dropping, the birth rate was falling, and greater numbers of young women, especially, were working outside the home, often through the first years of their married lives. Significantly, it was also a time when the conventional split between the public and the private was tenaciously maintained by the free-market system, through reinforced and new structures of gender segregation and discrimination and the failure of big business to accommodate adequately the changing needs of families in a dynamic social order.[119] For men, work life and home life were deemed wholly compatible by existing forms of socioeconomic organization. For women, however, reconciling the two was largely construed as a matter of internal conflict and personal choice, as a problem to be resolved at the level of the individual female psyche. Psychologists played an important role in this construction. The developmental mandate for adolescents to seek self-support had much to do with preserving the role of the white middle-class male as the primary self-sufficient provider for his family and very little to do with mapping the possibilities for broad-based female economic independence. Indeed, based as it was on a presumption that economic circumstances were entirely self-determined, it had even less to do with the conditions of necessity governing the realities of most working-class people's lives, man or woman, adolescent or adult. Child guidance experts were, in fact, at the forefront in providing scientific justification for the white middle-class girl's efforts to realize vocational achievement in this period. However, they were also leaders in construing love and ambition as a vying dualism in the developing female psyche, as well as in rendering the relationship between the public and the private a psychological, as opposed to a social, problem, to be resolved through the developmental process of each individual adolescent girl by and for herself.

With both marriage and employment rates of young women on the rise, the architects of female adolescence offered up a reinterpretation of the relationship between work and girlhood appropriate to modern times. They took part in a larger longstanding cultural discussion and debate about women's economic roles. In the 1920s, this conversation was joined by three sets of voices. The first group was composed of feminists in the National Women's Party who proposed the Equal Rights Amendment in 1923. ERA advocates drew on the language of liberal individualism to argue that women's political and social equality depended on their achievement of economic independence. According to ERA proponents, women had the same rights and responsibilities as men to self-support, irrespective of their marital ties or family relationships. Women's ability to work at whatever jobs suited their talents and inclinations, to be compensated for their labor equally to men, and to retain control of the wages they earned were the foundation for both their development as full human beings and for the contributions they were to make to social progress.

The second position on women's economic roles was voiced by middle-class women labor reformers affiliated with such organizations and government agencies as the National Consumers League, the Women's Trade Union League, the Women's Bureau, and the Children's Bureau, who continued the fight begun in the Progressive era to secure protective labor legislation for working women. Working-class women comprised half of the female labor force and two-thirds of employed wives in the 1920s. They performed low-paying, low-status, gender-segregated, and often dangerous jobs in the manufacturing, agriculture, and service sectors of the economy. From the many studies labor reformers conducted of their lives, they concluded that working-class women pursued wage earning out of necessity to ensure their families' survival. Although labor reformers supported equal pay and vocational training for women, they also endorsed the family wage, which idealized a sexual division of labor based on the roles of the male provider and female homemaker. Labor reformers were vehemently opposed to the ERA because it threatened laws that limited the hours or kinds of jobs women worked. They denied that working women were capable of competing on an equal basis with men because of their roles as mothers or because of the pernicious gender discrimination structuring the economy and society. Labor reformers were joined in the debate over the ERA by male trade unionists, who supported protective laws to preserve their own dominance within the traditionally male sectors of the economy, and by many working women themselves. Although some women trade unionists spoke of their work in terms of rights, personal satisfaction, and independence, many saw their wage earning in terms of the contributions they made to the family economy and feared the hardships they would incur if even the modicum of protection they received from labor laws was taken away.

The third voice in the conversation over women's economic roles was that of "career-marriage advocates." These were middle-class white women who pursued jobs in the white-collar sectors of the economy, which ranged from clerical work to creative and professional employment. Although cleaner and less physically taxing than working-class women's jobs, women's white-collar work, too, was marked by gender stratification, unequal pay, and few opportunities for advancement. Middle-class married women who worked outside the home garnered widespread attention in the popular press throughout the 1920s. Commentators weighed in on why women sought to combine marriage and career and how men, women, children, and society were affected. Those who supported women who combined career and marriage did not speak of women's work in terms of economic independence or necessity but rather in terms of personal fulfillment, individual choice, and social enhancement. Some feminists in the 1920s recognized that women could achieve economic independence only if domestic labor were socialized or professionalized or if men's roles in the private sphere also changed. Career-marriage advocates believed otherwise. Like women labor reformers, they believed that women were primarily responsible for homemaking. The individual working woman, they optimistically advised, should draw on her family's private resources to secure help with domestic labor. As keen supporters of companionate marriage, career-marriage advocates believed that women's work outside the home contributed to the happiness and well-being of those within it, without providing any critique of institution of the family.[120]

Child guidance experts' fashioning of the adolescent mandate to seek self-support made important contributions to the career-marriage advocacy position. Educational, vocational, social, and developmental psychologists alike overwhelmingly attested to the imperative for the girl to cultivate a "life-career motive" during her adolescence. Every girl, these experts proclaimed, was to pursue some personally satisfying and socially useful endeavor "over and above

the normal one in acquiring a husband." As Leta Hollingworth explained, because the modern industrial economy had moved production from the home to
the factory and had so diversified the workforce both boys and girls were now
able to select specialized occupations on the basis of their individual abilities and
interests, rather than sex or caste. For both, then, adolescence was a period of
discernment of individual capacities (made possible with the aid of mental tests),
of gathering and assessing information about vocational options, and of beginning the course of advanced education and training required to pursue a particular type of work.[121] Whereas experts directly correlated the boy's efforts to
navigate the dizzying world of vocational choices with the achievement of economic independence, most explications of the female adolescent mandate to
seek self-support maintained strong links to conventional female roles in the
private sphere. The girl's preparation for and engagement in work outside the
home was now understood to be a necessary component of her development but
only insofar as it allowed her to bide time efficiently before marriage, made
early marriage more economically feasible, contributed to the status of her new
family, kept her engaged enough in personal pursuits so as not to become a
domineering mother, and enabled her to respectably postpone childbearing, for
which, during adolescence, she was deemed physically, psychologically, and economically unprepared. Willystine Goodsell associated the girl's imperative to
work less with women's domestic roles than with the cultivation of the female
virtues of self-sacrifice and community service. "Every year," she lamented,
"there are graduated from high schools and colleges thousands of young women
who are actuated by the selfish desire to make a place for themselves in the
world, even at the cost of others, or to continue their agreeable intellectual
pursuits untroubled by the necessities and strivings of their fellow beings." Critical as she was of the "crude materialism" driving the life motives of modern
boys and girls alike, her hopes that girls would take up vocations that would
enable them to bring their "intelligence, goodwill, and working ideals" to the
ends of ameliorating the epidemic of selfishness in the modern world nonetheless distinctly disassociated female labor from any economic motive at all, domestic or otherwise.[122]

Another key difference between boys' and girls' experiences with the mandate to seek self-support lay in the sorts of "mental-hygiene problems" it engendered as they moved through adolescence. For the boy, struggles arose over
selecting a vocation from a vast array of choices and, perhaps, over the necessity
of reconciling his personal ambitions or preferences with his innate abilities. The
girl faced these problems as well, but her imperative to seek self-support also

conflicted with another of the fundamental mandates of the adolescent life stage itself. As Leta Hollingworth explained: "The vocational desires of intelligent girls are increasingly for work suited to their mental abilities. They do not wish to take up housework as a life-occupation, but the difficulty of reconciling mating with differentiated work is obvious. To organize the self as worker with reference to the self as mate successfully and harmoniously is one of the chief psychological problems of the adolescent girl of our day." Hollingworth accepted the inevitability of this conflict for all girls and put her faith in individual rationality and wherewithal to resolve it. Thus, even as she advocated for professional child care, birth control, and changes in public opinion about working mothers, she focused most of her attention in her discussion of the mandate to seek self-support on the capacity of a small, elite group of girls to make the key adolescent adjustments to vocation and mating, largely on their own accord. "This policy [of educating gifted girls according to their own unique talents and abilities]," she lauded, "is to give girls a chance to solve their own problems on their own initiative, by . . . permitting them to work out the relation between work and love by intelligent experimentation."[123]

Like Hollingworth, Blanchard and Manasses also inscribed the "relation between work and love" in psychological terms. They, however, were careful to note that the girl's rational capacity to navigate her choices was compromised by the "inner compulsions and drives" that influenced her decision-making process. "The young woman who is confronted with the choice of a career versus marriage may believe that she has given thoughtful consideration to her decision," they warned. "But her attitude toward these questions is seldom rational; rather it is motivated by the memories of her early experience." The adolescent girl had reached the age when she could begin to recognize the effects of those early experiences on her desires and behavior, but she often required psychological guidance to do so successfully. Furthermore, whereas Hollingworth focused on the ongoing experimentation by a small group of girls in resolving their psychological conflict, Blanchard and Manasses argued that for most girls, the struggle would be worked out through the exigencies of development itself. In her youth, they explained, the girl was expected to cultivate her unique capacities and talents and even to draw on them to contribute to the economic success of her new family during the early years of her marriage. Then, gradually, her egoistic ambitions would normally give way to the primary psychological motivation of her adult life, the desire to sacrifice her self-interest for the needs of her husband and children.[124]

Lorine Pruette was not so sure. In 1924, she published *Women and Leisure: A*

Study of Social Waste, a highly critical assessment of the middle-class woman's social position in advanced civilization. Women presided over smaller families in homes with dwindling requirements for physical maintenance, Pruette argued, but they had not proved themselves worthy of the leisure now bestowed on them. Instead, they had become the "mid-wives of a mediocre culture." The first part of *Women and Leisure* constituted an exhortation against female idleness and an endorsement of women's work outside the home. The latter part was primarily concerned with the analysis of the results of a survey of the "home interest and the career interest" of 347 adolescent girls, which Pruette claimed to be "a field study of the feminine psychic and emotional life more intensive and extensive than any which has [yet] been undertaken." In reporting on her investigation of girls' own accounts of their unique developmental challenge, she countered predominant expectations for either an essential female adolescent psychic conflict between desires for love and ambition or the reconciliation of the two wishes solely by way of individual effort, rational decision making, or the processes of individual development.[125]

What Pruette found to be significant about her findings was a "considerable vacillation of desires" predominating in the psyches of at least one-half of the girls she questioned. When girls were asked what they would most like to become if they could choose from "anything in the world," 61 percent chose occupations outside the home, while 39 percent said they would most like to be a wife and mother. When put to the test of actually choosing *between* home and career, however, 64 percent of those surveyed now replied that they would prefer a "home of [their] own, with husband and children to care for but without independent income and outside work."[126] When Pruette asked the girls she interviewed to record one of their daydreams, she found a similar "dual-motif" of desire. For Pruette, such seeming vacillation was not, however, a sign of the girl's inherent psychological inconsistency but rather of "the presence in the girl's life of two elements both of which deserve consideration" in a modern social order. As the recordings of their daydreams revealed, many girls managed satisfactorily to realize both categories of yearning on the level of the imagination. It was society, Pruette argued, not individual female psychology, that created the struggle in the growing girl and that ultimately thwarted the satisfaction of her sustained yearnings for *both* achievement and romantic love:

> An apparent inconsistency of desires, or fluctuation between desires, is compelled by the industrial and social conditions of the day. No such inconsistency is demanded of the man who expects both an active life outside the home and a

satisfactory home life as well . . . With the same fundamental desires, the same vital urge, the one appears because of social tradition and industrial exigencies inconsistent, the other consistent.

Once the social organization was restructured to better accommodate these two desires that were fundamental to women as well as men, Pruette maintained, girls and women would exhibit a greater measure of psychological health, all female social unrest would be rendered obsolete, and a major form of social wastage would be virtually eliminated.[127]

For Pruette, the normative seeker of self-support was the white middle-class girl. Indeed, despite her focus on the influence of social conditions in creating and potentially mitigating conflicts between love and work, she paid no attention to the role of economic necessity in shaping the girl's experiences with the two developmental mandates. Neither did Hollingworth or Blanchard and Manasses. Hollingworth expressed confidence that mental testing would put all adolescents on an equal playing field, with intelligence, not caste, determining whether a girl would work outside the home or what kind of work she would do. How "nonintelligent" girls who had no choice but to perform both manual and domestic work would deal with conflicts between them, she did not say. For Blanchard and Manasses, "fundamental attitudes" were far more important than "economic pressure[s]" in giving rise to the struggle between love and ambition. Girls from all social stations, then, would benefit more from psychological insight and self-understanding, rather than a change in economic circumstances, in pursuing a satisfactory resolution.[128]

Despite Pruette's doubts about the purely psychological and developmental nature of the home / career conflict, she was nonetheless careful to point out that the majority of girls she questioned "removed themselves from the suspicion of belonging to the mythical 'third sex' by their definite . . . desires for love and home."[129] Blanchard and Manasses also acknowledged that every girl most likely wanted "a judicious mixture of love and work and play" in her life. However, in a survey of their own, they, too, took pains to highlight their finding that 82 percent of 252 girls surveyed expressed a decided preference for marriage. "We are undoubtedly safe in saying that the modern girl seems to want marriage most of anything in life," they confidently asserted.[130] Such confidence flowed from several sources. It arose first from these psychologists' faith in empirical social science and their conviction that by objectively determining what girls said they wanted they had accurately measured what they wanted. It was also born out of their hope, rooted in their particular feminist inclination, that the new girl would be able to both cultivate her individuality and honor her femininity, to pursue a vocation appropriate to her unique inclinations and to fulfill the desire for home life that every normal girl was ultimately unable to ignore. Finally, their assertion derived from their more critical feminist understanding of the persistent social impediments to women's emancipation. "[T]he modern girl is at last in a position where she must face reality," they conceded, "and summon all the reserves of courage and energy which she may possess in order to adapt to it." Her challenge, then, was to figure out a way "to compromise with life situations, and to make the best possible adjustment to difficult circumstances."[131] Whereas Pruette took a primary interest in the "difficult circumstances" that rendered love and ambition a psychological conflict for the girl in the first place, Blanchard and Manasses remained more concerned with the individual girl's psychological "adjustment." During this period in which social and individual problems were increasingly cast into a psychological framework, it was Blanchard's and Manasses' focus that held greater sway among the architects of modern female adolescence. Every normal adolescent girl faced the mandates to seek self-support and to assume a domestic role and identity. Where conflict arose, psychological explanations and remedies would offer her the most compelling solutions.

"Achieving a Point of View": Intellect, the Self, and Sexual Difference

The fourth and final problem of adolescent adjustment was the mandate for the developing individual to cultivate a systematic explanation of the self in the universe. At adolescence, Leta Hollingworth contended, every child needed to arrive at "some satisfactory answer" to questions about "the meaning of existence" and the meaning of his or her "own existence in particular." She recognized at the outset of her discussion of this mandate that pondering such questions often included the formation of a religious sensibility. The linking of religiosity and adolescence had a long history in philosophical thought and had been prominently revived at the turn of the century by G. Stanley Hall's fascination with adolescent conversion. Hall's assertions about the connection between the two were premised on the claim that a religious impulse first manifested in the growing child as the sex instinct emerged at puberty. His more quixotic associations between sexual ardor and unity with the divine were, however, largely abandoned by subsequent thinkers in favor of a focus on the development of mental capacity. "Religion becomes a problem of adolescence," Hollingworth asserted, "not because there is at that period a development of religious instinct, but because *intelligence* develops during the teens to a point where question and answer arise as manifestations of growth in mental power."[132]

This shift in explanation of the adolescent's tendency toward religious fervor reflected a larger preoccupation in the psychological sciences during the early years of the twentieth century with the quality of human intelligence and, especially, with the differing quantities of mental capacity determined to be possessed by various social groups and discreet individuals. The injunction for the adolescent to "achieve a point of view" and thereby to "find" the individual self and locate that self in relation to the wider world was thus significantly informed by current formulations of the meaning of "intelligence." These, in turn, were related to wider intellectual and cultural debates concerning sexual difference. The relationship between femininity, intellect, and selfhood was in many ways remade during the 1920s. In these years, the results garnered from newly developed intelligence tests scientifically justified the girl's claims to individual achievement in educational and vocational endeavors. The displacement of mental capacity as a marker of femininity, however, resulted in a vacuum that almost immediately was filled by an array of psychological explanations that attempted to set new parameters around the self and its place in the social world based on alternative components of essential sexual difference.

The rise of intelligence testing in the United States was influenced by the French psychologist Alfred Binet, who in 1905 created a test to measure mental deficiency in school children. Together with Théodore Simon, Binet also developed a classification scheme that grouped test takers by "mental age." In 1908, Herbert H. Goddard brought the Binet-Simon tests to the United States. The Binet-Simon test was revised in 1916 by Stanford University psychologist Lewis Terman and became the Stanford-Binet IQ test. Initially, the IQ test was used to identify the mentally defective in the population, resulting in the "feeble-minded" category and the determination of a causal connection between abnormal intelligence and criminality. The IQ test was also used to classify and rank racial and ethnic groups, serving as the newest scientific tool for determining the relative worth of various human subjects. During the First World War, Terman, along with Herbert Goddard and psychologist Robert Yerkes, for the first time conducted intelligence testing on a massive scale, overseeing the systematic measurement of the mental abilities of almost 2 million draftees. The Army experiment led both to a greater enthusiasm among some psychologists for the social applications of intelligence testing, as well as to the first widespread public debate over the reliability and implications of the tests' findings. The 1920s was marked by sustained contention among social scientists regarding the initial applications of IQ testing. Many scientists acknowledged the complexity of the notion of intelligence and questioned the existence or measurability of a single trait known as general intelligence. Many also sought to discern the role of education and the environment in shaping inherited mental capacity.[133]

Even before the Army tests were conducted and debated, the links between low intelligence and antisocial behavior in children and adolescents had begun to be loosened, an effort that was spearheaded largely by the work of William Healy. In the wake of Healy's work on juvenile delinquency, sociologists and psychologists sought alternative explanations for misbehavior among the young beyond deficient mental capacity. The slackening of the bond between feeble-mindedness and delinquency also rendered the relationship between intelligence and psychological and social adjustment in the developing child more complex. Thus, Phyllis Blanchard assured that the majority of feeble-minded individuals could be guided into vocational pursuits suitable to their limited level of intelligence, so they would become relatively productive and, most certainly, un-threatening members of society.[134] As a corollary, Leta Hollingworth contended that it was actually children with *superior* mental capacity who had the most difficult time achieving a point of view and finding the self during adolescence; these children were, therefore, acutely susceptible to rebellious, antisocial, and self-destructive behavior. "It is in fact true," she asserted, "that a very high degree of intelligence is required in order to entertain the abstract thoughts that lead to anxiety about the future and to ethical considerations." Harry Hollingworth agreed. Like Leta Hollingworth, he posited that the "storm and stress" of adolescence was induced not by biology but by the complexities of modern life, which the "intellectually superior adolescent" was most poised and challenged to make sense of and adjust to. In contrast, among primitive peoples, whose simple environments were not conducive to the highest development of the intellect, youths were "supplied with a point of view upon life and death." They were, therefore, spared the difficulty and the privilege of having to wrest one out on their own. With these claims, the Hollingworths joined many of their fellow psychologists in valorizing adolescent angst among white middle-class youths (who were as a group coded as having average or above-average intelligence) as a wellspring of social progress and even a sign of the superiority of western civilization. Harry Hollingworth, however, also echoed the accompanying concern that young people would not be able to live up to such grand expectations. "Religious dogmas, literary standards, political methods, social institutions may all go into the discard until youth finds better ones or becomes discouraged in the search," he opined. "In due time the youth becomes a conservative, a radical grown beyond adolescence."[135]

Along with the complication of the association between intelligence and psychological adjustment or socialization, social scientists also leveled challenges at longstanding assertions about woman's inferior mental capacity. During the early twentieth century, a small but effectual group of female social scientists continued to steadily chip away at the notion that the female reproductive organs had anything to do with any individual woman's cognitive (dis)abilities.[136] At the forefront of this effort were psychologists Helen Thompson Woolley and Leta Hollingworth, who in addition to conducting their own research and publishing their own monographs, sequentially authored a series of reviews of the psychological literature on sex differences in mental traits for *The Psychological Bulletin* between 1910 and 1919. In these reviews, Woolley and Hollingworth charged their male colleagues with bias in interpreting psychological data; advanced the notion that socialization and not natural female deficiency accounted for those sex differences in mental traits that could be identified by sound scientific method; and highlighted the body of empirical evidence rapidly accumulating that revealed girls' performances on tests of general intelligence to be equal or superior to that of boys.[137] In 1918, Hollingworth admitted her difficulty at arriving at the title for her review, "Comparison of the Sexes in Mental Traits," thereby expressing the degree of confusion marking her discipline's shift away from assumptions about inherent psychological differences between the sexes:

> To entitle it "Sex Differences in Mental Traits" would lead the reader falsely to infer that all or most of the comparisons have shown differences. To call it "The Mental Traits of Sex" would imply that it discloses mental traits which are sex-limited. On the other hand, a title like "Sex Identity in Mental Traits" would be unfair, especially to such expressions of opinion as are to be included, which take the time-honored view that there are, and must be notable, inherent psychological differences between the sexes. Simply to adopt for a title "The Psychology of Sex" would give the erroneous impression that the review treats of literature pertaining to the sexual instinct. The title finally chosen seems to circumvent most of these difficulties.[138]

By 1919, Hollingworth rendered such confusion, and the consistent return of contradictory findings about the relative intelligence of males and females, a sign of certainty that the search for difference in mental traits in the sexes was a futile and pointless exercise. "Perhaps the logical conclusion to be reached on the basis of these findings," she resolved, "is that the custom of perpetuating this review is no longer profitable and may as well be abandoned."[139] By 1922, Terman himself asserted that results garnered from modern mental testing determined that women's claims to intellectual equality with men could no longer be ignored. "[A]mong psychologists," he stated, "this issue is as dead as the ancient feud as to the shape of the earth."[140]

In the literature on normal child development, what emerged during the 1920s was a consensus that intelligence was indeed inherited but individually, not through group affiliation. Nurture was deemed to have an important influence on the manifestation of mental facility, but the focus was largely on its potentially negative effect. Environmental factors, according to Blanchard, had "a limiting rather than an accelerating power over intellectual development." The native capacity that the IQ test claimed to measure was reified as the most reliable indication of each individual's chances for educational and vocational success, so long as inherent ability was met with commensurate forms of social opportunity. When it came to the adolescent mandate of "achieving a point of view," then, any individual boy or girl of a certain intellectual capacity, at least, had the same chance of facing and resolving the crisis of the self that portended both the distress and the promise of youth that were its hallmarks in modern society.[141]

Even as inherited individual intelligence became in some ways more important to those concerned with understanding and guiding the development of the child, it also became less less important. "Intelligence, however high," Blanchard cautioned, "cannot be used to the best advantage by the persons thus improperly socialized, because it is comparatively powerless to control the emotional responses or to alter their early conditionings."[142] Having been revealed as an unreliable indicator of (anti)social behavior, or of sexual difference, intelligence increasingly became only one of several components that shaped the development of the individual self and was frequently deemed to be the least significant among them. Personality and temperament, inclination and desire, and many nonintellectual qualities such as industriousness, perseverance, flexibility, confidence, ambition, sociability, and cooperativeness, which psychologists recognized as more pliable than cognitive capacity, overshadowed mental prowess as the keys to both fulfillment and happiness for the individual self, as well as harmony and progress for the wider society. "There are . . . many factors involved in success in life," Helen Thompson Woolley succinctly stated in an early attempt to devise a "new scale" for measuring the physical and mental capacities of adolescence. This was as true for girls as it was for boys. "So far as present studies show," Ruth Shonle Cavan and Jordan True Cavan explained, "intelligence is not closely linked to other traits or characteristics, and whether a girl develops a well-adjusted personality does not depend upon her degree of mental capacity so much as upon the emotional and social habits which are instilled in her from childhood on."[143]

As Leta Hollingworth explained, "a person's Self," was composed of the "sum total" of how he or she "appears, thinks, feels, and acts." The final goal of adolescent development was for the child to become aware of the many traits, characteristics, thoughts, emotional responses, experiences, and habits that shaped who he or she was up to that point in the life cycle, judiciously select among them, and unify them into a coherent, stable whole. For many psychologists of the 1920s, this imperative for self-determination was as much a right and responsibility for the growing girl as it was for the boy. "This unifying element [the girl] will find in those aspects of her home, her vocation, her social and her community life which come to be considered worthwhile," Grace Loucks Elliott declared, "and which determine what she avoids and to what she gives herself with abandon."[144] Yet, at the same time that some psychologists were envisioning the possibility that the girl might join with her brother in this most consequential task of discovering and shaping the self on her own accord, others set out to delimit the contours of the self she would find when she went looking for it. With the mental equality of women firmly established, these psychologists now looked to the extraintellectual qualities of the self to offer insight into the meaning of essential sexual difference. Until the age of adolescence, Hall conceded in his depiction of the flapper, intelligence tests constituted an accurate measure of the girl's individual capacities and offered a useful guide to the appropriate directions in which her educational and vocational endeavors might move. At that point, though, when puberty established the terms for biological sexual difference, these tests "baulk, stammer, and diverge," and remained unable to chart the vast, and vastly important, territories of the girl's self that were determined by an enduring femininity. "Common observation seems to tell us that boys and girls are different," concluded psychologist Beth Wellman a decade later. "Perhaps the aspects of mental life that have been measured are not the crucial ones in which the real differences may be expected."[145]

With this query in mind, psychologist Joseph Jastrow, who throughout his influential career never wavered from his search for differences in the masculine and feminine mind, took on both the socialization argument and the results from IQ testing to make a case for the importance of sexual difference to the continued advancement of modern civilization. In a 1929 article titled "The Implications of Sex," Jastrow placed himself firmly within the social scientific consensus of the late 1920s regarding the influences over human development to affirm that girls and boys, men and women, were the product of both nature and nurture. Too often missing from this dictum, he pointed out, was the more compelling truth that nurture was best understood as "man's interpretation of nature's intentions." "It is itself natural," he explained, "that men should find in nature the suggestions for their direction of human traits to what becomes established custom and career. Sociological trends may seem unnatural because they misinterpret or distort natural differences of men and women, and they certainly can both exaggerate and suppress them; but in some measure they follow their clew."[146]

In regard to the problem of determining the significance of intelligence, Jastrow attempted to establish a hierarchy of psychological traits, in which certain attributes were determined to be more reliable than others for revealing nature's differing plans for the two sexes. Concerning IQ testing, "[w]e may concede both the validity of the measurements and the sufficient accuracy of the yardstick, uncertain as it is," he allowed. "What is questioned is the assumption that for reaching a significant comparison of efficient masculine and feminine traits, one trait is as good as another, is as pertinent as another." Traits were, in fact, to be "weighed as well as measured," with the rather prosaic form of intellectual proficiency gauged by the IQ test earning a low score in terms of its indication of the nature or significance of sexual difference.[147]

Once he determined the comparable cognitive capacity of men and women to be a similarity that made no difference, Jastrow then identified a whole range of diverging masculine and feminine "supporting qualities" that were to better serve as the guideposts for any man or woman's search for selfhood, purpose, happiness, or achievement. These "supporting qualities" in the male included exploration, invention, inquiry, mastery, and objective interest; in the female, they were sympathy, esteem, good will, gentility, and the "refinements of the arts of living." "The resulting perspective of the masculine mind and the feminine mind grows in contrast on a nearer approach, a closer analysis," he wrote. "The elements, the features of the composition, are much the same, for so are men and women by virtue of a common heredity and endowment; but what is emphasized, central, and of major import in the one recedes in the other in favor of a different set of major factors, each in turn in closer relation to what is biopsychologically male or female."[148]

Although Jastrow's argument clung tenaciously to past formulations of essential sexual difference that emphasized both the complementary and hierarchical nature of that difference, it also exhibited a careful awareness of the current scientific and social terrain on which it maneuvered. Just as Hall had been compelled to allow for the economic, sexual, and political "emancipation" of the flapper by the beginning of the 1920s, so Jastrow accounted for the role of nurture, conditioning, and socialization—the touchstones of current developmental thinking—at decade's end. Yet both psychologists nonetheless managed to chart the qualities and proclivities of a female self that continued to locate girls and women firmly within the domestic sphere. Harking back to the past, firmly located in the present, such arguments were also harbingers of things to come. In the early 1920s, convinced that the significant mental differences between the sexes did not lie in the realm of intelligence, but in other aspects of the self, Lewis Terman began work on the quantifiable measurement of the traits of "masculinity" and "femininity," qualities he associated not with cognition but with personality and temperament. Terman and his associates' *Attitude Interest Analysis Survey* was first published in book form in 1936 and served as a tool to diagnose the normalcy of the psychological subject's gender identification well into the 1960s.[149]

As it had begun in the nineteenth century, the intellectual transformation of "old" girls into "new" continued to be a highly contested process during the early decades of the twentieth century as social scientists described and prescribed the development of the "normal" adolescent girl for modern times. In the 1920s, psychologists named and interpreted the four "life-problems" of the adolescent stage of development to allow for both challenge to and retrenchment of the ideologies of gender that had structured ideals of girlhood in the past. Thus, prior efforts to identify the qualities and consequences of essential sexual difference were dampened, if not defeated, over the course of the decade by the reification of the universality of developmental mandates and the individuality of the human subject, by reigning environmentalist assumptions in child guidance theory, and by empirical determinations of female intellectual equality. A widespread acceptance among psychologists that the mandates of female development now included some measure of autonomy, sexual expression, vocational achievement, and self-determination especially benefited the white middle-class girl. Yet, the conceptual problem remained regarding how she would exercise these masculine privileges of adolescence within the parameters of her femininity.

Some social scientists in this period saw adolescence and femininity as utterly incompatible, as with those who deemed the girl an essential failure at psychological weaning. Others assured that the enduring qualities of the feminine psyche could not help but dull the edge of each of the four mandates, as with those who deemed the girl's sexual expression to be more a sign of her eagerness to please than a reflection of autonomous desire. For feminist psychologists, the problem of reconciling femininity and adolescence proved to be especially thorny. As biological explanations of child development were complicated by environmentalist approaches, the "feminized" qualities of adolescence gave way to a focus on the problems and opportunities of the adolescent to exercise the capacities for autonomy, choice, and self-determination. Leta Hollingworth and Lorine Pruette at times responded by seeking to challenge those conditions in the family and society that rendered the mandates of femininity and adolescence incongruous in the first place. Unwilling to sacrifice the essential female "love ideal" to the masculine imperative for "egoism and self-aggrandisement," Phyllis Blanchard took a different approach and reiterated the possibility of a "feminized" adolescence, though now for the girl alone. One point on which these scientists did agree was that in modern times, girls, like boys, were indeed "supposed to catch" adolescence. They also collectively affirmed that once the girl caught it, and took on those struggles "universal" to the stage and peculiar to her sex, the psychological sciences would provide much of the insight necessary to explain her to herself, her parents, and the society in which she came of age.

CHAPTER FIVE

Adolescent Girlhood Comes of Age?

*The Emergence of the Culture Concept
in American Anthropology*

In his *Studies in the Psychology of Sex,* Havelock Ellis carefully considered the development of the female sexual instinct during puberty. The girl, he concluded, was especially prone to a volatile adolescence because of the manifestation during the teenage years of a sexual instinct that, while rarely spontaneously experienced, once aroused fluctuated unsteadily between greater ardor and more intense inhibitions than the uniformly aggressive impulse of the boy. Quoting one commentator on the gendered difference in the experience of adolescence, Ellis noted, "Adolescence is for women primarily a period of storm and stress, while for men it is in the highest sense a period of doubt."[1] For Ellis, this gendered bifurcation of the qualities of adolescence helped to explain the nature and significance of sexual difference, one of his primary aims in the *Studies.* For other contemporary social scientists, however, the comment encapsulated a shift in the overarching terms organizing the concept of adolescence during the early decades of the twentieth century from naturally occurring stress to culturally specific doubt.

Key to the shift in focus was what scholar and reformer Miriam Van Waters referred to as "the anthropological point of view."[2] The anthropological work that articulated its terms most assertively was Margaret Mead's *Coming of Age in Samoa.* Published in 1928, following the young anthropologist's first trip to the South Sea Islands, *Coming of Age* attracted significant scholarly attention and enjoyed widespread public appeal. Simultaneously rooted in the Victorian ethnographic tradition's approach to studying the "other" so as to better understand and appreciate "ourselves" and tied to what historian Susan Hegeman describes as the modernist anthropologist's posture as "delicately poised between social worlds," Mead's work was aptly subtitled *A Psychological Study of Primitive Youth for Western Civilization*. Were the difficulties associated with adolescence, Mead wanted to know, "due to being adolescent or to being adolescent in America?"[3] "Being adolescent in America" was the oft-quoted conclusion she drew, with her cross-cultural comparison serving as both critique and reaffirmation of the norms of western culture.[4] Mead argued that American youth paid a price that their Samoan counterparts did not, in the form of greater psychological mal-adjustment, for the seemingly boundless life choices offered those coming of age in a highly complex society. Her final recommendation was for social scientists, teachers, and parents to find ways to reduce and manage such socially induced stress among adolescents, without sacrificing the benefits of individuality and freedom that constituted their birthright in modern America.[5]

In its attention to the relative influences over development of the forces of nature and nurture, Mead's work contributed to the wider interest among social scientists from across the disciplines during the 1920s in discerning the relation-ship between biology and culture, particularly as it pertained to issues of race, gender, and child development. Placed within this intellectual context, Mead's early work has been read by historians as constituting a defiance of nineteenth-century scientific racism, an expression of early-twentieth-century feminism, and a challenge to Victorian organicism in ideas about the child. Each of these contributions is most often assessed separately, though, with little attention paid to the ways in which the categories of race, gender, and age informed one another in Mead's anthropology. Thus, historian of anthropology George W. Stocking, Jr., writes that "the biological determinism Mead confronted in her Samoan ethnography was a generically human rather than a racially specific one," without considering the ways in which race and gender figured into late-nineteenth- and early-twentieth-century conceptualizations of puberty and ado-lescence.[6] Recent scholarship *has* brought together the categories of race and gender to expose the white liberal bias of Mead's feminism; but here, there is little interest in the relationship of this problem to the questions about child development that Mead brought to the forefront.[7]

This chapter explores the interrelationships among Mead's concerns with race, gender, and child development in her early work by focusing on the figures at the center of *Coming of Age in Samoa*: adolescent girls "in primitive and modern society."[8] To formulate her conclusion that it was the social environment and not biological necessity that determined the quality of the adolescent experience, Mead's field study focused on the life cycle of the Samoan girl. She justified her choice of subject with practical and theoretical reasons. As other female ethnologists had discovered before her, being a young woman herself meant that she could achieve a greater access to and intimacy with girls than with boys.[9] As important for the development of her scientific argument as this pragmatic consideration, however, was her attention to the history of ideas about female adolescence. Mead acknowledged that the girl had long been more closely associated with the dangers of puberty than the boy. Mothers, she pointed out, were constantly warned that girls posed a special problem during this age. They were told that "as your daughter's body changes from the body of a child to the body of a woman, so inevitably will her spirit change, and that stormily."[10] Claims about the natural stressfulness of adolescence had been established on Victorian conceptions of the civilized girl's growing body. Mead's analysis of primitive girlhood authoritatively refuted those claims. Skillfully refashioning notions of gender, race, age, and culture, she rejected earlier associations of adolescence with somatic vulnerability and the attendant feminine qualities of dependence, emotionality, and altruism in favor of the masculine attributes of autonomy, rationality, and individualism. In this way, Mead's study of primitive adolescent girlhood ironically moved the concept of adolescence closer to the ideals of white, western male culture. To be sure, Mead touted adolescence as an exportable product of that culture, within the reach of both western girls and primitive "others" who were able and willing to adapt to its norms. She also recognized that some fine-tuning of those norms was necessary, noting that there were things that could be learned from other cultures' approaches to sexuality and the family, for example. Even so, as formulated in *Coming of Age in Samoa*, Mead's concept of adolescence was consolidated around a set of expectations for development that, although making gestures toward female and racial "others," largely ignored the differences in their experiences of growing up that might have posed alternatives to the mandates of adolescence and also left unexamined the potential of various groups of young people to lay full claim to the stage's privileges.

Mead made the case for the challenging and original nature of her contribution, recalling that she wrote *Coming of Age* at a time when puberty and adolescence were "firmly equated in everyone's thinking." She also contended that ethnologists' knowledge of primitive girls was "far slighter" in the mid-1920s than their knowledge of boys.[11] The next sections attempt to assess these assumptions, by examining ideas about the girl in "the anthropological point[s] of view" leading up to Mead's work on Samoa. We begin with Victorian anthropology and its core intellectual paradigm, cultural evolutionism. According to the cultural evolutionists, all cultures were organized into the hierarchical stages of savagery, barbarism, and civilization, with a combination of innate racial characteristics and environmental conditions determining whether a culture evolved to a higher state or remained fixed in a primitive condition.[12] The problems of race, gender, and child development intersected in this paradigm— on the child/savage equation and again on the fascination with puberty rites. These intersections were articulated in G. Stanley Hall's work with greatest effect. Therefore, we briefly revisit Hall's particular incarnation of the child/ savage analogy and then focus on his analysis of the ethnographic literature on female initiation ceremonies. Next, we turn to the reassessment of Victorian interpretations of female puberty rites by scholars Elsie Clews Parsons and Miriam Van Waters. As members of the generation of college-educated women coming of age during the turn of the century, Parsons and Van Waters joined psychologists Phyllis Blanchard, Leta Hollingworth, and Lorine Pruette in positing alternative interpretations of the many changes taking shape in modern girls' and women's lives from those of more conservative thinkers like Hall.[13] Intellectually, they were influenced by the work of Franz Boas, the single most important figure in leading American anthropology away from scientific racism and toward a pluralistic concept of culture that interpreted human differences as the product of dynamic social and historical conditions, as opposed to inherent racial (in)capacities.[14] The new idea of culture was incipient, rather than fully articulated, in Parsons's and Van Waters's writings on female adolescence.[15] Nonetheless, more than a decade before Mead, each of these scholars followed the general intent of its Boasian meaning to pose the first anthropological challenges to the notions of natural racial hierarchy and essential gender difference that lay at the heart of evolutionary accounts of adolescent development. Finally, we examine Mead's *Coming of Age in Samoa*. With its primary counterpoint Hall's notion of natural storm and stress, Mead's first ethnography barely acknowledged those works in psychology and anthropology that by the mid-1920s constituted a substantial tradition of thought in focusing more attention on the social, rather than the biological, aspects of the adolescent stage of development.[16] More emphatically than anyone had before her, however, Mead articulated a relationship between "adolescence" and "culture" that managed to undermine the gender and racial hierarchies inscribed in evolutionary theory *and* to valorize goals for adolescent development associated with the virtues of both masculinity and western civilization.

"SAVAGE CHILDHOOD" AND "ADOLESCENT RACES":
RACE, GENDER, AND CHILD DEVELOPMENT IN
VICTORIAN ANTHROPOLOGY

In *Coming of Age in Samoa*, Mead offered her analysis of female adolescence in part as a counterbalance to what she saw as the comparatively greater fascination with pubertal boys by nineteenth- and early-twentieth-century ethnographers. In recalling her purposes for writing *Growing Up in New Guinea*, published just two years later, she made the more general case for the necessity of studying primitive children as scientific subjects in their own right: " [I]f, as Freud, Piaget and Levy Bruhl maintained, adult savages . . . were like civilized children, what were primitive children like? This sounds like an obvious question, but no one had asked it. There was an assumption that somehow primitives—child and adult—could be thought of together, while for our own children, the most meticulous attempts to trace their psycho-social development were in order. So, I set out to find out what a set of primitive children were like"[17] Although the lives of primitive children that Mead sought to illuminate *were* largely overlooked in Victorian anthropology, the equation of adult savages and civilized children was one of its central tenets. Indeed, it was an analogy that proved to be equally useful for those expounding on the nature of the savage as for those describing the growth of the child.

For those interested in developmental questions during the second half of the nineteenth century, whether they were of a phylogenetic, ontogenetic, or cultural nature, the comparative method was used as a key explanatory device. The comparative method allowed that in the absence of actual evidence from the historical record, earlier stages of physical, psychic, or cultural evolution could be filled in by observing those currently exhibiting primitive attributes in each of these respective areas.[18] Only half-critically referring to its use as "wild work," the psychologist William James made note of the various natural and social categories metaphorically related by the comparative method: "So it has come to pass that instincts of animals are ransacked to throw light on our own; and that the reasoning faculties of bees and ants, the minds of savages and infants, madmen, idiots, the deaf and blind, criminals, and eccentrics, are all invoked in support of this or that special theory about some part of our own mental life."[19] What all of these groups were thought to have in common were mental characteristics that marked them as inferior on the single linear scale of intellectual and moral development prescribed by evolutionary theory. Chief among these shared characteristics was the inability to subordinate instinctual urges to human rational control. Notable anthropological thinkers such as E. B. Tylor and Herbert Spencer made ample use of the comparative method as they endeavored to ascertain the stages of sociocultural and psychic evolution and to explain the reflective and causal relationships between culture and mind.[20]

Of all of the analogies maintained by the comparative method, perhaps none was more frequently employed than that of the civilized child to the savage adult. According to the British eugenicist Francis Galton, the infancy and early childhood of all races were congruous; then, the conditions of civilization so prolonged individual development that civilized men continued to evolve toward an ever superior level of mental capacity, while their savage counterparts remained "children in mind, with the passions of grownup men."[21] Galton's statement, however, suggested that the dual defining characteristics of the savage—mental inferiority and sexual potency—could not entirely be squared with prevailing Victorian notions of the innocent child. Hence, some adjusted the comparison to liken savage adults more accurately with civilized adolescents, in whom was recognized an often inept, largely unconscious, and frequently overwhelming sexual passion. In his work *Savage Childhood*, ethnographer Dudley Kidd, for example, positioned the Kafir culture of South Africa perpetually "at the dawn of puberty": "Not a few observers have pointed out that the imagination in the Kafirs runs to seed after puberty; it would be truer to say that it runs to sex."[22] Comparisons of either sort could encompass a certain romantic primitivism that celebrated the innocent child, the carefree youth, and the noble savage. With far greater effect, however, the child/savage equation served to organize and maintain hierarchical and exploitative social relations among the races and to justify imperialistic racial policies of civilized governments that ranged from the paternalistic to the punitive.[23]

Conversely to ethnographers like Kidd, who used notions of childishness to characterize the adult savage, G. Stanley Hall drew on assumptions about savagery as a way to describe and prescribe the development of the civilized male child. The influences of Victorian anthropology loomed large in Hall's work, and he drew prominently from the cultural evolutionists' use of the comparative method and their Lamarckian explanations about the interrelationships among individual, biological, and cultural development.[24] Hall's enthusiasm for recapitulation theory, in particular, vitalized the child/savage equation in a new way for audiences interested in the development of the child. As the highest stage of cultural evolution attainable, civilization involved a fundamental paradox: only white men were capable of achieving it, yet its realization threatened to undermine the very manliness that made it possible, by sapping men of their vital nerve force and requiring their undue compliance to existing social norms and conditions. Hall attempted to resolve this contradiction by reorienting the dualistic relationship between savage virility and civilized restraint in developmental terms. Growing boys should, he argued, be allowed to express the more primitive aspects of their masculine natures throughout their childhood and early adolescence. Then, by the mid-teenage years, such primitiveness was to give way to the more refined manly qualities of civilization, with the experiences of boyhood savagery safely inoculating male youth against civilization's effeminizing tendencies. This combination of primitive prowess and civilized self-control linked the white adolescent boy simultaneously with the past and the future, and as such was to facilitate human evolution's gradual culmination in a race of the "super-man."[25]

Hall, then, took the "anthropological point of view" primarily as a way of accounting for and directing the development of the civilized boy. However, in the final chapter of *Adolescence*, entitled "Ethnic Psychology, or Adolescent Races and Their Treatment," he also turned his attention to the "savage" half of the child/savage equation and to the current status and future prospects of the world's primitive races. Given his valorization of childhood and adolescence, Hall's ascription of a child/adolescent nature to primitive peoples opened the way for a critique of certain attributes of modern life, including a modest challenge to some of the more virulent racism of his day. Condemning imperialistic racial violence, Hall warned that the extirpation of primitive peoples could rob all of humanity of some of those qualities necessary for the achievement of evolutionary perfection. "If unspoiled by contact with the advanced wave of civilization, which is too often its refuse, and in which their best is too often unequally matched against our worst," he averred, "[savages] are mostly virtuous, simple, confiding, affectionate, and peaceful among themselves, curious, light-hearted, amazingly religious and healthful, with bodies in nearly every function superior to ours and frequently models for the artist, and the faults we see are usually those we have made . . . The best of the lower races represent that most precious thing in the world—stocks and breeds of men of new types and varieties, full of new promise and potency for our race, because heredity so outweighs civilization and schooling."[26] In the decade following the publication of *Adolescence*, Hall fashioned himself an expert on racial pedagogy and advocated for the protection and education of "adolescent races" to both academic and popular audiences. From 1906 to 1908, he served as president of the National Congo Reform Association, which petitioned the United States government to intercede on behalf of the Congo against their colonial oppressors. In speaking out about the treatment of racial minorities in his own country, Hall similarly appealed for reform on the basis of a mix of romantic primitivism, optimistic meliorism, and condescending paternalism. Nonwhite races the world over, he maintained, ought to be dealt with in the same "enlightened" fashion as civilized children, with attention to their needs for nurturance, understanding of their developmental inferiorities, and respect for all of the potential, individual and racial, that lay buried within them.[27]

The child/savage equation thus foregrounded the figures of the civilized boy and the primitive adult. And yet, Mead's "obvious question" in *Growing Up in New Guinea*—whither the primitive child?—and her related question in *Coming of Age in Samoa*—whither the adolescent girl?—did also receive some attention in the Victorian anthropological tradition. As a figure distinct from his or her savage parents, the primitive child appeared in the ethnographic literature most often as a signifier of the cultural inferiority of a particular group of people. Just as the savage woman was almost uniformly presented by the cultural evolution-ists as a victim of sexual and economic exploitation, so the primitive child was depicted as appallingly deprived and degraded, whose social and emotional treatment fell far below the child-rearing standards set by the civilized families of the white middle class.[28] In the nineteenth-century anthropological tradition's relentless rendering of difference to establish cultural hierarchy, one group of amateur ethnographers featured the primitive female child, in particular, as a prominent figure. Historian Joan Jacobs Brumberg shows that in the years fol-lowing the Civil War, the evangelical women of the women's foreign-mission crusade made the case for the superiority of western Christian civilization by focusing their ethnographic efforts on the female life course. For these female missionaries, one of the most important measures of "heathenism" was the sexual degradation of girls through the early loss of virginity due to child mar-riage. In a body of widely disseminated missionary literature, these evangelical women provided lurid descriptions of "girlless villages" throughout the Middle East, Asia, and Africa, places in which barely pubertal girls were forced into sexual relationships years before they were physically, socially, or emotionally ready. "The 'girlless village,'" Brumberg explains, "was the ethnologists' way of reporting that among the heathen there was no observable period of depen-dency in which young women capable of reproduction were sheltered and pro-tected by family or community."[29] Thus, the superiority of Christian civilization rested, at least in part, on its allowance of female adolescence.

The discussions of puberty rites in the Victorian anthropological literature likewise advanced the view that the degree of a culture's recognition of adolescence measured its place on the cultural hierarchy. Along with this notion emerged a tension between the universality and particularity of adolescence that influenced the work of psychologists and anthropologists interested in child development throughout the 1920s. As one of the first to articulate the terms of this tension, Hall tried to explain the relationship between puberty rites and the adolescent stage of development. "The universality of these rites and their solemn character testify impressively to a sense of the critical importance of this age almost as wide as the race," he declared. "Here education began and extended up toward more mature years and downward toward infancy almost in exact proportion as civilization and its luggage of cultures and skills increased. The functions of the teacher began genetically with the rude regimentations, tortures, mutilations, instructions often most antihygenic, and immoral ceremonies of these initiations to manhood, womanhood, and often at the same time to nubility, with almost no interval after the first physical signs of puberty, for the slow processes of maturation of body and soul. The progressive increase of this interval is another index of the degree of civilization."[30] Hall attended to "savage pubic initiations" both to reveal the universal psychobiological aspects of the adolescent stage of development and to rank the cultural differences between civilized and primitive peoples. He represented puberty rites as concentrated versions of the essential tasks of adolescent development and as mere rudiments of the prolonged transition between childhood and adulthood that made modern civilization possible.

In the case of the girl and the meaning of female puberty rites, the tension between the universal and particular aspects of adolescence was worked out along the fault lines of gender and race. To Hall, female puberty rites signified the universally determining influences of biology on the girl's development. Thus, in his discussion in *Adolescence* of the initiation rites of primitive peoples from around the globe, he narrowly consolidated meanings of female maturity around the biological moment of the first menstruation. The onset of the menses, he maintained, was recognized the world over as the key determiner of the destiny of the girl. Elements of isolation, confinement, and taboo common to so many female initiation rites testified to the uniformly powerful influence of the girl's periodic function on her life, as well as to the potential danger to girl and to community if such power was not managed properly. And the rites' virtually ubiquitous emphasis on preparing the girl for marriage surely indicated that the roles of wife and mother constituted a female biological fate unchanged, and in Hall's vision, unchangeable, across time and space. Indeed, he referred to the primitive girl presented for marriage as a "debutante" at her "coming-out feast." He did so not to make a case for the "survival" of primitive behaviors among civilized peoples who had advanced through the same early stages of cultural evolution where current savages now rested but, rather, to present the meanings and mandates of femininity in an unbroken line of continuity from past to present and place to place. It was thus less significant that debutante balls were similar to primitive initiation ceremonies than that those ceremonies were like debutante balls. According to Hall, comparative evidence from primitive cultures revealed that despite, or more accurately because of, the variety of cultural forms used to express it, girls everywhere shared a singular biological destiny, which could be ignored by themselves or their communities only at great biological, psychological, and social peril.[31] This was not to say, however, that all of these cultural expressions of the essence of femininity served women equally well. Hall was highly critical of what he saw as a uniformly adverse view of menstruation among primitive peoples, and he faulted civilized societies for replicating such ancient repulsions in the form of fear and ignorance. Continued cultural evolution, he hoped, would one day enable the full recognition of the fundamentally positive personal and social power of the menstrual woman. Then, female initiation ceremonies of all kinds would unequivocally celebrate the unique contributions of the girl's biological destiny to evolutionary progress.[32]

If in his explanation of female puberty rites, Hall relied on the evidence of race to reveal the essence of gender, he also used ideologies of gender to uphold hierarchies of racial difference. In his review of the comparative ethnographic literature on the pubertal girl, he acknowledged that there had long been some debate regarding the correlation between the onset of the menses and geography and climate. Precocity, he affirmed, was as much the product of "nerve strain" induced by the complex conditions of civilized life as it was of the racial characteristics inherent in peoples living in the tropical regions of the world. The real differences between primitive and civilized girls lay less in the timing of menarche than in the degree of ease or difficulty with which the reproductive function was established and, most crucially, in the amount of time a society allowed for the development of sexual difference at this stage of the life cycle. "Menstrual phenomena," Hall noticed, "seem to be more and more marked as we pass up the scale" of the cultural hierarchy. Savage female initiation rites did reveal a certain reverence for women, he conceded. However, the fact that "after the almost universal pubescent initial seclusion practiced among primitive people" sexual difference was "so commonly ignored" served as both sign and cause of "tribal arrest or decline" in Hall's mind. The menstruation of civilized girls was thus more painful and more productive of emotional volatility than among their savage counterparts. Such difficulties were, however, ultimately redeemed by the accompanying physical, mental, and social moratorium thought to enable the refinement of the civilized girl's complex biological organization and to provide the opportunity for her distinctive development into a preeminent class of woman. According to Hall, then, stressful adolescence was one of the crosses of racial superiority that civilized females were simply meant to bear.[33]

The female initiation rites of primitive peoples interested Hall for their crude indications of the invariably precarious nature and paramount importance of female puberty in the girl's development. They also illuminated the distinctive and more refined biocultural adolescent stress produced by modern civilization, as well as alluded to the possibility of an advancing western civilization's potentially more humane ways of dealing with this crucial life stage. Conceiving of female adolescence as a stage of development with universal and particular elements that could be illuminated through cross-cultural comparison, Hall presented that which was uniformly shared among girls around the world to be the product of the biocultural evolution of gender and that which was distinct the result of the biocultural evolution of race. Two scholars, Elsie Clews Parsons and Miriam Van Waters, in the years between the publication of Hall's *Adolescence* and Mead's *Coming of Age in Samoa*, drew on Franz Boas's ideas about culture to offer up the first anthropological alternatives to such cultural evolutionary conceptualizations of female adolescence.

OF "DEBUTANTES" AND "DIVERSITY": ANTHROPOLOGY RECONSIDERS THE EVIDENCE ON FEMALE PUBERTY RITES

Margaret Mead set out for Samoa in 1925 under the tutelage of Franz Boas, the founding father of cultural anthropology in the United States.[34] Boas proposed that Mead investigate the problem of the relationship between individual development and cultural distinctiveness for her first ethnography, which had antecedents in Boas's own work as an anthropologist. From his liberal German Jewish family, Boas imbibed the values that would inform his life's work: equality of opportunity, political and intellectual freedom, commitment to the search for scientific truth, and dedication to furthering the progress of all of humankind.[35] Trained in geography and physics, he gradually formulated his ideas about the role of historical processes in shaping and transmitting what would come to be understood as cultural phenomena, as well as his conceptualization of those phenomena in relative terms, during his first ethnographic field trips to Baffinland and British Columbia in the 1880s. Following his immigration to the United States in 1886, Boas combined his ethnological orientation with the problems and methods derived from physical anthropology, which had long been primarily concerned with the physical measurement and classification of immutable racial types. Although bound by many of the same assumptions about heredity and evolution held by scientific racists, his efforts as a physical anthropologist were motivated by an interest in understanding the processes of racial formation and undertaken with a critical stance toward arbitrary classification.[36]

It was from this particular intellectual vantage point that Boas focused some of his attention on the physical growth of the child. Some of the original impetus for this work most likely derived from G. Stanley Hall's interests in child development. As president of Clark University, Hall named Boas docent in anthropology in 1889, the first such appointment at an American university. Hall's own considerable occupation with charting the norms of physical growth notwithstanding, in a letter some months before making his appointment, he expressed to Boas his prediction that "the physical part of anthropology is a little stagnant and that the myth customs and belief side is the next to grow."[37] Nonetheless, during his brief tenure at Clark and for the next two decades, Boas developed his theoretical orientation in part under the aegis of physical anthropology.[38] He challenged some existing assumptions about biological growth and also laid the broad foundation for cultural anthropology's subsequent interest in the relationships among race, gender, culture, and child development.

Conceptualizations of the growth process had long been implicated in theories of gender and racial difference and hierarchy. During the second half of the nineteenth century, the widely acknowledged early and rapid growth spurt of the pubertal girl, along with the tremendous physical energy thought to be expended on its realization, justified limiting the civilized adolescent girl's pursuit of educational opportunity. At the same time, the commonly accepted notion of the arrested growth of primitive peoples at the cusp of adolescence accounted for racial inferiority. For both groups, deficient physical growth held serious consequences for mental development. In the case of the girl, the growth of her mind and body were thought to compete fiercely for vital energy, with the needs of her body prescribed to prevail over her educational ambitions. In the case of savages, it was argued that the cessation of the physical growth of the brain in early adolescence determined a lack of mental complexity in adulthood.

Boas first questioned these assumptions in a series of studies on the growth of school children conducted during the 1890s, with the findings reported in the journal *Science*.[39] His starting premise was that influences operative during the growth process accounted for the differences among various groups of adults. Thus, he proposed, the study of growth during childhood and adolescence constituted "a most important branch of anthropology," and, conversely, "a study of the anthropology of children" was "of the greatest importance for a knowledge of the conditions and laws of growth."[40] Challenging the customary "cross-sectional" approach to the study of growth, which relied on simultaneous measurements of children of different ages, he pioneered in the use of longitudinal analysis of the growth of the same individuals over several years.[41] He discovered a variation in the growth rates of individual children that were not seamlessly correlated to differences in nationality, social class, or gender. Introducing the concept of the "tempo of growth," a fundamental contribution to growth theory,[42] Boas concluded that each child's growth represented a distinct pattern of acceleration and retardation that was the result of three factors: heredity, the life history of the individual, and environmental conditions. He also took to task one of the "most fundamental deductions" of growth theorists— that favorable or deficient physical development "establish[ed] a basis of precocity and dullness" in mental development. What all the measurements of the development of children's bodies and minds revealed, he pointedly qualified, was "that mental and physical growth are correlated, or depend upon common causes; not that mental development depends upon physical growth." He further asserted that children's intelligence could not be categorically characterized as "bright" or "dull." Maintaining that the measurement of a child's intellect at any particular moment did not serve as an indication of the child's essential capability, Boas argued instead that it reflected his or her individual pace of development, which if "retarded" could be caught up by the time of maturity.[43] In an address given before the American Association for the Advancement of Science in 1894, he suggested the implications of his growth studies for the question of racial differences. The meaning of the purported arrested growth of primitive peoples, he argued, depended on the rate of growth during the growing period. If that growth were more rapid, the fact that it ended early should have no effect on mature physical form, or by connection, on adult mental capacity. His findings from the growth studies were also worked into the argumentation of his best-known work, *The Mind of Primitive Man*, published in 1911.[44]

Following the growth studies,[45] Boas turned to what would be his most important examination of the effect of the environment on physical growth, a study of immigrants and their children conducted for the U.S. Immigration Commission between 1908 and 1910, the results of which were published in 1912 as *Changes in Bodily Form of Descendants of Immigrants*. The study involved the analysis of various bodily measurements of some eighteen thousand individuals, drawn from the populations of Eastern European Jews, Bohemians, Sicilians, and Neapolitians, and in smaller numbers, of Poles, Hungarians, and Scots, in New York City. From this data, Boas made comparisons between foreign-born and native-born children, between both of those groups of children and their parents, and between individuals born in Europe in a particular year and children of immigrants who left Europe in that same year. He discovered what he considered to be surprising changes both in the rate of the development of children growing up in the United States, as well as in the physical "type" of each immigrant group. Most important, he found changes in the cephalic index, or ratio of the length to the width of the head, of the children of immigrants, thereby undermining one of physical anthropology's most cherished assumptions—that particular head forms corresponded to specific racial types and that such forms were fixed by heredity and passed on unchanged across generations.[46]

In the growth studies and the immigration report, Boas made the case for the plasticity of human types and for some measure of influence of the social environment on mental and physical phenomena then widely accepted as entirely the products of biological heredity. The critique of racial formalism propounded by these investigations provided some of the basic ammunition for his own understanding and elaboration of the culture concept. These studies also paved the way for the relationship between cultural anthropology and child development studies that would begin to unfold in earnest during the third decade of the twentieth century. Boas returned to the study of growth curves during the 1930s, when he was largely concerned with the timing and intensity of the adolescent growth spurt and the effect of these factors on adult stature.[47] The explicit problem of the relationship between physical growth and culture, however, would not be taken up until the 1940s, when Margaret Mead brought cross-cultural analysis to bear on the work of psychologist and growth theorist Arnold Gesell.[48] Meanwhile, as Boas and his students formulated innovative methodological approaches and theoretical orientations toward the "myth customs and belief side" of their discipline, they entertained new possibilities for the ways in which the culture concept and the developing child might offer insight into one another. Mead would again ultimately engage in the most avid effort to understand and explicate this relationship, in both the scholarly and popular realms. Before her initial pursuit of such questions in *Coming of Age in Samoa*, however, two scholars influenced by Boas revisited the ethnographic evidence on female puberty rites to offer the first anthropological critiques of Victorian assumptions about female adolescence.

The first of these scholars, Elsie Clews Parsons, showed signs early on that she might be inclined to take on the ideas about female adolescence propounded by nineteenth-century ethnographers. Born in 1874 to wealthy New York parents, whose marriage was both bitterly unhappy and doggedly conventional, the young Elsie Clews questioned and defied the expectations for conservatism and quiescence placed on her by her family and social class. Despite her parents' disapproval of higher education for girls, her determination and intelligence reluctantly convinced her father to allow her to enroll in the newly founded Barnard College in 1892.[49] Her primary intellectual influence was Franklin Giddings, a founder of academic sociology in the United States. A proponent of evolutionary explanations of social organization and social change, Giddings nonetheless rejected Spencer's valorization of unbridled individualism as the driving force of evolution. Instead, he posited that social advancement was propelled by human beings' ability to identify with one another, with the contact among heterogeneous groups fostering the highest evolutionary ends of social cohesion and individual self-realization. Although Giddings did not expect women to reach the maximum potential for individual freedom, even in a fully evolved society, his theories held great appeal for the iconoclastic and socially conscious Elsie, as did his emphasis on the special role to be played by the social scientist in diagnosing and transforming society's ills. Inspired by her mentor, Elsie married her enthusiastic study of sociology with involvement in the fledgling settlement house movement in New York City, establishing a chapter of the College Settlements Association at Barnard during her junior year. In these years, she was also profoundly intrigued by the ideas of the French sociologist Gabriel Tarde, whose bold challenge to nineteenth-century notions of biological determinism went further than Giddings in emphasizing the importance of respect for individual differences and the individual's capacity for freedom and invention as the foundation for social harmony and the impetus for social change. Following her graduation in 1896, Elsie went on to do graduate work in sociology and education at Columbia University. On receiving her doctoral degree, she accepted a fellowship that supported her work as the director of student fieldwork for Giddings's sociology courses and also began teaching her own course on "Family Organization."[50] In 1900, she married the lawyer and political and social reformer, Herbert Parsons. With the hopes of creating a modern marriage founded on sexual egalitarianism, she continued her teaching and settlement work after the wedding and again following the birth of her first two children, until her husband's election to Congress moved the family to Washington in 1905.[51]

During her tenure as a teacher, Parsons laid the groundwork for her subsequent careers as both professional ethnographer and public intellectual. Viewing the family as "a particularly well-chosen subject for the elementary student of society," she directed her students in amassing data on the customs, beliefs, practices, and organization of immigrant families living in New York's West Side.[52] In 1906, she published her guidelines for leading students in such fieldwork, along with her extensive lectures from her course on the family, as a textbook intended for use by college lecturers and directors of home-reading clubs. Although Parsons began *The Family* by warning her readers that the passion for classification in evolutionary thought "may lead to non-scientific just as well as to scientific results," she nonetheless conceived of the changes in family life from the "primitive simple family," to the "matriarchate," the "patriarchiate," and finally the "modern simple family," largely in linear and progressive terms. Where she parted company from previous ethnographic renditions of the evolution of the family was in her rejection of the view that that the modern family represented a state of near perfection. Instead, she argued that the civilized family had much evolving still to do, most particularly in shedding its vestiges of patriarchy and the release of control by husbands and fathers over women and children. She thus explicitly cautioned her students against conflating *"what is"* and *"what ought to be"* in family life and offered her own set of prescriptions for the treatment of women and children in modern society, which, while often corresponding to biological "facts," were also derived from deliberate "ethical interpretation," consideration, and choice.[53]

Like previous ethnographers, Parsons saw the family as the crucible of evolution, with the extension of infancy and the prolonged education of children a primary cause of individual development and social advancement. She also similarly recognized initiation ceremonies as "extremely significant in the study of culture," affirming that they functioned largely to inculcate "the personal traits thought to be necessary to successful manhood or womanhood in the given group." She did document some variation in the expectations for womanhood that were passed on through these ceremonies—such as the somewhat unusual emphasis placed on the girl's physical strength and productivity in the puberty ceremonials of the Thomson River Indians. In what would become a hallmark of her later work, though, she was also careful to note that the parental power that accompanied filial dependence almost universally weighed more heavily during this period of life on girls than on boys. Indeed, one of the most formidable ways parents the world over exerted control over children was in their efforts to preserve the chastity of their unmarried daughters.[54] In the "most advanced type of family," which was, according to Parsons, yet to be completely evolved, children were to be afforded a prolonged period of dependency that furthered not only the needs of parents for familial preservation and of society for group cohesion, but their own interests as independent individuals as well. "[I]mmature off-spring," she asserted, "must be supported, protected, and educated throughout the period of immaturity in such a way that they will be perfectly adapted to their total environment, and will also be able to avail themselves of whatever opportunities for progressive individual variations may spring from their own natures and *be tolerated in their environment.*"[55] For Parsons, the "ethical fitness of given traits of family structure" were to be judged by how well those traits promoted the "equal opportunity for all for the development of personality." Such a "golden rule of democracy" was, she further insisted, to apply especially to the growing girl, who if she was to become a "fit" mother of an advancing generation of offspring required even greater "opportunities for personal de-velopment" than her brother. Such opportunities included an equal chance to become "a producer as well as a consumer of social values," along with the right to sexual information, enjoyment, and expression, which were to be real-ized through work outside the home, sex education, contraception, and trial marriage.[56]

The Family was immediately and vociferously denounced by the clergy and the mainstream press for its deviant ideas about sex and marriage, which were deemed all the more blasphemous because they were advocated by the osten-sibly respectable wife of a congressman. In a reaction in part to the criticism, Parsons wrote little in the years following its publication, devoting her intellec-tual energies instead to reading widely in the ethnographical literature and observing the manners and mores of the her own social class in Washington, D.C., which to her constituted as fascinating a subject for anthropological field-work as any primitive culture.[57] When she returned to New York City in 1911, she developed a relationship with Franz Boas, his students, and supporters, most notably Alexander Goldenweiser, Robert Lowie, Paul Radin, A. L. Kroeber, and Edward Sapir, who were on their way in transforming anthropology around a newly conceived concept of culture. Especially important to their thinking was the critical positivism espoused by the Austrian physicist, mathematician, and historian of science Ernst Mach, who rejected abstract theorizing and posited that all knowledge was constructed out of provisional experience and need. At the same time, Parsons also joined in other currents of intellectual radicalism that flowered in Greenwich Village in the years leading up to World War I, sharing ideas about feminism, socialism, and psychoanalysis with the likes of Walter Lippmann, Max Eastman, Margaret Sanger, and Randolph Bourne.[58]

Guided by these intellectual influences, Parsons became increasingly suspicious of evolutionary sociology's speculative schemes and relentless classifications, as well as its underlying faith in progress. She now considered what an anthropology rooted in empiricism and critical positivism might reveal about the varied customs and beliefs shaping the meaning and experience of femininity that were such a central part of every culture, including her own. One thing such an approach laid bare, she discovered, was the sheer constancy of the limitations circumscribing female development across widely different cultural milieus. Cross-cultural comparisons of femininity were not, of course, new to the anthropological tradition. Victorian anthropology had manipulated the ethnographic evidence on woman to remark on both the naturalness of femininity and the evolutionary distance the civilized woman had traveled away from the savagery of her primitive ancestors. Hall's descriptions of female puberty rites in *Adolescence* followed this same pattern, reinforcing the cultural evolutionists' claims about the biological basis of both femininity and racial hierarchy. Parsons fundamentally challenged this sort of cross-cultural comparison. Locating primitive and civilized customs on the same analytic plane, she rejected evolutionary theory and anticipated subsequent functionalist approaches to the study of human behavior. As she explained it, human nature was shaped less by the distinct biologies of sex, race, and age than by a constellation of universally shared unconscious psychological motivations. It was two of these motivations—the fear of change and the desire for control—that invariably led to the sort of invidious classifications and exertions of power responsible for the universally oppressive character of the female experience.[59]

Parsons's first work to explore this set of themes was *The Old-Fashioned Woman: Primitive Fancies about the Sex*, a walk through the female life cycle that drew together existing ethnographic evidence on primitive cultures with her own original observations of middle- and upper-class American society. In her examination of the customs of femininity governing the early stages of the woman's life, Parsons's emphasis was on the ubiquitously repressive character of all female initiation ceremonies. Hall, too, had acknowledged the harsher elements of female pubic rites, as well as what he saw as civilization's foolish replication of them in the modern negative attitude toward menstruation. Significantly, though, he also procured these rites as evidence of what he saw as the essential significance of the girl's coming of age, the initial personal and social recognition of the inevitable biological destiny of woman. Parsons's scathing review of ancient cloisters and modern debutante balls would brook no such interpretation. Instead, she viewed the customs of girlhood worldwide as indication of a universal social investment in stifling female potential and in channeling whatever remained toward limited purpose. Thus, Japanese taboos restricting the girl's use of her brother's possessions; the custom in New Ireland of shutting girls up "in dark cages for four or five years"; and the preference of the Mexican nobility to sequester their girls in convent schools were all of a piece with this universal tendency.[60] Yet these customs were no more barbaric, Parsons insisted, than the West's modern approach to sex education, which kept the girl cloistered in mind as in body and which based her social desirability on her remaining "truly a virgin in her soul." In fact, in acknowledging that "among many savage tribes adolescent girls are without any hesitation carefully instructed about marriage and maternity," Parsons described the valorization of female sexual ignorance as a uniquely modern phenomenon. All the more insidious, she seemed to suggest, were the shared unconscious motives of human behavior when exhibited by a society ostensibly devoted to the exercise of rationality and the democratic sharing of knowledge.[61]

Following the period of seclusion that uniformly attended the onset of the first menstruation, the girl was required to "come out" of her confinement and embrace the narrow role society prescribed for her adulthood. This was the moment in the girl's life when "a rigid line is drawn," a line which "determines her dressing, her hygiene, her occupations, her friends, her name, her behaviour, her point of view." Every culture, Parsons noted, treated the "debutante" in similar fashion. Everywhere attention was paid to her looks and to the modification of the girl's body to meet her particular culture's idealized standards of femininity. Everywhere girls were put on display, judged against one another by themselves and others to determine their suitability for the narrow role now expected of them.[62] Despite these overwhelming cross-cultural similarities, however, Parsons did acknowledge that modern debutantes tended to be older than those girls coming out in primitive cultures. "More primitive *debutantes* also seek the reputation of being a *belle*—sometimes in ways more convincing and consequential than ours; but whatever their plan, they are unable to devote much time to it because they invariably do what is expected of them and—get married," she noted. "Realising that the object of their coming-out, of their finery and 'make-up,' of their dancing, parading, and visiting is to advertise themselves as marriageable and to allure suitors, they know better than to linger on in that limbo of the 'older girls' in which impatient parents and a disappointed public find it so difficult to take an interest."[63] Counter to Hall's claim that an extended adolescence served as a measure of the virtues of civilization, Parsons viewed modern female adolescence simply as a longer period of preparation for more of the same. Primitive girls may have been more compliant than their modern sisters in accepting the limitations placed on them, but, to their credit, they were also considerably less naive. Civilized girls, on the other hand, may have been able to maneuver through the mandates of youthful femininity with slightly more agency and capacity for choice, but they also remained woefully ignorant of their complicity in their own oppression. Modern girls, Parsons lamented, blissfully engaged in the frivolity of the adolescent years with little awareness of the restrictions that invariably followed from the period of female initiation in all times and all places.[64]

Missing from *An Old-Fashioned Woman* was any substantive explanation about why the experience of female development, and the ceremonials that marked its various stages, should be so similar across cultures. Over the next several years, Parsons more fully explicated her ideas about the universal motivations that governed the human mind. *Fear and Conventionality, Social Freedom,* and *Social Rule* were her major works on this theme, each exploring a different set of psychological dynamics that helped shape the customs, institutions, and relations that composed social life in all cultural milieus. Sex and age, Parsons asserted, were "the two greatest sources of difference between its members society has to apprehend." She explained that because human beings both dreaded innovation and longed to connect with one another, and also because they possessed an instinctive "will to power," whatever "natural differences" existed between men and women or children and adults came to be managed in all cultures through the creation of arbitrary group identities. The result was the regulation of the expression of authentic personalities and the restriction of truly intimate relationships between individuals.[65] The "ceremonial of growing up" functioned as one key cultural tool for this sort of hegemonic classification. As Parsons explained it, both the severity of primitive puberty rites and the difficulties of modern adolescence were products of the generic human "reluctance . . . to meet change." All adults, she further contended, approached adolescence as a period of crisis not because of any characteristics essential to biological puberty but because they were reluctant to relinquish any of their own power to the next generation. Thus, Parsons insisted that the 'ordeals' entailed in the puberty rites of primitive girls and boys were "not so much tests as expressions of social control." For girls and boys in "less primitive cultures," schools fulfilled the function of categorizing and containing the "junior age-classes" in strikingly similar ways. "Apprenticeship or school discipline," she comparatively noted, "is maintained through the usual social instruments of subjection—through ridicule, privations, compulsory labour, imprisonment, and in some school systems through torture."[66]

Parsons's early body of work challenged previous anthropological interpretations of female adolescence. Most significant was her suggestion that the difficulties marking the girl's adolescence the world over revealed less about the biology of female puberty than about the essential nature of the human mind. In her hands, then, the ethnographic evidence on female adolescence moved some distance away from the evolutionary paradigm that had long contained it. For Parsons, the counterpoint to biological distinctions of sex, race, and age was a psychological unity that bound all human beings together. The product of universally shared fears and desires, female adolescent stress was likely, but not necessarily inevitable, with remedies to be found in a collective change of mind about both growth and gender. While Parsons acknowledged that the achievement of such change would not be easy, she also noted that the early signs of an "unconventional society" were already beginning to show themselves.[67] "No more segregated groups, no more covetous claims through false analogy, no more spheres of influence, for the social categories," Parsons prophesized about such a society in her conclusion to *Social Freedom*. "And then the categories having no assurances to give to those unafraid of change and tolerant of unlikeness, to those of the veritable new freedom, to the whole-hearted lovers of personality, then the archaic categories will seem but the dreams of a confused and uneasy sleep, nightmares to be forgotten with the new day."[68]

Where Parsons saw only relentless similarity in the cross-cultural treatment of the adolescent girl, Miriam Van Waters drew attention to the highly complex and variable ways in which primitive peoples approached the phenomenon of female puberty. "Recognition . . . of the diversity of the problem," she wrote in her meticulously researched article "The Adolescent Girl among Primitive Peoples," "is the first needful conclusion." No single theory, she argued, could adequately interpret the range of ethnographic evidence on girlhood without "do[ing] violence to the conditions of primitive life." And the least effective of such singular theories was that which conflated biological puberty with social adolescence. "The human element, as distinct from the purely sexual," she admonished, "has not been sufficiently emphasized in the study of primitive concepts of puberty."[69] Hereby rejecting the reductive biologism of Victorian anthropology, Van Waters became the first scholar to highlight the multiple and overlapping social, religious, pedagogic, esthetic, and physical functions served by the customs of adolescent girlhood. Her early scholarly work on the adolescent girl from "the anthropological point of view" was never as popular as that of Hall, Parsons, or Mead. She did, nonetheless, hold true to the general conclusions she drew here as she reached out to wider audiences in her later reform efforts as a leading administrator of Los Angeles's juvenile justice system and as a writer of popular works on juvenile delinquency, which did earn her national recognition.[70] Even more significant for the history of anthropological ideas about female development, however, was Mead's recognition of Van Waters' influence on her own approach to the topic. " 'The Adolescent Girl among Primitive Peoples' had been the only existing material in the field when I started my work," Mead admitted, for which, she acknowledged, she had "always been grateful."[71]

Van Waters's attention to cultural diversity and social environment in her study of female adolescence grew out of a family background quite different from Parsons's. She was born in 1887 into a solidly middle-class family and was raised by loving, liberal parents and the close-knit community of the Portland, Oregon, evangelical Episcopal church where her father served as pastor throughout her childhood and adolescence. From her father, especially, and from the array of visitors who passed through the rectory of St. David's Church, from church officials to theological students, to itinerant workers of all kinds, the young Miriam was encouraged toward, and eagerly embraced, a love of learning, ecumenicalism, a respect for social scientific inquiry, and a commitment to social justice. Growing up was not without its stresses and strains, however, as she also assumed a heavy load of domestic and child-care responsibilities during her mother's frequent bouts of physical and nervous exhaustion.[72] In 1905, she enrolled at the coeducational University of Oregon, where her professors further fostered both her academic ambitions and her devotion to social service. As a philosophy major, she took courses in literature, social history, economics, and psychology, hereby receiving ample exposure to the evolutionary theories of Darwin and Spencer. Her own interpretation of these theories, as expressed in her commencement address of 1908, however, emphasized the importance of both individualism and social responsibility to the process of social change. Van Waters also increasingly pondered and spoke out about the "woman question" in these years; as editor of the campus magazine, she urged her fellow female students to take a more active role in asserting their right to equal treatment in the university's sponsoring of extracurricular activities.[73]

In 1910, Van Waters left Oregon for Clark University, where she intended to pursue doctoral work in psychology under Hall's direction. In Worcester and its environs, she became acquainted with more radical social theories, including socialism and cultural relativism. She also developed an ardent enthusiasm for woman suffrage. One of the most important influences for Van Waters at this stage of her academic career was Jane Addams, whose work inspired her thinking about the possibilities for an active role for women in public life and also directed her attention to the problems of youth in modern society. Despite these many opportunities for intellectual stimulation and growth, however, Van Waters's match with Hall proved to be an incompatible one. During the first two years of her graduate study, she found herself increasingly at odds with Hall's autocratic teaching methods and his biologically based theories. Particularly troublesome were his efforts to push her intellectual endeavors toward what she considered the narrow arena of experimental psychology and away from broad philosophical questions and social reform. After an aborted attempt at a thesis topic on female juvenile delinquency designed to meet Hall's methodological demands, Van Waters switched both fields and advisors, turning instead to the study of anthropology with professor Alexander Francis Chamberlain. Chamberlain had been a student of Franz Boas at Clark, receiving the first doctorate in anthropology offered in the United States, and succeeded his mentor when Boas resigned from the university in 1892. His scholarship on folklore, childhood, and Native American peoples emphasized the environment and learning to account for racial differences. Also appealing to Van Waters were Chamberlain's progressive politics and his avid effort to combine scholarship and social engagement in his own life and work. With the supportive guidance of her new mentor, Van Waters combined research on delinquent girls from the Portland Municipal Court with a study of the ethnographic literature on puberty rites. "The Adolescent Girl among Primitive Peoples" was the condensed version of this dissertation, a two-part installment piece published under the coeditorial direction of Hall and Chamberlain in the *Journal of Religious Psychology*.[74]

At the outset of her analysis, Van Waters declared her break with evolutionary theory. In this exploration of the diverse customs of adolescent girlhood, Hall's recapitulation theory, she stated, "is not, of necessity, assumed." Thus, her method of cataloguing native attitudes toward puberty in girls was not intended to provide evidence of the origin and development of puberty customs but was meant to furnish "insight into the present significance of these customs in the mind of the people who practice them."[75] Van Waters's functionalist approach challenged the notion that any one theory could serve to explain female puberty rites. Predominant among such totalizing explanations, she pointed out, was the theory that all customs surrounding the girl's coming of age were derived from physical necessity and revealed the essential biological nature of femininity. Contrary to widely accepted belief, she argued, all primitive cultures did not mark the onset of the menses with a special ceremony and of those that did, there was by no means a consensus about what this physical phenomenon meant to the girl or her community. Drawing attention to numerous cultures that did not maintain the alleged universal association between menstruation and impurity, she unmoored female puberty rites from their conceptual grounding in the biological limitations of femininity. "Sex does not dominate here completely, as so often asserted," she claimed. "The human element is supreme." For Van Waters, complex social, religious, economic, educational, and physical priorities shaped the meanings and experiences of female adolescence across cultures. In every culture it was the interaction of a specific set of such priorities, not the inherent nature of sex or race, that determined the treatment of the adolescent girl and the significance of her coming of age.[76]

Van Waters acknowledged that some previous theories of puberty rites did take into account the broad social functions of these customs. Most such interpretations, however, were generally only applied to ceremonies for male adolescents. Indeed, the boy's coming of age had long been seen as signifying more than the essential nature of male physiology; rather, male puberty rites were recognized for their role in individuating an adult self and in situating that self in appropriate relationship to society and the supernatural. Largely ignored in these interpretations, Van Waters argued, was the presence of such concerns displayed in female puberty rites as well. Thus, the element of seclusion that in many cultures attended the girl's arrival at puberty was generally assumed to derive from universal acceptance of the essential impurity of menstruation. In Van Waters's reading, such seclusion was as likely to mark the girl's arrival at social maturity and her right to personal freedom as to signify the limitations of her sex. She noted that like the boy the girl's confinement was often self-chosen, and like the boy her period of isolation frequently involved important pedagogic instruction and rigorous tests of her capacity for independence. Indeed, from a painstaking examination of some three hundred constituent elements of puberty ceremonies from around the globe, Van Waters discovered that only 13 percent of the total number were modes of treatment that applied exclusively to girls. Her data showed that the vast majority of puberty customs was not only shared by both sexes but also had much in common with those rites that attended other groups in the community occasionally receiving special attention, such as warriors, shamans, hunters, mourners, and pregnant women. Drawing on the work of French ethnographer Arnold Van Gennep, who argued that each type of primitive ceremony had to be studied as part of the whole sequence of a community's rites of passage, Van Waters insisted that "customs relating to the adolescent girl cannot be studied apart from their relation to other social groups and classes." She did, nonetheless, admit that "sex discrimination," by which she meant both the differential and unequal treatment of girls in a society, was organized through and reflected in many of the ceremonies that marked the girl's coming of age. What was important for Van Waters, however, was that the marking of sexual difference not be uncritically accepted as an inevitable, and hence "natural," component of female puberty ceremonies, especially at the expense of pursuing understanding of the social and religious values that produced such differentiation and gave it specific form in each culture.[77]

Despite her focus on the unrecognized diversity in primitive attitudes toward female puberty, Van Waters nonetheless felt compelled in her conclusion to "reduce the facts to their simplest general terms" and discern some common tendencies from among her data. She discovered that three concerns emerged as pervasive in all female puberty ceremonies: attention to periodicity, to individuation, and to socialization. Those trying "to envisage clearly the problems centering in the adolescent girl in modern society," she argued, would do well to acknowledge these "universal" facets of adolescent development and strive to accommodate them as best as a complex civilization would allow. Relating her anthropological findings to two case studies of girls in trouble with the Portland Municipal Court, Van Waters advocated that those undertaking work with the delinquent girl, in particular, ought to take their first cue from primitive societies and provide appropriate outlets for the "essential traits" of "energy, activity, [and] independence" manifested by all adolescents.[78] Admittedly, one normal manifestation of such adolescent energy was sexual desire. While keeping with her ecumenical approach of cataloguing the range of sexual behaviors, attitudes, and customs of primitive peoples from around the globe, Van Waters also clearly registered her preference that modern civilization learn from those cultures that rejected the sexual double standard, provided for sex education, and even accepted homosexuality.[79]

Like Hall before her, then, Van Waters used cross-cultural comparison to account for both the particular and the universal aspects of the adolescent stage of development. Her analysis of that which was shared by all adolescents and that which was distinct to certain groups resisted the reductive biologism of previous cultural evolutionists' accounts of female development, however. Van Waters's overwhelmingly dominant concern with the *dis*similarities in the form and function of female puberty rites around the globe, along with her interpretation of these particulars as the product of dynamic cultural influences, allowed her to challenge the cultural evolutionists' notions of natural racial hierarchy and essential gender difference. Likewise, her broad application of the "universal" attributes of adolescence as facets of a common human nature undermined the exclusivities of race and gender that prevailed in earlier interpretations of puberty rites. Thus, Van Waters noted that the recognition of "periodicity" in these ceremonies referred not only to the establishment of the female menstrual cycle but also to the "alternating levels of adolescent mental growth" characteristic of both male and female psychology at this stage of development. Her most significant contribution was her explicit extension of the task of individuation, with all of its associations with individual freedom and competent maturity, to girls as well as boys. "The tendency of primitive observance of puberty to mark and set forth the individuality of the girl," she stated, "is very apparent."[80] Van Waters *did* recognize the role played by such ceremonies in initiating the girl into sexual life and in assimilating her to the social life of her gender, virtually the only aspects of female adolescence attended to by the cultural evolutionists; but she was also especially careful to emphasize the ways in which even these seemingly inevitable facets of the girl's development were variably shaped by the "human element" in every culture.

Elsie Clews Parsons's and Miriam Van Waters's anthropological accounts of female adolescent development thus differed from one another in significant ways. While expressing some hope that human beings had the capacity to overcome their fear of change and desire for control, Parsons determined the passage from childhood to womanhood thus far in the history of human societies to be mostly a uniformly grim affair, with adolescent girls depicted primarily as victims of deeply held psychological attitudes toward age and gender. In contrast, Van Waters's analysis of female adolescence emphasized diversity of experience and complexity of meaning. She sanguinely recognized in existing female puberty ceremonials the possibility of paths of development that affirmed the growing girl's femininity, as well as her wider humanity, even amid the prevalence of various cultures' enactments of "sex discrimination." Despite these differences, however, both scholars drew on the work of Franz Boas to counter the prevailing evolutionary explanations of female adolescent development. In neither of these scholars' works was "culture" a fully articulated concept. Nonetheless, each offered a compelling counterpoint to biological reductionism by reanalyzing female development in terms of both human universals and social differences. Beyond these initial publications, however, neither Parsons nor Van Waters continued on with this particular line of research. Parsons remained in anthropology but turned her attention to detailed expositions of Pueblo and New World African cultures. Van Waters maintained a focus on the problems of adolescent development, but left academic anthropology for a career as a professional reformer in the field of juvenile justice.[81] It would, then, be left to yet another of Boas's students to build on the broad implications in these works and to elaborate more fully the relationship between female adolescence and the culture concept.

ADOLESCENCE AND CULTURE IN MEAD'S
COMING OF AGE IN SAMOA

Like Parsons and Van Waters, Margaret Mead experienced her own mix of inspiration and limitation during her formative years that helped to germinate her thinking about race, gender, and child development and pave the way for her life's work as a scholar and public intellectual devoted to understanding and to educating others about these themes. As biographer Lois W. Banner explains, Mead's autobiographical writings optimistically recall a childhood, adolescence, and young adulthood marked by strong familial support, good luck, and boundless opportunity, while recounting a fair measure of personal insecurity and suffering, as well.[82] Born in 1901, Mead spent her early childhood moving between numerous homes in the Philadelphia area, where her father was a professor at the Wharton School, and the town of Hammonton, New Jersey, where her mother conducted research for her doctoral thesis on the lives of the Italian immigrants in that community. Emily Fogg Mead hailed from a privileged Chicago family, received an undergraduate degree from the University of Chicago, and pursued graduate work in sociology at that institution as well. The demands of marriage, motherhood, and social reform work, including efforts on behalf of the American Association of University Women and the Women's Trade Union League, prevented her from finishing her dissertation. She did, however, complete a master's essay carefully documenting the social organization and culture of Hammonton's Italians. Mead admired her mother for her intellectual acumen, her progressive child-rearing techniques, and her liberal ideas about racial and ethnic difference. She also rejected her mother's elitism and her tendency toward emotional restraint and sexual repression.[83]

Mead responded to and characterized her father with similar ambivalence. Brilliant and liberal-minded like his wife, Edward Mead avidly encouraged the intellectual prowess of his oldest child. However, he was also selfish, demanding, contemptuous, aggressive, controlling, acquisitive, and unfaithful. By Mead's

account, the most important influence in her young life was her paternal grandmother, Martha Mead, who lived with the family for some twenty-seven years.
Mead unabashedly praised her grandmother's loving and generous manner, but
even she provided complicated fodder for Mead's thinking about gender and her
efforts to fashion an identity and role for herself as a woman in the modern
world. Martha Mead had attended college and worked as a teacher for many
years (with decidedly progressive inclinations), supporting herself and her only
son after the untimely death of her husband. Her devotion to education and her
independence were, however, complemented by her enthusiasm for all things
feminine and domestic, aspects of a woman's life that Emily endured but did not
actively embrace.[84]

Despite some arbitrary threats by Mead's father that he would not support
her educational endeavors, the women in her life fully expected that she would
go to college. Mead spent one mostly difficult year at DePauw University in
Indiana, her father's alma mater, where she undertook Herculean (and ultimately successful) efforts to fit in to the sorority and fraternity student culture
that dominated the place. She then transferred to Barnard, a college that by the
1920s was renowned for its intellectually stimulating atmosphere, emphasis on
career achievement for women, and first-rate undergraduate curriculum in the
social sciences.[85] Mead took courses in psychology from Harry Hollingworth, in
sociology from William Fielding Obgurn, and in anthropology from Franz Boas,
all of whom encouraged her to think about new ways to integrate cultural and
psychological approaches to the study of human behavior and social phenomena. She graduated in 1923 with a bachelor's degree in English and, with half-
hearted hopes of becoming a high school psychologist, immediately began work
toward a master's degree in psychology at Columbia, where Boas provided the
key direction for her work. Conservative social scientists in control of the National Research Council had condemned Boas's pacifism during World War I.
With the advent of intelligence testing during the war, these social scientists
faulted him for failing to attend adequately to biology in explaining human
differences. In this climate of hostility, he encouraged Mead to focus her master's
thesis research in psychology on the mental capacity of the Italian immigrants of
Hammonton, New Jersey. Challenging the assumptions of IQ testers that English proficiency made no difference in measurements of immigrant groups'
mental abilities, she argued that some combination of biological and environmental factors accounted for various groups of children's performance on intelligence tests.[86]

As she completed her thesis, Mead realized that the life of an educational psychologist was not for her. Inspired by Boas and incited by his teaching assistant, Ruth Benedict, she began advanced graduate work in anthropology. To support herself, Mead worked as a research assistant to William Fielding Ogburn, who was joining with such liberal sociologists as William I. Thomas in issuing critiques of the nuclear family. For her doctoral thesis, written under Boas's tutelage, she examined whether Ogburn's concept of "cultural lag," or the notion that technological change preceded change in traditional attitudes and beliefs, was applicable to the primitive peoples of the South Pacific.[87] Even before she completed the thesis, Mead began planning her first trip to the field, which was arranged as a compromise with Boas. Mead was determined to visit Polynesia intending to continue the research into the processes of cultural change she had begun in her dissertation. Boas was satisfied that enough work had been done about cultural borrowing. Engaged as he was with directing anthropology toward new questions about the workings of human culture, Boas wanted Mead to explore individual development, preferably among the native peoples of the United States. Mead was reliant on Boas for intellectual and institutional support, and Boas was ultimately sympathetic to Mead's ambitions. She therefore acquiesced to the subject matter; he agreed to the location, and Mead made her way to Samoa to conduct an anthropological study of the adolescent girl.[88]

Samoa, a U.S. protectorate with a naval base, met Boas's standard as a field site where Mead's health and safety would not be unduly at risk. She began her nine-month stint of fieldwork in the fall of 1925 on the main island of Tutuila and then moved to the more remote island of Ta'u. Her research consisted of a combination of observation of and interviews with inhabitants on three villages on Ta'u, including detailed interviews with some fifty girls ages 8 to 20.[89] In a letter he wrote to Mead before her departure, Boas identified the specific components of female adolescent life he hoped she would consider. Most importantly, he wanted to know how young girls in Samoa "react to the restraints of custom." He also wanted her to investigate the "excessive bashfulness of girls in primitive society," as well as and the "interesting problem[s]" of "crushes among girls" and "the occurrence of romantic love."[90] With Boas's overarching intellectual problem and particular concerns as her guide and Hall's evolutionary biologism as her foil, Mead laid out the fundamental questions of her fieldwork: "In the course of development, the process of growth by which the girl baby becomes a grown woman, are the sudden and conspicuous bodily changes which take place at puberty accompanied by a development which is spasmodic, emotionally charged, and accompanied by an awakened religious sense, a flowering of idealism, a great desire for assertion of self against authority—or not? . . . Can we think of adolescence as a time in the life history of every girl child which carries with it symptoms of conflict and stress as surely as it implies a change in the girl's body?"[91]

On the basis of her findings of the smooth, uncomplicated development of Samoan girls, the answer was decidedly "no," although that, of course, begged her next question: "If it is proved that adolescence is not necessarily a specifically difficult period in a girl's life—and proved it is if we can find any society in which that is so—then what accounts for the presence of storm and stress in American adolescence?"[92] Culture, and not the exigencies of growth, race, or sex, was responsible for the prevalence or absence of difficulties among the young, Mead maintained, and her proposal hereby placed the culture concept and the developing child in reciprocal relationship with one another. An understanding of culture, she promised, would illuminate the process of ontogeny far better than a race- and gender-based biology had in the past. Likewise, respectful appreciation of the sundry processes of child development among different peoples would provide insight into both the fundamental "personality" of any culture under investigation, as well as the mechanisms by which such a culture was reproduced.

As with Parsons and Van Waters, Mead's use of the culture concept challenged the racism and sexism inscribed in evolutionary theory and endemic in American society. At the time of her writing, African Americans faced continued political, social, and economic discrimination, maintained by the constant threat of physical violence, in the South, along with dashed expectations in those northern cities to which they had migrated in the years surrounding World War I with hopes for greater racial equality. Anti-immigrant sentiment resulted in a series of quota laws aimed at limiting the numbers of southern and eastern European immigrants entering the United States. While a new generation of new girls (particularly those of the white middle class) were beginning to take for granted opportunities for education, work, political participation, self-expression, and even sexual pleasure, they still bucked up against a gender-stratified economy, a persistent sexual double standard, and the increasing commodification of female sexuality.[93]

Coming of Age made the case for racial tolerance and gender equality on several grounds. Mead's choice of subject matter demanded that the lives of nonwhite children—and girls, no less—were worthy of attention, detailed exposition, and respect. Going even further than Van Waters, who sought to complicate earlier interpretations of female puberty rites, Mead was the first anthropologist to take seriously the significance of all aspects of girls' lives, primitive and modern, including their work, play, education, friendships, family and community relationships, and sexual attitudes and behaviors.[94] Further-more, Mead asserted, as a unique and valuable social organization in its own right, and not as a mere representative of an earlier stage of cultural evolution, Samoa had things to teach modern civilization. Although an initial chapter describing the languid rhythm of "A Day in Samoa" smacks of romantic primitiv-ism, Mead was, for the most part, interested in documenting those specific Samoan child-rearing attitudes and strategies that might be useful in enabling modern children to adjust to the complexities of their lives, not escape them. The first area where the Samoan way stood to improve on modern life was in the realm of family relationships. Mead found that Samoan society was organized into large households composed of near and distant relatives and presided over by chiefs, in which age, not relationship, bestowed authority. In such an arrange-ment, she explained, the adolescent girl "stands virtually in the middle with as many individuals who must obey her as there are persons to whom she owes obedience." Overseeing the activities of her younger relatives in the household provided her with an "ample outlet for a growing sense of authority." Equally important, the authority adults had over her was diffuse. Indeed, if a girl was dissatisfied with her home environment, she simply moved to another residence within the household group. Samoan girls were thus freed from the intensely emotional relationships of the "tiny, ingrown biological family" experienced by modern girls. Whether one ascribed to the psychoanalytic view that such rela-tions entailed sexual conflict, Mead declared, they surely had a tendency toward pathological dependence. The larger point, though, was that counter to the psychoanalytic proposition about the inevitability of parent-child discord, the Samoan case held out the hope that the emotionality of adolescence that was the product of intensive family connections could be mitigated through the adjust-ment of the child's social environment.[95]

The second aspect of Samoan society better attuned to optimum develop-ment was its laid-back attitude toward sexuality in general and female sexuality in particular. As several of her reviewers noted, Mead's depiction of "free love" among primitives was not so new either to ethnographers or to the general public, both of whom were accustomed to interpreting the sexual license of primitive peoples as both sign and cause of their inferior position in the evolu-tionary hierarchy. Building on the work of British anthropologist Bronislaw Malinowski, Mead reversed this precept, now holding up Samoa as a model for a far more repressive American culture to follow.[96] In childhood, she described, Samoan boys and girls were amply exposed to the range of bodily functions related to birth, sex, and death. In addition to witnessing numerous postmortem

Cesareans, miscarriages, and autopsies, Samoan children made a game out of searching the palm groves for lovers and practiced masturbation with abandon. Adults interfered with children's observation of sexual activities not because they deemed the behaviors wrong in themselves or children to be essentially innocent but because spectatorship of any "emotionally charged" event by a nonpartici- pant was considered unseemly. None of this, Mead assured, resulted in pre- cocious heterosexual activity because of the taboo proscribing close relations between girls and boys of the same age. Rather, early sex education inoculated children against shocking revelations about sexuality and the neuroses that ac- companied them later in life.[97]

Although by no means a matter of cultural consensus, Mead's call for sex education and her acknowledgment of the importance of sex adjustment for the healthy development of the personality trod ground made familiar by a range of reformers and intellectuals since the end of the nineteenth century, from social purity advocates to anarchists. Bolder still were her claims about the insignifi- cance of female puberty and her endorsement of casual, experimental sexual activity for adolescent girls. As Van Waters had found to be the case for many of the societies whose puberty ceremonies she examined, Samoa was a place where the girl was not singled out at her first menstruation for any special rite or ritual. In comparing the treatment of the prepubertal and pubertal girl, Mead observed, "No ceremony had marked the difference between the two groups. No social attitude testified to a crisis past." In the Samoan experience, the dynamics of female biological development did not debilitate the girl or threaten the commu- nity and so did not require that she be secluded, protected, policed, or limited in her activities in any way. Rather, the several years following her unremarkable arrival at puberty constituted "the best period" of the her life, when her responsi- bilities for baby tending ended and the routine tasks now expected of her such as weaving and cooking exacted neither great ambition nor concentrated effort.[98]

With so few demands on her, much of the teenage Samoan girl's energy could be directed toward clandestine sexual adventuring. Mead found that the majority of Samoan adolescent girls and boys, for whom social rank was not an issue, engaged in numerous brief, often simultaneous affairs conducted surreptitiously "under the palm trees" or in their own homes, which their parents for the most part ignored. A public elopement, in which a couple openly consummated their relationship by running to one set of relatives or the other, was motivated by "vanity and display," conducted when "the boy wishe[d] to increase his repu- tation as a successful Don Juan, and the girl wishe[d] to proclaim her con- quest." Admittedly, the virginity of daughters of chiefs and village princesses was guarded more carefully than average girls. However, even they shared with their sisters of lesser rank a "definite avoidance of forming any affectional ties" with their sexual partners, along with a broader acceptance of what constituted "normal" sexual behavior, both of which made sexual neuroses almost unheard of among Samoans. Indeed, some Samoan girls and boys also engaged in "casual homosexual relations," although these "never assumed any long-time importance," since Samoans regarded them merely as "imitative of and substitutive for heterosexual" activity. The downside was that Samoan girls never experienced the sort of romantic love prized by their western counterparts. Yet, Mead wondered whether Samoans' treatment of sex as "something which is valued in itself" better enabled them "to count it at its true value." Whereas some old-fashioned American girls discounted the importance of sexuality in their lives and other more modern ones naively conflated sexual desire with the all-encompassing love relationship, Samoan girls experienced their sexuality as more than nothing and less than everything and were far better adjusted as a result.[99]

By arguing that the shared human process of ontogeny was variously managed by culture, holding up certain aspects of Samoan child rearing for western emulation, refusing to see girls as determined by biological puberty, and claiming for them sexual pleasure and autonomy, Mead undermined some of the assumptions about race and gender long figured into the concept of adolescence. Yet her Samoan ethnography actually did as much to further earlier evolutionary associations between optimal adolescent development and the attainments of western civilization. Through the culture concept, Mead authoritatively organized the adolescent stage of development around the opportunities and problems associated with individual freedom. Modern, heterogeneous societies, she argued, offered the young a dizzying array of choices of how to be and act and what to think and believe. Such conditions could present enormous difficulties for youth, which could be lessened somewhat by drawing on some of the child-rearing methods practiced by primitive societies. Cultural complexity, however, also guaranteed the incomparable opportunity for modern adolescents to question, to weigh, to challenge, and ultimately to choose ways of life that would bring them individual success and personal satisfaction, as well as poise them to maintain a fortuitous balance between cultural continuity and cultural change. In making such contentions, though, Mead also failed to recognize that "storm and stress" had long been identified as a problem peculiar to civilized youth. Indeed, despite Victorian anthropology's emphasis on the brutal trials imposed on the young during pubertal ceremonials, it was certainly not novel for Mead to characterize a primitive people as unencumbered, or to find that the children among them had easy access to sexual knowledge and that their daughters experienced few difficulties at the onset of menstruation. Nor was it new to consider American youth as besieged by the complexities of modern life or to deem the challenges faced by such adolescents paradoxically as threats to personal and social stability and as sources of individual and cultural renewal. For Mead, the contrast between carefree Samoan youth and troubled American adolescents did not imply an inherent hierarchy of capacity or worth. Nonetheless, her analysis continued to privilege the cultural complexities and developmental difficulties independently negotiated by modern youth and to equate such negotiation with individual and social superiority.[100]

According to Mead, it was largely because Samoans grew up in a "simple, homogeneous primitive civilization" that they were not called on to exercise the capacities for autonomy, individuality, and rational choice that were mandated of their western counterparts during the adolescent stage of development. Instead, Samoan children and youth were uniformly trained in "traditional" tasks appropriate to their gender and continuous with previous generations, exposed to a single worldview complicated at most by the presence of the Christian pastor in the village, discouraged from displaying ambition or precocity, and channeled into friendships based on household relationship as opposed to voluntary association.[101] While Mead did praise the sexual freedom Samoan adolescents were able to realize, she saw this to be a necessary and desirable, though not sufficient, condition for the highest development of the personality. Thus, her reading of the role of the dance in Samoan culture revealed it to be the exception that proved the rule. The dance was, Mead averred, the only activity in which Samoans displayed the kind of competitive self-cultivation promoted in the rearing of American children. "Each dancer moves in a glorious individualistic oblivion of the others, there is no pretense of co-ordination or of subordinating the wings to the centre of the line," she explained. "Often a dancer does not pay enough attention to her fellow dancers to avoid continually colliding with them. It is a genuine orgy of aggressive individualistic exhibitionism." Mead acknowledged that as the dance progressed, it often became "flagrantly obscene and definitely provocative in character." Such a likelihood was not as significant, however, as its function "in the development of individuality and the compensation for repression of personality in other spheres of life."[102]

Vestiges of cultural evolutionism in Mead's work were reinforced by her suggestion that Samoan adolescents lacked the biological wherewithal to sustain the prerogatives of individualism that the dance briefly held out to them. The accounts of child development in Mead's early work explicitly rejected evolutionary theory's notion of innate racial differences and argued for the importance of environmental influence on all manner of growth. Advocating for the contributions the study of the "primitive child" could make to the field of social psychology in the first edition of *A Handbook of Child Psychology,* she clearly asserted her position: "Assuming then that the primitive child starts life with the same innate capacities as the child of civilized parents, the startling differences in habit, emotional development, and mental outlook between primitive and civilized man must be laid at the door of a difference in social environment."[103] The reductive interpretation of this guiding assumption by some readers of Mead's work notwithstanding, however, Mead never sought to replace biological with cultural determinism.[104] Indeed, biology maintained an important hold on her work through her belief in both a common human nature and innate individual differences that took form in each person's unique temperament.[105]

In *Coming of Age*, Mead argued that it was the combination of social conditions and the capacity for temperamental variation within a group of people that accounted for the presence or absence of adolescent difficulty in any culture. Adolescents in western culture had such an array of life options available to satisfy their varied temperamental inclinations, she admitted, that they should actually have a *better* chance at adjustment than Samoan youth who came of age within a much more limited field of social opportunity. A measure of hypocrisy in American culture helped to explain the prevalence of distress and rebellion among its young. America promised its adolescents extraordinary freedom of lifestyle, Mead accused, without providing the proper means of "education for choice" and while continuing to insinuate a "single standard" for thought and behavior.[106] Comparable lack of Samoan youthful maladjustment Mead credited to the particular alchemy of a simple, static society and a people without the range of temperamental variation that would lead to either individual or collective upsetting of traditional mores:

> Samoa's lack of difficult situations, of conflicting choice, of situations in which fear or pain or anxiety are sharpened to a knife edge will probably account for a large part of the absence of psychological maladjustment . . . Furthermore the amount of individualization . . . is much smaller in Samoa. Within our wider limits of deviation there are inevitably found weak and nonresistant temperaments. And just as our society shows a greater development of personality, so also it shows a larger proportion of individuals who have succumbed before the complicated exactions of modern life.[107]

Mead therefore was neither a cultural determinist nor a cultural relativist. Thus, her reinterpretation of adolescence through a cultural lens in no way precluded her from valorizing one society's set of values or way of life over another. The values she endorsed were associated not only with modern civilization but with masculinity as well. Indeed, one of the primary contributions of *Coming of Age* was to normalize for modern girls the same claims to independent selfhood long enjoyed by boys of the white middle class. As their behaviors attested, Mead declared, modern girls had the same developmental needs and rights as boys to experiment with sexual partners, exert autonomy from their parents, ascribe to and reject belief systems, and weigh vocational options. Yet her furthering these goals also perpetuated the notion of western cultural superiority. For it was not only the prevalence of choice in western society but the prospect that the modern girl might share in her brother's capacity for choice for the first time in history that was one of the surest signs of the ultimate promise of western civilization.[108] Mead granted that the development of the Samoan girl was made easier by her culture's casual approach to sexuality. However, her development was also more severely limited by that same culture's lack of capacity to offer girls a broader range of opportunities for individuality and choice. As Mead described her, the Samoan girl grew up "slowly and quietly like a well-behaved flower," maintaining "a very nice balance between a reputation for the necessary minimum of knowledge and a virtuosity which would make too heavy demands." Expectations for her development were clear and relaxed, with the outcome, marriage, "to be deferred as long as possible," but "inevitable" nonetheless.[109] In contrast, nothing was inevitable about the development of the modern American girl. Her unlimited choices to think, to feel, and to behave frequently rendered her baffled and neurotic, but ultimately established her as a claimant to individualism and rationality, the most valuable gifts, according to Mead, any culture could bestow.[110]

Although Samoan girls were reared to be passive and compliant and to give little effort and less thought to cultivating a unique self, they were also, paradoxically, noticeably individualistic; they were less capable of engaging in cooperative effort or of experiencing specialized affection, particularly in friendships with fellow females. Mead admitted that this led to fewer of the "crushes" that were so prevalent in American girls' experiences, as well as a decided absence of female adolescent "gangs." But it also resulted in a female sphere held together neither by individual preference nor a sense of social solidarity, but by "regimented associations" of relationship groups, over which the girl herself exercised little control.[111] Both the lack of female independence and of voluntary female solidarity, Mead concluded, were symptomatic of a society in which too great a pronouncement of sexual difference prevailed. A "rigid sex dichotomy," characterized by antagonism and avoidance from childhood on, meant that no Samoan, male or female, ever regarded the other "simply as individuals without relation to sex."[112] In registering her distaste for such a social arrangement, Mead turned earlier anthropological accounts of gender difference on their head. In Victorian anthropology, the less sexually differentiated a people, the less advanced the culture. For Mead, however, exacting adherence to sex roles precluded recognition of individuality and was to be rejected out of hand by modern society. "The strict segregation of related boys and girls, the institutionalised hostility between pre-adolescent children of opposite sexes in Samoa are cultural features with which we are completely out of sympathy," she stated. "For the vestiges of such attitudes, expressed in our one-sex schools, we are trying to substitute coeducation, to habituate one sex to another sufficiently so that difference of sex will be lost sight of in the more important and more striking differences in personality." Unlike Van Waters, then, who sought out evidence in female puberty rites of nonwestern cultures' encouragement of the girl's individuality and independence, Mead declared female autonomy to be a vital sign and measure of the advancement of modern civilization.[113]

Mead's rendering of autonomy as a normal attribute of modern female adolescence made important contributions to the liberal variant of early-twentieth-century feminism, which made claims for women's equality on the basis of their common humanity with men. Her assertion that a society's achievement of gender equality was directly proportional to the attention it paid to the development of personality, however, compromised the egalitarian potential of that contribution by depending on a kind of cultural comparison that recalled earlier anthropological claims to racial hierarchy. It also served to undermine that current of ideas about adolescence that had valued, if always ambivalently, the "feminized" attributes of this stage of development. Hall conceived of adolescence as the one time of life when boys were as inclined as their sisters toward selfless service and religious devotion. He described in loquacious terms how the reproductive potential of the establishment of the sex instinct awakened all adolescents to their connection with the past and future of the human race, as well as with the wider mysteries of the universe. By the 1920s, such a notion seemed quaint, at best, as the dominant cultural ethos came to stress the importance of individual success, self-fulfillment, and personal pleasure. While feminist psychologist Phyllis Blanchard held out for an alternative set of values that esteemed altruism over egoism, even she pulled back from deeming these to be universal to girls and boys. More developmental thinkers of the 1920s rendered the feminine "love ideal" subordinate to the masculine goal of self-realization.[114] For her part, Mead enthused that modern adolescents were, indeed, disposed toward selflessness, idealism, and religiosity but determining what cause to support or church to join were seen as more opportunities for the adolescent to exercise choices in the process of fashioning a unique personal identity. So, a modern girl looking to her family members as sources of inspiration for her own beliefs, values, and activities might turn to a father who was "a Presbyterian, an imperialist, a vegetarian, a tee-totaler, with a strong literary preference for Edmund Burke," an aunt who was "an agnostic, an ardent advocate of woman's rights, an internationalist who rests all her hopes on Esperanto, is devoted to Bernard Shaw, and spends her spare time in campaigns of anti-vivisection," and so on.[115] Girls—and boys—were no more or less inclined toward service than what they might choose to eat or to read. Most important was that all of these possibilities for doing and being existed as coequal, if sometimes contradictory, choices for the personality in the making.

Mead's analysis also sought to qualify the emotional experience of adolescence. Adolescent girls and boys were emotionally volatile, she argued, because of the difficult choices they faced in a complex society, not because of the exigencies of biological puberty. That established, she also wanted to free American adolescents from the specific emotional poignancy that arose out of the "evils inherent in the too intimate family organization," the effects of which the large, heterogeneous Samoan household protected against. While not willing to relinquish the specialized affection between parents and children (and especially between husbands and wives) that enabled the expression of the unique personality, Mead warned that it had to be acknowledged that this came "at the price of many individuals' preserving through life the attitudes of dependent children, of ties between parents and children which successfully defeat the children's attempts to make other adjustments, of necessary choices made unnecessarily poignant because they become issues in an intense emotional relationship." Mead herself did not identify this as a distinctly female problem, nor did she ascribe it to an unfortunate weakening of the patriarchal family. Indeed, she was as concerned with the effects on the girl of the "domineering, dogmatic" father as with the protective and solicitous mother.[116] In this way, she was able to draw on psychoanalytic assumptions about the centrality of family dynamics in the life of the child while also rejecting sexist remedies for family conflict. Nonetheless, the point drew on a long history of negative associations between emotionality, (inter)dependence, and femininity. Perhaps more importantly, it fed into the larger trend currently being advanced by both psychoanalytic and behavioristic interpretations of child development about the emotionally crippling nature of motherhood and the failures of girl children, in particular, to achieve "psychological weaning" from their families.[117]

In deliberately seeking to include modern girls among the adolescents who were to enjoy the opportunity to realize autonomous selfhood, then, Mead also contributed to the theoretical exclusion of developmental processes and outcomes that resonated with alternative ways of knowing and being during the experience of coming of age. She also failed to consider adequately the ways in which the American adolescent girl bore resemblance, in terms of limited cultural expectations for her development, to Samoa's "well-behaved flower." Mead briefly reflected on the ways in which "our American theory of endless possibilities" and "our myth of endless opportunity" might produce "bitter rebellion" in those individual adolescents for whom the choices they had to make were "in contrast to the opportunities which they are told are open to all Americans."[118] However, in *Coming of Age*, the problem of girls, and other social groups, facing restricted choices in a choice-oriented society was only minimally addressed in favor of an ultimately optimistic vision of the equal potential of all modern adolescents to shape themselves as they would.

Mead's observation that the Samoan girl faced "neither revelation, restriction, nor choice" in her coming of age allowed for appreciating and scrutinizing both primitive culture and western civilization.[119] Samoan culture was to be emulated for its gradual approach to child development that avoided any sudden revelations about sexuality at the onset of biological puberty and the cusp of social adolescence. However, lack of opportunity for individual choice in Samoa, which helped to account for the absence of conflict among Samoan adolescents, Mead rejected as a wholly undesirable solution to the challenges modern youth faced. In this trio of comparative cultural characteristics, "restriction" presented the most thorny problem for Mead. In Samoa, the combination of few restrictions and fewer choices worked well to produce ease of adjustment among its young and to assure the continuity of a simple, homogeneous society. In Manus, the subject of Mead's next study, a rigid set of social expectations, largely enforced by the inducement of shame, also guaranteed cultural continuity, but at the expense of frustrated youth who grew into profoundly unhappy adults.[120] In the case of American adolescents, Mead maintained that neither a strict set of social expectations, which mired the young in tradition, nor a lax set of social expectations, which never required them to impose larger meaning on disparate choices, would preserve and further the most valued attributes of modern society. Needed above all else, she concluded, was for modern youth to become aware of themselves as cultural beings, to learn to recognize the ways in which every culture, even ones premised on freedom of choice, demarcated certain limits *and* the ways in which purposeful individuals could affect their social environment to make a better world:

> . . . we have one great superiority over . . . all primitive peoples. To them their customs are immune from criticism—given, ordained, immutable. They move unselfconsciously within the pattern of their homogeneous, self-contained societies. We, caught almost as completely in a far more complex pattern, have acquired the ability to think about it. Our young people pay the price of heterogeneity in the choices which they must make . . . But it is possible for us to give them in some slight measure the benefit of that heterogeneity also. It is choice which makes us culture-conscious, which makes it possible for us to see our society as a complex of possible courses.[121]

Deeming the edge obtained by modern adolescents to be a function of their "culture-consciousness," Mead coded her own versions of the concepts of "adolescence" and "culture" with a remarkably similar set of expectations. She declared the ideal adolescent to be capable of exercising autonomous choice and the ideal culture to be able to sustain possibilities for freedom through its heterogeneity and dynamism. In the same way, according to Mead, just as adolescent deviance was never to languish as utter aimlessness, culture was not to devolve into absolute relativism. Indeed, purposeful adolescent rebellion was, she proposed, one vital impetus to progressive cultural change. Even as Mead echoed Hall's vision of the felicitous relationship between a plastic adolescence and the evolution of civilization, she updated her predecessor's Lamarckianism with an "anthropological point of view" that presented the promises and difficulties of the adolescent stage as a product of culture rather than biology. As Parsons and Van Waters had begun to do, Mead too undercut reductive race- and sex-based theories of ontogeny by emphasizing the overriding importance of the "human element" in shaping the meanings and experiences of child development among different peoples. She also, however, renewed the idea of a privileged relationship between optimum adolescent development and the distinctly masculine achievements of western civilization.

Epilogue

Defining and explaining adolescence and reconciling the concepts of adolescence and femininity have engaged biomedical and social scientists from the mid-twentieth century to the present. This epilogue will highlight some of the most important contributions to the scientific discourse on adolescence and adolescent girlhood from 1930 onward. Recent aspects of the history of adolescence are, of course, worthy of full studies and have become a vibrant area of scholarly interest. The intent here is to show that the contested meanings of adolescence and female adolescence formulated during the nineteenth and early twentieth centuries continue to inform scientific knowledge and broader cultural common sense about the teenage child, even as such meanings have been reworked in particular historical contexts marked by distinctive forms of social relations, cultural practices, and intellectual concerns.

As in the period from 1830 to 1930, the decades from the mid-twentieth century to the present have been marked by social, political, cultural, and economic changes that have shaped and been shaped by the experiences of boys and girls growing up. Many professionalized scientific experts have sought to understand, to explain, and to prescribe these experiences in various ways. Certainly, an important influence in young people's lives throughout this period has been the growth of a mass consumer culture. As early as the 1920s, marketers and advertisers identified the "teenager" as a social category, with discrete needs and desires that could be satisfied by purchasing particular consumer goods. Historian Kelly Schrum demonstrates that white middle-class high school girls were the first young people marketers catered to; high school girls were not only models for certain attributes of adolescence but also were the "first teenagers."[1]

The commercial category of "teenager" and the scientific category "adolescence" intersect with and inform each other in ways scholars have just begun to probe. Marketers initially appealed to girls on the basis of contradictory attributes and behaviors that girls had been experimenting with for some time and that social scientists had endeavored to account for, contain, and legitimate: girls' concern for their appearance, their interest in cultivating sex appeal, their susceptibility to influence by their peers, their longing to express themselves as individuals, their emotional volatility and vulnerability, and their intellectual capability. Juliet B. Schor, an economist and researcher of current commercial trends, argues, "Marketers took the psychology [of development] and reconceptualized the process of growing up as a process of learning to consume." As marketers used the developmental paradigm throughout the twentieth century, they also further refined developmental stages, with such creations as "preteen" and "tween," each with distinct physical, psychological, and behavioral characteristics. Marketers have also succeeded at undermining certain core tenets of the developmental paradigm in ways children's rights advocates or critical developmental psychologists have not. Marketers have challenged the notion of the child's growth into maturity as a slow, gradual process marked by innocence and dependence. They have pushed expectations for autonomous decision making (around consumer purchasing), self-expression, and worldly knowledge and pleasure into earlier stages of the life cycle. Marketers claim to offer children and teenagers new opportunities for independence and empowerment. What girls and boys have made of such opportunities, and what they have gained and lost in the process, interested scholars in both the sciences and the humanities during the second half of the twentieth century and has been a subject of ongoing cultural and political debates.[2]

From the 1930s through the 1950s, as the commercialized teenager came into its own and as attending high school became more widespread, the category of adolescence received renewed attention by scientists working in the interdisciplinary field of child development studies. Arnold Gesell and his colleagues at the Yale Clinic of Child Development and the Gesell Institute of Child Development (founded in 1950) further elaborated on the stage theory of growth, which continued to influence some child development experts and the public. Following his profiles of the normal infant, preschool, and school-aged child, Gesell arrived at the detailed exposition of the "intricate transitional years" of adolescence late in his career, with the 1956 publication of *Youth: The Years from Ten to Sixteen*.[3] As in his work on the earlier stages of the child's development, Gesell acknowledged the importance of each adolescent's uniqueness and individuality, which he claimed was largely the product of genetic endowment, and of the role of environmental factors in "support[ing], inflect[ing], and modify[ing]" developmental progress. He also reasserted his predominating interest in the "underlying ground plan" of growth, which he deemed to be inherent in the species and the "prime causative force" of the "maturity traits" marking this period of life. On the one hand, Gesell's meticulous accounting of the physical, mental, emotional, and social changes young people experienced during each year of this phase in the life cycle undermined some of the "loose, sweeping generalizations" about the adolescent epoch, particularly notions of its overwhelming turbulence and rebellious orientation. At certain moments during these years, Gesell assured, young people were capable of controlling their emotions and exhibiting a purposeful, rather than impulsive "spirit of independence." On the other hand, by parsing out the qualities of adolescence in such an exacting way, Gesell further solidified their "characteristicness" and intensified expectations for their proper manifestation, in "obedience to [the] deep-seated laws and cycles" of development.[4] With his emphasis on biologically driven maturation and standardized criteria for adolescents' physical growth and mental development, Gesell both heightened and alleviated the anxiety of parents and teenagers concerned with measuring up to the normative characteristics of the adolescent stage. In an era in which deviance of any kind was highly suspect, his ideas appealed to parents seeking definitive answers about their adolescent children's normality and explanations for their adolescents' misbehavior in genetics or the vagaries of the developmental stage, rather than their own poor parenting.[5]

The findings recorded in *Youth* were based on cross-sectional and longitudinal analyses of interviews with and examinations and measurements of eighty-three boys and eight-two girls from white middle-class New Haven families. In commenting on the homogeneity of the sample, Gesell averred that he and his colleagues "assumed that these sequences [of development] are not fortuitous and that they would show a significant relation to maturity levels and chronological age despite diversities in individuality." The universality of such maturity levels was, nonetheless, compromised by "sex variations," which Gesell deemed to be the "most far-reaching constitutional factor" distinguishing adolescents and "the whole of humanity." Although Gesell cautioned that gender differences could "easily be overstated" and suggested that girls and boys "meet most stages of development in a highly comparable manner," he also ascribed to adolescent girls and boys distinct masculine and feminine characteristics that comfortably reinforced dominant expectations for cold war gender roles, which emphasized male bread-winning and female homemaking.[6] According to Gesell, girls matured faster, were more oriented toward domesticity, marriage, and motherhood, and were "more sensitive to moral and personal issues" than their brothers. Boys were more concerned about careers than girls and, most notably, emerged from their sixteenth year with a "sense of independence" and "self-assurance," aptly poised to take on their "share of the world's work." Born as he was "with certain inalienable traits which are inherent in the very patterns of his development," the boy achieved what the girl never quite managed to—becoming "an individual personality in his own right." In keeping with his conformist sensibilities, Gesell also reminded his readers of the complex and paradoxical nature of individuality in a democracy: It could only be made manifest and preserved within the context of "obedience" to both natural laws and social rules and regulations. As an ideal primer for the 16-year-old boy, he recommended the "Code of Conduct for American Servicemen While Prisoners of War." More than a military manual, Gesell contended, this "remarkable document" would acquaint adolescent boys with the "ideals and traditions" of the United States, serve as a "bulwark against enemy political indoctrination," and help to fashion them into the sort of "loyal Americans" that their progression through the stages of development guaranteed for and required of them as they came of age.[7]

Gessell's ideas were both challenged and reinforced by Lawrence K. Frank, another important figure who shaped the concept of adolescence in this period. Trained in economics, Frank served as associate director of the Laura Spellman Rockefeller Memorial in the 1920s and 1930s and played a crucial role in facilitating research in the science of normal child development and in popularizing this body of scientific knowledge among a wider public. His interest in adolescence was prompted in part by new social concerns about juvenile misbehavior at home during the Great Depression and World War II, as well as by the participation of adolescents in totalitarian movements abroad in the same period. During his tenure at the Rockefeller Memorial and other philanthropic organizations, a spate of longitudinal studies of adolescent growth and development were launched at Harvard, Yale, the University of Iowa, the University of Chicago, the University of California at Berkeley, and other child development institutes. These studies began with and reinforced many of the assumptions about adolescence promulgated by scientists in the nineteenth and early twentieth centuries, most notably that adolescence was a particularly consequential stage of the life cycle that was important both to the individual and society and that the attributes and difficulties of the age resulted from some mix of biological and environmental influences.[8]

Frank's own writings on adolescence emphasized anew that the primary mandate of the adolescent stage of life was individualization and also furthered the links established in both Hall's evolutionary theory and Mead's cultural anthropology between adolescent independence and the ongoing advancement of modern society. Departing from Gesell, Frank insisted that adolescents not be forced into blind obedience of their parents' values or way of life because parents' values would be rendered obsolete in a society advancing toward greater democracy and expanded scientific knowledge. He therefore praised the peer culture for enabling adolescents to rebel against such conventions, while also keeping the centrifugal forces of individualism in check. Nonetheless, he recognized that the pressure for adolescents to fit into the peer group, which was often sanctioned by parents, posed problems as well. His hope was that parents, teachers, and medical and psychological professionals would encourage adolescents to honor their individual differences and to adjust, without obliterating, their unique capacities and perspectives to the spectrum of thought and behavior considered "normal."[9]

Both Frank and the child development researchers involved in the longitudinal studies modeled the individualizing adolescent after the white middle-class boy. These were the adolescents, in Frank's scheme, who were to be given the most latitude in rebelling against stagnant adult conventions and who would guarantee the ongoing success of American democracy and scientific innovation. Like Gesell, Frank and the child development researchers set gendered parameters around such boys' ability to deviate from the norm, worrying over adolescent boys who did not conform to masculine standards of physical prowess and flagging delayed puberty as a distinctly male developmental problem. However, for girls, the clash between the individuated adolescent ideal and compliance to female gender roles was more pronounced. In the face of a new round of threats to women's domestic roles during the Depression and the Second World War and under the growing influence of a particularly virulent antifeminist strain of Freudian theory, Frank and the child development researchers worried about girls who went through puberty too early and who did not meet what they deemed to be appropriate physical and behavioral standards of femininity.[10] Such standards were, indeed, modern, in that adolescent girls were expected to be physically appealing, sociable, popular, and exuberant. Lost to Frank's and Gesell's analyses were the links made by such thinkers as Parsons, Hollingworth, and Mead between the adolescent girl's capacity for individualization and the potential for fundamental social change or, conversely, by Hall, Addams, and Blanchard between such change and the transformative possibilities of the more traditional virtues of femininity.

Erik H. Erikson advanced a complicated gender perspective and arguably made the most important contributions to ideas about adolescence during the mid-twentieth century. Erikson's biographers connect his early life with the concept of the identity crisis, with which he is so prominently associated. Erikson was born in 1902 in Germany to a single mother from a prominent Jewish Danish family and a biological father, most likely a Danish Gentile, whom he never knew. He was somewhat ambivalently adopted by the German Jewish pediatrician his mother married when he was 3 years old. As his biographer Lawrence J. Friedman notes, Erikson eventually recognized that "his whole childhood involved learning to navigate borders—between Judaism and Christianity; Denmark and Germany; mother, stepfather, and biological father." Following his years as a markedly unhappy student of classical literature and languages at the Karlsruhe gymnasium, where he graduated in 1920, he spent the next seven years studying art and "wandering" through Germany and Italy. In 1927, at the behest of his former classmate and friend Peter Blos, he traveled to Vienna, where he trained in psychoanalysis with Anna Freud, a pioneer in the field of child analysis; became acquainted with some of the concepts of early ego psychology, as propounded by Anna Freud and fellow psychoanalysts Heinz Hartmann, Wilhelm Reich, and Paul Federn; completed a teaching degree in Montessori education; and taught at the Heitzing School, an experimental institution that brought psychoanalytic principles and work to bear on children's education.[11] This rich environment enabled Erikson to further develop a theme he had begun to formulate in his early twenties, before he was familiar with Sigmund Freud's ideas or studied with Anna Freud, and that would become the central intellectual concern of his career—the interaction between children's inner emotional life and outer social worlds as they moved through the stages of the life cycle. In the face of political turmoil in central Europe and his growing disillusionment with the Vienna psychoanalytic community, whose members were largely uninterested in what he would eventually refer to as his "configurational approach" to the understanding of the human personality, Erikson and his family immigrated to the United States in 1933. In the years before the publication of his first book, *Childhood and Society*, in 1950, he developed his career as a child analyst, researcher, lecturer, and writer and cultivated professional, intellectual, and personal relationships with the most important figures in child development studies in the United States, including William Healy, Lawrence Frank, and Margaret Mead.[12]

Childhood and Society comprises a collection of essays revised from Erikson's writings, published and unpublished, of the late 1930s and 1940s. Several of his chapters—including those on childhood among the Sioux and Yurok American Indian tribes, the "legends" of Adolf Hitler's and Maxim Gorky's childhood and youth, and "the American identity"—reflected the influence of the culture and personality school of thought on his work and contributed to the studies of national character that occupied many prominent behavioral and social scientists during and after World War II. Other chapters drew on psychoanalytic theory and Erikson's own clinical work to take thinking about child development in innovative new directions, most notably by recasting Freud's psychosexual developmental stages into biopsychosocial stages; extending these across the life cycle, from infancy into old age; and giving more attention to the role of the normally developing ego in autonomously facilitating connections between the inner emotions and external social conditions.[13]

These essays elaborate on the importance of establishing identity for the human personality, a task that Erikson depicted, somewhat contradictorily, as occupying the individual throughout the life cycle, while reaching its point of most crucial achievement during adolescence. Identity, as Erikson defined it, was "the accrued experience of the ego's ability to integrate . . . [childhood] identifications with the vicissitudes of the libido, with the aptitudes developed out of endowment, and with the opportunities offered in social roles." As with the other polarities that the ego struggled to resolve during development (trust vs. basic mistrust, autonomy vs. shame and doubt, etc.), identity was wrested out of a conflict with its opposite—role diffusion. In Nazi Germany, the inability of wayward male youth to resolve their identity crises successfully made them susceptible to Hitler's enticements toward militarism and racial hatred and helped to explain the Führer's rise to power. In the increasingly industrialized and standardized society of the United States, adolescents in Erikson's adopted country faced a similar danger of losing their individual identities and succumbing to conformist pressures, if not of authoritarian demagogues than of a technological and bureaucratic social organization. Mitigating such danger in Erikson's analysis, however, was the historic promise of America to sustain possibilities for choice, opportunity, dynamism, and tolerance in individual and social life. While Erikson did not always adequately recognize the intellectual influences that contributed to the formulation of his concept of identity, he also expressed some concern that popularizers too facilely credited him with originality. *Childhood and Society* garnered attention from midcentury cultural critics, socially concerned academics, and, following the highly successful publication of a revised paperback college textbook edition in 1963, college students who were worried about threats to individual selfhood in a technological and bureaucratic American society. We know, too, that Erikson's ideas resonated with a long history of ideas about adolescence in American thought—about the challenges and possibilities that made this stage unique in modern society, about this stage of life as a period of personality integration and individualization, and about the importance of adolescent adjustment to the social progress, freedom, and democracy of western civilization.[14]

Erikson expanded on the themes in *Childhood and Society* in subsequent works, including his biographies of Martin Luther and Gandhi, *Insight and Responsibility*, and *Identity: Youth and Crisis*. In these writings, he prominently associated the identity crisis, and the adolescent stage of the life cycle, with male development. He assumed the normativeness of the boy's development and, relying on Freudian orthodoxy, depicted the developing girl as different and deficient because she lacked a penis. Determined by her biological imperative to mother, he explained, the girl faced a narrower range of identity options that precluded her from fully engaging in autonomous, active self or social creation. Moreover, once the girl became a woman, she often acted as a foil to the boy's achievement of individualism, as mothers served as the primary vehicle through which the conformist expectations of the older generation were conveyed to children.[15]

Such indictments of femininity notwithstanding, however, Erikson also saw himself as construing a model of development that was congenial to girls and women and that valorized the essential female strengths of nurturance, generosity, and empathy. In his essay, "Inner and Outer Space: Reflections on Womanhood," in which he reported on his findings about sexual difference from observing the play of boys and girls participating in a child guidance study at the University of California, he outlined a "post-Freudian position" on female development that demanded that "feminine ego-strength be studied and defined in its own right." Here, Erikson grounded the establishment of female identity on "the existence of a *productive inner bodily space*," which he now claimed had "a reality [for the developing girl] superior to that of the missing organ" of the penis. In doing so, he perpetuated a long tradition of thought about the auspiciousness of female developmental difference that had been expressed by scientists and intellectuals in the nineteenth and early twentieth centuries. More immediately, he drew on the ideas of mid-twentieth-century psychoanalytic psychologist Helene Deutsch, who located women's capacity for sexual pleasure, psychological health, and personal fulfillment in the distinctive attributes of the female body and its unique role in reproduction. Longstanding struggles to reconcile expectations for femininity and adolescence and his own configurational approach to development, however, also required Erikson to qualify his biological determinism. He thus characterized adolescence as a period of "psychosocial moratorium" from adult activities in the girl as well as the boy, during which she was "relatively freer from the tyranny of the inner space" (compared with the stages of childhood or adult womanhood) and could pursue identities as an individual "*some*body" in political and economic life. In the case of male and female alike, he conceded, "anatomy, history, and personality are our *combined destiny.*" Even so, the "reality" of the "inner space" exerted an influence over the girl's development that in one way or another, she was ultimately unable to deny. " [A] true moratorium," Erikson remarked, "must have a term and a conclusion: womanhood arrives when attractiveness and experience have succeeded in selecting what is to be admitted to the custody of the inner space 'for keeps.' "[16]

Erikson's allusions to girls closed in by their inner space, at best, or tyrannized by it, at worst, conveyed the message that female development was both different from and inferior to male development. In addition to insisting that the girl's honoring of her inner space was essential for her own psychological well being, though, he also proclaimed that society needed the feminine virtues that female bodily experience made possible. In place of "ruthless self-aggrandizement," he declared, mid-twentieth-century America required "new kinds of social inventions and . . . institutions which guard and cultivate that which nurses and nourishes, cares and tolerates, includes and preserves." He also entertained the possibility that boys and men were not precluded by biology from adopting such maternal values.[17] Indeed, the fundamental premise of Erikson's developmental scheme was that the individual's sense of self across the life cycle arose out of reciprocity and mutuality with his or her immediate caretakers and the larger social world. This interdependence made possible the other overlapping achievements of development—intimacy, generativity, and ego integrity —from infancy to old age.

Whatever Erikson's intended meanings, the complexities and ambivalences in his thinking about gender exposed him to a range of (mis)interpretations by both supporters and critics. Many popularizers and synthesizers of his work ignored both his emphasis on the centrality of interdependence in development and his interest in female development to reduce the realization of identity at adolescence to the achievement of atomistic individualism, a capacity most often associated with boys. Meanwhile, in the context of the second wave of women's rights activism of the 1960s and 1970s, liberal feminists issued harsh indictments of Erikson's alleged Freudian-inspired biological determinism and his justifications for women's position of social inequality, emphasizing instead, as some feminists had in the past, the similarities in the developmental mandates and capacities of boys and girls. He fared little better with the emergence of the cultural feminists in the mid-1970s, whose valorization of female difference actually had much in common with his own rendition of the inner space. It was in part Erikson's purported conception of identity formation as a function of separation that spurred Carol Gilligan, his former teaching assistant and an architect of female adolescence in the late twentieth century, to formulate her own theory of female development that emphasized the role of relationship in the life of the adolescent girl.[18]

Other developmental thinkers contributed to mid-twentieth-century thinking about adolescence. Most notably, Swiss psychologist Jean Piaget, who garnered intense interest among American psychologists in the 1950s, proposed a biologically based theory of cognitive development, which asserted that at adolescence the developing child became capable of abstract thought and logical reasoning. He also offered a related theory of moral development, which posited that morality developed in phases and was a function of moral reasoning over behavioral response. American psychologist Lawrence Kohlberg further expounded a cognitive developmental view of morality. Kohlberg mostly focused on the phases of morality that developed with the capacity of hypothetical thought, that is, during or after adolescence.[19] Anna Freud, who was a frequent visitor, lecturer, and teacher in the United Stages from the 1950s until her death in 1982, offered a psychoanalytic account of the developing child that exclusively emphasized biology as the primary cause of the storm and stress of the teenage years. She maintained that because of the emergence of the genital drive at puberty, which entailed both quantitative and qualitative changes in the "libido economy," the equilibrium between the id, ego, and superego normally attained in early childhood was inexorably upset, causing emotional disturbances and dramatic changes in behavior. Rhetorically posing the "ever-recurrent question" about "whether the adolescent upheaval is welcome and beneficial as such, whether it is necessary, and more than that, inevitable," she answered affirmatively on all counts. In responding to the new anxieties raised by the change in sex drive, the ego experimented with new defenses, the most important of which was the adolescent's "breaking of the tie" with the love objects of childhood— the parents. The adolescent accomplished this by affiliating with peers who offered alternative ways to think and behave and provided appropriate partners with whom to establish heterosexual adjustment. Freud's view was challenged at midcentury by several studies that revealed that many "normal" young people did not reject parental values at adolescence and also that many parents encouraged the development of independence in their adolescent children. Even so, she certainly had the weight of history behind her. Freud echoed Hall's acceptance of the adolescent sensibility as essentially "inconsistent and unpredictable." She also joined numerous other thinkers in the past and present, including Parsons, Mead, Hollingworth, and Erikson, in asserting that wresting out independence from members of an older generation who were not especially eager to confer it was the primary challenge of adolescent development and, indeed, the only way for the child to attain maturity and truly "grow up."[20]

These ideas continued to be synthesized, interpreted, elaborated on, and applied during the mid-twentieth century by an array of professional experts in mental hygiene, child guidance, and child psychology. Beginning in the 1950s, they were joined by another group of scientific professionals devoted specifically to the care of the adolescent body and mind—physicians in the newly formulated adolescent medicine specialty. As historian Heather Munro Prescott explains, adolescent medicine emerged out of heightening scientific and cultural concern about the adolescent stage of life, as well as the efforts of mid-twentieth-century pediatricians to expand the scope of their field. The first institution devoted to adolescent health was the Adolescent Unit at Boston's Children's Hospital, founded in 1951. The unit was headed by J. Roswell Gallagher, M.D., who before his appointment to Boston Children's Hospital served as the director of health services at the Phillips Academy in Andover, Massachusetts, for fourteen years. At both institutions, Gallagher provided clinical treatment for his patients, conducted research into their growth and development, and published his findings in both major scientific journals and popular magazines and advice manuals, hereby establishing his reputation as a midcentury national authority on adolescence.[21]

Claiming that physicians such as himself were better attuned to the needs of the whole child than the child psychiatrists who headed up child guidance clinics, Gallagher marketed the Adolescent Unit to parents by lauding its attention to the emotional, behavioral, *and* physical problems—and the complex relationship among them—of their adolescent children. Such an approach appealed to parents who sought an "uncomplicated biomedical solution for their children's problems" that did not indict their child-rearing capabilities. Yet, Gallagher's clinical approach and intellectual contributions resisted both biological reductionism and the primacy of parental needs and points of view. Along with Mead, Frank, Erikson, and Anna Freud, he endorsed the predominant view among developmental thinkers at midcentury that individualization (whether necessitated by biology or culture) was the most important task of the adolescent stage of life. Like Piaget, he also believed that adolescents had the mental capacity to understand and to decide their own medical care. Even so, Gallagher's critique of conformity and support of adolescent independence, as with other developmental thinkers of the period, was limited by his concomitant reinforcement of cold war era gender roles. Thus, when he treated female patients and when he wrote about female development, he was mostly concerned with helping adolescent girls achieve appropriate "feminine role identification," which entailed that girls accept their "natural" inclination toward protecting their sexual virtue, caring for their physical appearance, and assuming the roles of wife and mother in the private home.[22]

With the second wave of feminism in the 1960s and 1970s, such buttressing of conventional expectations for femininity came under heated attack by feminist intellectuals, both within and outside the purview of the biomedical and social sciences.[23] Psychologist Carol Gilligan issued the most important challenge to midcentury conceptions of female adolescence. Gilligan was born in 1936 into a family that cultivated her interests in music, language, and literature. Her childhood home was also one that fostered awareness of the importance of human connection and of social responsibility. "I grew up . . . during the Holocaust," she explains," with refugees from Europe constantly through the house, and so the sense of what you have to value, and how people help one another, and how that's key to survival, and people don't live alone—that was all part of my childhood." Gilligan studied literature at Swarthmore College and graduated with a doctorate in psychology from Harvard in 1964, where she was schooled in midcentury experts' ideas about female development. After a brief respite from academia, she went on to teach at Harvard and eventually became one of a small number of women faculty members to be tenured at that institution.[24] Her first book, *In a Different Voice: Psychological Theory and Women's Development,* examined the moral "judgment and action" of males and females at various points in the life cycle in several situations of "moral conflict and choice." Taking on both Erik Erikson and her mentor Lawrence Kohlberg, who deemed adolescent boys and men more capable of exercising the supreme moral power of abstract judgment, she argued that girls and women made moral decisions and achieved a sense of self "through [their] relationship with others." Furthermore, the "quality of embeddedness in social interaction and personal relationships" that differently marked girls' and women's lives was, Gilligan proposed, as important in human development, if not more so, as men's achievements of separation, individuation, and sense of impartial justice.[25]

Following the feminist sociologist Nancy Chodorow, Gilligan argued that the distinctive developmental paths of boys and girls emerged out of the differences in each sex's pre-Oedipal relationship with the mother. At adolescence, she went on to explain, "when identity and intimacy converge in dilemmas of conflicting commitment," the boy's orientation toward "self-expression" and the girl's toward "self-sacrifice" gave rise to quite distinctive developmental problems—for the former, "a problem of human connection" and for the latter "a problem of truth." Gilligan asserted that the association she was making between women and relationship was "not absolute" and that the "contrasts between male and female voices" that she observed in her research were presented "to focus a problem of interpretation rather than to represent a generalization about either sex." In addition, she also identified as the central paradox of human development the "truths" that "we know ourselves as separate only insofar as we live in connection with others, and . . . we experience relationship only insofar as we differentiate other from self."[26] These caveats and nuances notwithstanding, however, Gilligan's text played a pivotal role in launching the sensibility, intellectual standpoint, and politics of a cultural feminism that above all else valorized girls' and women's distinct and superior capacity for care, compassion, and connection. It also revitalized a scientific and cultural debate about female adolescence that had not been so vigorously engaged since the turn of the twentieth century.

Gilligan continued to develop the themes of a distinctive and valuable relational female morality and sense of identity through her role as a founding member of the Harvard Project on the Psychology of Women and Girls' Development during the 1980s and 1990s. This feminist collective carried out several research studies employing a "relational approach to psychological inquiry" that entailed listening to the voices of girls from varying socioeconomic backgrounds—ranging from "at-risk" girls at an urban public high school to privileged girls attending the prestigious Emma Willard School in Troy, New York. The result was a spate of publications by Gilligan and her colleagues that focused on the problems and possibilities that the girl's distinctive developmental orientation posed for her at adolescence. Gilligan was already well known inside of and outside of academic and scientific circles because of the phenomenal success of *In a Different Voice*. She now achieved national celebrity, with *Time* magazine naming her one of the twenty-five most influential people in the United States in 1996.[27] In *Between Voice and Silence: Women and Girls, Race and Relationship*, Gilligan and fellow researchers Jill Taylor McLean and Amy M. Sullivan summed up the Harvard Project's fundamental conclusion about female adolescence. "At adolescence," they argued, "a shift takes place for many girls as they experience a relational impasse and a developmental crisis. To be in relationship at this juncture often jeopardizes 'relationship.' " Faced with the challenge of both conforming to the standard of adult femininity for "selflessness" *and* achieving the standard of maturity for "separation and independence," adolescent girls found themselves in a lose-lose situation. On the one hand, the authentic relationships they knew in childhood were compromised by new requirements of adult femininity that they please and serve others. On the other hand, the possibility for any relationship was undermined by expectations that they become autonomous individuals "entire unto themselves." Being in this position potentially gave rise to tremendous stress, confusion, and unhappiness in adolescent girls, putting them at risk for various psychological and social difficulties. Yet, McLean, Taylor, and Gilligan insisted, if psychological researchers such as themselves and society were to attend to girls' capacity for resilience and resistance in the face of their unique developmental challenge, a more humane society might at last be born. "The margins," they asserted, "can also offer a potentially transformative perspective on institutional and social norms, beliefs, and practices—an epistemological privilege that allows for an awareness and critique of standards of behavior or attitudes that diminish, demean, or disempower individuals or groups." In reflecting on the girls' voices she listened to at the Emma Willard School, Gilligan pondered how such girls might have, more simply, conceived of their relationship to fundamental social change. "What would happen," she imagined them wondering, "if what was inside of us were to enter the world?"[28]

Certainly, the work of Gilligan and her colleagues took a pioneering approach to the study of female adolescence. Indeed, no other scientific researchers in the past expressed such a deeply moral respect for their subjects, reflected so critically on their own role as producers of psychological knowledge, or made such forceful connections between the problems and possibilities of girls' development and the psychological and social empowerment of women. Even so, certain continuities with the past are worth noting as well. Most obviously, members of the Harvard Project built on a long tradition in scientific thought about girls' and women's greater capacity for connectedness and their corresponding superior morality, a perspective that has been employed by as many conservative thinkers as progressive ones. More subtly, perhaps, Gilligan and her colleagues were not the first to discover that girls faced a conflict between "love and ambition" at adolescence. Nor were they even the first feminist psychologists to suggest that an interdependent female self was a model for human development that had the potential to remake the world.

Much of the criticism of Gilligan's work has fallen out along some familiar lines, as well.[29] One group of critics includes feminist psychologists and intellectuals who take issue with both Gilligan's research methods and her conclusions. Their concern is that her conception of female development is essentialist in that it posits the existence of an "authentic" self before the influence of social forces and that it presumes women's inherent and unchanging difference from men. "Imagine, if you will," Carol Tarvis proposes, "that femininity is fragmented, messy or haphazard rather than coherent and authentic. Or that the expression of identity is a contingent and temporary business rather than a matter of finding one's true self. Or that femininity is a kind of ongoing project which has to be ongoingly constructed through social interaction rather than an object to be discovered, suppressed, or lost. In other words, femininity is a social construction rather than a psychological entity."[30] These critics also contend that despite some attention to the categories of race and class, Gilligan's work reifies gender difference as the fundamental distinguishing characteristic between human subjects. Moreover, these critics claim, although Gilligan attributes the greater female capacity to care to the different ways girls and boys are mothered in early childhood, she does not adequately attend to the larger context of social conditions and power relationships within which the different female voice, insofar as it exists, emerges. Like Hollingworth, Pruette, and Mead, such feminist researchers and thinkers maintain that boys and girls are more similar than not in their developmental capacities and mandates and that what differences do exist are the product of socialization.

Meanwhile, other critics charge that Gilligan's portrait of girlhood is overly optimistic and sentimentalized. Indeed, some of her colleagues have begun to investigate girls' capacity for anger, aggression, emotional cruelty, and pleasure-seeking sexuality and to study how, why, and with what effects such inclinations develop during childhood and adolescence. Although such tendencies certainly did not go unnoticed in the past, by thinkers ranging from Hall and Freud to Blanchard and Pruette, the current round of research offers up a decidedly critical feminist analysis of them, variously interpreting them as a dangerous product of or an empowering challenge to the values and relations of patriarchy shaping contemporary society.[31] Finally, many of Gilligan's critics worry that the far-reaching effect that her paradigm (or more reductive popular versions of it) has had on ideas and practices in psychotherapy, education, the law, and business will ultimately hurt girls and women's opportunities to achieve full economic, political, and social equality in the twenty-first century.

Another group of critics claims that Gilligan's research has done a deep disservice to boys, by eclipsing the particular developmental challenges they face in growing up, ignoring their own capacity and need for connectedness with others, devaluing their uniquely male qualities and capacities, and requiring them to conform to a "feminized" standard for moral, emotional, and social development. Such concerns resonate with earlier developmental thinking. In the 1870s,

Mary Putnam Jacobi and other critics of Clarke sought to counterbalance the exclusive focus on the girl's development by turning their attention to the physical, mental, and emotional vulnerabilities the boy experienced during his own "epoch of development." The closing decades of the nineteenth century then witnessed a heightened interest in the civilized boy's growth into manhood, a process most thoroughly explored by G. Stanley Hall. By valorizing the masculine traits of virility and self-control, decrying against the effeminate influences pervading civilized society, and holding girls up as a model for some of the moral and emotional qualities he admired, Hall actually presaged the range of contributions that composes the contemporary discourse on male development.

As in the past, what literary scholar Kenneth B. Kidd calls the "new boyolgy" owes much to broader cultural anxiety about an imperiled white middle-class masculinity, as expressed in the contemporary men's movement, and to current conceptions of female development. Some architects of the "new boylogy," including psychologists William S. Pollack, Dan Kindlon, Michael Thompson, and Eli H. Newberger, acknowledge their indebtedness to Gilligan's work and proclaim an interest in taking it in new directions, rather than rejecting it, or its feminist principles, outright. By doing so, they end up reinforcing the notion that boys and girls are fundamentally different (with these differences rooted in both nature and nurture), while also claiming that boys have a similar need to have their voices heard and to develop their own moral and emotional capacities for relationship with others. Drawing on Chodorow and Gilligan, these psychologists maintain that boys' capacity for intimacy is threatened long before their adolescence, when in early childhood the imperative to establish a masculine self requires them to separate from the close emotional bonds established with their mothers. Thus, they contend, helping boys understand and express the range of feelings that make up their emotional lives, especially their longing to care for and be cared for by others, must be addressed from an early age.[32]

Other contemporary boyologists, such as psychologist Michael Gurian and cultural critic Christina Hoff Sommers, virulently denounce this feminist perspective on boys' development, rejecting any claims that deny essential differences between boys and girls. Boys must not be raised or educated in the same manner as girls, they argue, but rather must be allowed to express their innate capacities for energy, independence, and wildness, qualities that place them in a position of opposition to and power over all that is feminine within themselves and in the broader social world.[33] The important differences among them notwithstanding, all of these boyologists deem white middle-class boys to be in a state of crisis, at risk of academic failure, low self-esteem, identity confusion, depression, suicide, drug abuse, and delinquency. If ignored, they warn, such a condition can only portend the direst consequences for such boys and the society in which they are coming of age. The literature on male development intimates that such consequences are worse than the effects experienced by boys living under conditions of racism and poverty, whose experiences boyologists largely ignore. The literature also suggests that such boys' problems are worse than the developmental crisis faced by their white middle-class sisters, which boyologists once again envision alternatively as a model for and foil to the boy's own, while also finally determining it to be one of less personal and cultural significance.[34]

A final response to Gilligan's work has arisen from a group of developmental psychologists influenced by postmodern critical theory. The task of these scholars, psychologist Erica Burman explains, is to expose the developmental paradigm as "a modern, Western construction," whose theory and practice is suffused with "inequality and differential treatment on the basis of class, culture, gender, age and sexuality." In their scheme, Gilligan's reconceptualization of female development does not go far enough because it continues to operate within the terms of the paradigm that it seeks to challenge. As Valerie Walkerdine insists, "Other stories which have been the object of domination through their incorporation into schemes of pathologization cannot be spoken within a developmental framework." The solution she is proposing, then, is not merely the "tacking [of] difference onto a developmental model," for "that ship is holed below the water line." Rather, it is to provide "specific concrete and local analyses" of how subjects are constituted through "discursive practices," resulting in "[s]tories about change and transformation" that are not reductive, essentialist, universalizing, hierarchical, or relativist. Such an approach, Walkerdine and Burman promise, offers up a radical alternative to what they deem to be the oppressed and oppressive modern subject, which has been constituted through a developmental paradigm that privileges individualism, reason, progress, and the nature / nurture dualism at the expense of what they consider to be more morally sustainable ways of being and knowing in the world.[35]

This study of the early invention of the category of female adolescence draws from this critical approach to provide some further historical explanation for how the norms for adolescent development were established and authorized by physicians, psychologists, and anthropologists working within and participating in the formulation of the developmental paradigm in the nineteenth- and early-twentieth-century United States. One of the goals here has been to show that ideas about the white middle-class girl were quite important for that process.

Beginning with the nineteenth-century physicians involved in the health reform movement through the contributions of Margaret Mead in the field of cultural anthropology, thinkers and scientists took up the problem of female development as an object of particular concern, hereby giving rise to the figure of "the adolescent girl" and helping to formulate the modern concept of adolescence. Efforts by these thinkers to reconcile the concepts of femininity and adolescence went a long way in contributing to the constitution of the modern subject as described by contemporary critical developmental psychologists. It was also, however, at that very juncture where expectations for adolescence and femininity did not quite fit together that the hegemony of the developmental paradigm was sometimes questioned and the limits of developmental orthodoxy probed. Whether the ongoing project of issuing such a critique ought to come from within the developmental paradigm or only can only be effective from "beyond" or outside of it is unclear. However, Erica Burman is correct: It is worth pursuing one of the promises of the developmental framework held out by several of the thinkers explored in this book—that we can sustain a mode of thinking through which we might imagine "how we can create, and can become, the people who can bring about and can inhabit a very different world."[36]

Notes

INTRODUCTION

1. F. Scott Fitzgerald, "Bernice Bobs Her Hair," in *F. Scott Fitzgerald: Novels and Stories, 1920–1922* (New York: Library of America, 2000), 365. Fitzgerald's story was published in 1920. *Little Women* was published in 1868 and 1869.

2. Paula S. Fass, *The Damned and the Beautiful: American Youth in the 1920s* (Oxford: Oxford University Press, 1977); Nancy F. Cott, *The Grounding of Modern Feminism* (New Haven, CT: Yale University Press, 1987), Chapter 5; Carolyn Kitch, *The Girl on the Magazine Cover: The Origins of Visual Stereotypes in American Mass Media* (Chapel Hill: University of North Carolina Press, 2001), Chapter 6; Kelly Schrum, *Some Wore Bobby Sox: The Emergence of Teenage Girl Culture, 1920–1945* (New York: Palgrave Macmillan, 2004).

3. Phyllis Blanchard and Carlyn Manasses, *New Girls for Old* (New York: Macaulay, 1930); Fitzgerald, "Bernice Bobs Her Hair," 365.

4. William Kessen, "The American Child and Other Cultural Inventions," *American Psychologist* 34, no. 10 (Oct. 1979): 815–820; William Kessen, *The Rise and Fall of Development* (Worcester, MA: Clark University Press, 1990); Ross D. Parke, Peter A. Ornstein, John J. Rieser, and Carolyn Zahn-Waxler, "The Past as Prologue: An Overview of a Century of Developmental Psychology," in *A Century of Developmental Psychology*, ed. Ross D. Parke, Peter A. Ornstein, John J. Rieser, and Carolyn Zahn-Waxler (Washington, DC: American Psychological Association, 1994), 1–70; Richard M. Lerner, ed., *Developmental Psychology: Historical and Philosophical Perspectives* (Hillsdale, NJ: Lawrence Erlbaum Associates, 1983).

5. For key works by Gilligan and her colleagues on female development, see the "Essay on Sources." Among the American Association of University Women's publications are *The AAUW Report: How America's Schools Shortchange Girls* (1992); *Hostile Hallways: The AAUW Survey on Sexual Harassment in America's Schools* (1993); *Growing Smart: What's Working for Girls in School* (1995); *Girls in the Middle: Working to Succeed in School* (1996); *Separated by Sex: A Critical Look at Single-Sex Education for Girls* (1996); *Gender Gaps: Where Schools Still Fail our Children* (1998). Mary Pipher, *Reviving Ophelia: Saving the Selves of Adolescent Girls* (New York: Grosset-Putnam, 1994).

6. Carol Gilligan, "Preface: Teaching Shakespeare's Sister: Notes from the Underground of Female Adolescence," in *Making Connections: The Relational Worlds of Adolescent Girls at Emma Willard School*, ed. Carol Gilligan, Nona P. Lyons, and Trudy J. Hanmer (Cambridge, MA: Harvard University Press, 1990), 9, 10.

7. Barbara Hudson, "Femininity and Adolescence," in *Gender and Generation,* ed. Angela McRobbie and Mica Nava (London: Macmillan, 1984), 31, 51, 47, 53. See also Sue Lees, *Sugar and Spice: Sexuality and Adolescent Girls* (London: Penguin Books, 1993), 4, 15–17.

8. Joseph F. Kett, *Rites of Passage: Adolescence in America, 1790 to the Present* (New York: Basic Books, 1977), "Introduction" and Chapters 5–8; John R. Gillis, *Youth and History: Tradition and Change in European Age Relations, 1770–Present* (New York: Academic Press, 1981), Chapter 3.

9. Kett, *Rites of Passage,* 137, 6.

10. Joseph M. Hawes, "The Strange History of Female Adolescence in the United States," *The Journal of Psychohistory* 13, no. 1 (Summer 1985): 51–63. See also Elizabeth Lunbeck, *The Psychiatric Persuasion: Knowledge, Gender, and Power in Modern America* (Princeton, NJ: Princeton University Press, 1994), 189.

11. See especially Carroll Smith-Rosenberg, "From Puberty to Menopause: The Cycle of Femininity in Nineteenth Century America," in *Clio's Consciousness Raised: New Perspectives on the History of Women,* ed. Mary S. Hartman and Lois Banner (New York: Harper & Row, 1974), 23–37; Smith-Rosenberg, *Disorderly Conduct: Visions of Gender in Victorian America* (New York: Alfred A. Knopf, 1985), 53–76; Joan Jacobs Brumberg, "Chlorotic Girls, 1870–1920: A Historical Perspective on Female Adolescence," in *Women and Health in America,* ed. Judith Walzter Leavitt (Madison: University of Wisconsin Press, 1984), 186–195; Brumberg, "'Ruined' Girls: Community Responses to Illegitimacy in Upstate New York, 1890–1920," *Journal of Social History* 18 (Winter 1984): 247–272; Brumberg, *Fasting Girls: The Emergence of Anorexia Nervosa as a Modern Disease* (Cambridge, MA: Harvard University Press, 1988); Brumberg, *The Body Project: An Intimate History of American Girls* (New York: Random House, 1997); Jane H. Hunter, *How Young Ladies Became Girls: The Victorian Origins of American Girlhood* (New Haven, CT: Yale University Press, 2002); Kathy Peiss, *Cheap Amusements: Working Women and Leisure in Turn-of-the-Century New York* (Philadelphia: Temple University Press, 1986); Mary E. Odem, *Delinquent Daughters: Protecting and Policing Adolescent Female Sexuality in the United States, 1885–1920* (Chapel Hill: University of North Carolina Press, 1995); Ruth M. Alexander, *The "Girl Problem": Female Sexual Delinquency in New York, 1900–1930* (Ithaca, NY: Cornell University Press, 1995).

12. In addition to Hudson and Kessen, cited in notes 4 and 7, see Julian Henriques et al., *Changing the Subject: Psychology, Social Regulation, and Subjectivity* (London: Methuen, 1984); John R. Morss, *The Biologising of Childhood: Developmental Psychology and the Darwinian Myth* (Hove and London: Lawrence Erlbaum Associates, 1990); Morss, "Making Waves: Deconstruction and Developmental Psychology, *Theory & Psychology* 2, no. 4 (1992): 445–465; Valerie Walkerdine, "Beyond Developmentalism?" *Theory & Psychology* 3, no. 4 (1993): 451–469; Erica Burman, *Deconstructing Developmental Psychology* (London: Routledge, 1994); Nancy Lesko, *Act Your Age! A Cultural Construction of Adolescence* (New York: Routledge, 2001); Claudia Castañeda, *Figurations: Child, Bodies, Worlds* (Durham, NC: Duke University Press, 2002); Sheila Greene, *The Psychological Development of Girls and Women: Rethinking Change in Time* (London: Routledge, 2003); Catherine Driscoll, *Girls: Feminine Adolescence in Popular Culture and Cultural Theory* (New York: Columbia University Press, 2002).

13. Walkerdine, "Beyond Developmentalism?" 452, 453–454, 455.

14. For explanations and examples of this approach, see Nancy M. Theriot, "Women's

Voices in Medical Discourse: A Step toward Deconstructing Science," *Signs* 19, no. 1 (Autumn 1993): 1–31; Martha H. Verbrugge, "Recreating the Body: Women's Physical Education and the Science of Sexual Differences in America, 1900–1940," *Bulletin of the History of Medicine* 71, no. 2 (1997): 273–304; Sally Gregory Kohlstedt and Helen Longino, "The Women, Gender, and Science Question: What Do Research on Women in Science and Research on Gender in Science Have to Do with Each Other?" *Osiris* 12 (1997): 3–15; and Evelyn Fox Keller, "Developmental Biology as Feminist Cause?" *Osiris* 12 (1997): 16–28.

15. For theoretical exploration of the "transnational circuits of exchange" through which the child and childhood are figured in culture, see Castañeda, *Figurations*, 5–7.

16. On this last point, see Burman, *Deconstructing Developmental Psychology*, 186.

17. For helpful reviews of this literature, see Gerald F. Moran, "Colonial America, Adolescence in," in *Encyclopedia of Adolescence*, 2 vols., ed. Richard M. Lerner, Anne C. Petersen, and Jeanne Brooks-Gunn (New York: Garland, 1991), 1:157–164; and Helen King, *The Disease of Virgins: Green Sickness, Chlorosis and the Problems of Puberty* (London: Routledge, 2004), 86–88. See also Driscoll, *Girls*, 28, 82. See the "Essay on Sources" for more on the works in European and American history that challenge the location of the "birth of adolescence" in the nineteenth-century United States.

18. In addition to Kett and Gillis, see John and Virginia Demos, "Adolescence in Historical Perspective," in *The American Family in Social-Historical Perspective*, ed. Michael Gordon (New York: St. Martin's Press, 1973), 209–221; John Demos, *Past, Present, and Personal: The Family and the Life Course in American History* (New York: Oxford University Press, 1986), 92–113; and John Neubauer, *The Fin-de-Siècle Culture of Adolescence* (New Haven, CT: Yale University Press, 1992), 3–6.

19. Lyn Mikel Brown and Carol Gilligan, *Meeting at the Crossroads: Women's Psychology and Girls' Development* (Cambridge, MA: Harvard University Press, 1992), 2.

20. Carol Gilligan, "Prologue," in *Making Connections*, 1. The edition of the *Handbook of Adolescent Psychology* to which Gilligan refers was published in 1980.

CHAPTER ONE. "LAWS OF LIFE"

1. Robley Dunglison, *Medical Lexicon: A Dictionary of Medical Science,* 11th edition (Philadelphia, 1854), 602.

2. Joan Jacobs Brumberg, *The Body Project: An Intimate History of American Girls* (New York: Random House, 1997), 3–5, 23–25; Heather Munro Prescott, *A Doctor of Their Own: The History of Adolescent Medicine* (Cambridge, MA: Harvard University Press, 1998), 16; Joseph F. Kett, *Rites of Passage: Adolescence in America 1790 to the Present* (New York: Basic Books, 1977), 44; John Demos, *Past, Present, and Personal: The Family and the Life Course in American History* (New York: Oxford University Press, 1986), 95–96; and Jane H. Hunter, *How Young Ladies Became Girls: The Victorian Origins of American Girlhood* (New Haven, CT: Yale University Press, 2002), 130–132.

3. Michael Rutter, *Changing Youth in a Changing Society: Patterns of Adolescent Development and Disorder* (Cambridge, MA: Harvard University Press, 1980), 5; Catherine Driscoll, *Girls: Feminine Adolescence in Popular Culture and Cultural Theory* (New York: Columbia University Press, 2002), 50, 81.

4. This point is made most convincingly by C. Dallett Hemphill in *Bowing to Necessities: A History of Manners in America, 1620–1860* (New York: Oxford University Press, 1999), 37–39. See also Demos, *Past, Present, and Personal*, 97; Ross W. Beales Jr., "In Search of the Historical Child: Miniature Adulthood and Youth in Colonial New England," in *Growing Up in America: Children in Historical Perspective*, ed. N. Ray Hiner and Joseph M. Hawes (Urbana: University of Illinois Press, 1985), 17–24; and Gerald F. Moran, "Colonial America, Adolescence in," in *Encyclopedia of Adolescence*, 2 vols., ed. Richard M. Lerner, Anne C. Petersen, and Jeanne Brooks-Gunn (New York: Garland, 1991), 1:159–160.

5. The term *semidependence* is Kett's, *Rites of Passage*, 29. The term *semiautonomy* is Michael B. Katz's in *The People of Hamilton, Canada West: Family and Class in a Mid-Nineteenth-Century City* (Cambridge, MA: Harvard University Press, 1978), 256. See also Demos, *Past, Present, and Personal*, 99.

6. Kett, *Rites of Passage*, 137–138.

7. Hunter develops this argument throughout *How Young Ladies Became Girls*. The quotation is on p. 6.

8. Bernard Wishy, *The Child and the Republic: The Dawn of Modern American Child Nurture* (Philadelphia: University of Pennsylvania Press, 1968), 3–78. For a particularly suggestive analysis of the intersections of literature and science in nineteenth-century conceptualizations of child development, see Sally Shuttleworth, "The Psychology of Childhood in Victorian Literature and Medicine," in *Literature, Science, Psychoanalysis, 1830–1970: Essays in Honour of Gillian Beer*, ed. Helen Small and Trudi Tate (Oxford: Oxford University Press, 2003), 86–101.

9. See, for example, William Buchan, *Advice to Mothers* (Philadelphia: John Bioren, 1804); George Logan, *Practical Observations on Diseases of Children* (Charleston, SC: A. E. Miller, 1825); and William P. Dewees, *Treatise on the Physical and Medical Treatment of Children* (Philadelphia: Carey & LEA, 1825).

10. For the relevance of the themes explored by the antebellum health reformers to developmental psychology in the twentieth century, see Ross D. Parke, Peter A. Ornstein, John J. Rieser, and Carolyn Zahn-Waxler, "The Past as Prologue: An Overview of a Century of Developmental Psychology," in *A Century of Developmental Psychology*, ed. by Ross D. Parke et al. (Washington, DC: American Psychological Association, 1994), 1–70.

11. Howard P. Chudacoff, *How Old Are You? Age Consciousness in American Culture* (Princeton, NJ: Princeton University Press, 1989), 9–10; Kett, *Rites of Passage*, 11–13; Demos, *Past, Present, and Personal*, 96; Hemphill, *Bowing to Necessities*, 37–38; Beales, "In Search of the Historical Child," 17–24.

12. The discussion that follows on changes in the experiences of youth from the seventeenth through the mid-nineteenth century focuses on the Northeast, in part because this is the region that historians have investigated most thoroughly and also because the discourse about developing youth that emerged in the 1830s was largely a response to the changing experiences of young people in urbanizing, industrializing America.

13. Judith S. Graham thoroughly examines the debates by historians over the nature of colonial childhood in *Puritan Family Life: The Diary of Samuel Sewall* (Boston: Northeastern University Press, 2000). See also Steven Mintz and Susan Kellogg, *Domestic Revolutions: A*

Social History of American Family Life (New York: Free Press, 1988), 14–16; Kett, *Rites of Passage*, 16; Demos, *Past, Present, and Personal*, 96–97.

14. Mintz and Kellogg, *Domestic Revolutions*, 16.

15. Moran, "Colonial America, Adolescence in," 159; Demos, *Past, Present, and Personal*, 98–99.

16. For references to this literature, see note 17 of the "Introduction" and the "Essay on Sources."

17. See Roger Thompson, "Adolescent Culture in Colonial America," *Journal of Family History* 9 (Summer 1984): 131–141; Thompson, *Sex in Middlesex: Popular Mores in a Massachusetts County, 1649–1699* (Amherst: University of Massachusetts Press, 1986), 83–96, 97–109; Beales, "In Search of the Historical Child," 17–24; Moran, "Colonial America, Adolescence in," 160–164; and Graham, *Puritan Family Life*, 154–155.

18. Demos, *Past, Present, and Personal*, 99. See also Susan M. Juster and Maris A. Vinovskis, "Nineteenth-Century America, Adolescence in," in *Encyclopedia of Adolescence*, 2 vols., ed. Richard M. Lerner, Anne C. Petersen, and Jeanne Brooks-Gunn (New York: Garland, 1991), 2:698–699; and Harvey J. Graff, *Conflicting Paths: Growing Up in America* (Cambridge, MA: Harvard University Press, 1995), 26–33.

19. E. Anthony Rotundo, *American Manhood: Transformations in Masculinity from the Revolution to the Modern Era* (New York: Basic Books, 1993), 18–20.

20. Anne M. Boylan, "Growing Up Female in Young America, 1800–1860," in *American Childhood: A Research Guide and Historical Handbook*, ed. Joseph M. Hawes and N. Ray Hiner (Westport, CT: Greenwood Press, 1985), 166–167; Jacqueline S. Reinier, *From Virtue to Character: American Childhood, 1775–1850* (New York: Twayne Publishers, 1996), 134–135; Nancy F. Cott, *The Bonds of Womanhood: "Woman's Sphere" in New England, 1780–1835* (New Haven, CT: Yale University Press, 1977), 19–62.

21. The literature describing and problematizing separate spheres is voluminous. See the "Essay on Sources" for scholarship that helped to frame this discussion.

22. Boylan, "Growing Up Female in Young America," 166–170. For the argument that young women's wage work promoted individual freedom, see Thomas Dublin, *Women at Work: The Transformation of Work and Community in Lowell, Massachusetts, 1826–1860* (New York: Columbia University Press, 1979); Dublin, *Farm to Factory: Women's Letters, 1830–1860*, 2nd edition (New York: Columbia University Press, 1993); and Dublin, *Transforming Women's Work: New England Lives in the Industrial Revolution* (Ithaca, NY: Cornell University Press, 1994). For the argument from European history that young women's wage work in early industrialization served family needs, see Louise A. Tilly and Joan W. Scott, *Women, Work, and Family* (New York: Holt, Rinehart, and Winston, 1978).

23. Nancy F. Cott makes the argument that antebellum young women experienced marriage as a loss of personal freedom, in *The Bonds of Womanhood*, 78–83. Catherine E. Kelly finds that for provincial New England young women, at least, marriage did not entail a loss of self but rather for many was their surest route "to adulthood, to full selfhood, to female excellence," in *In the New England Fashion: Reshaping Women's Lives in the Nineteenth* Century (Ithaca, NY: Cornell University, Press), 107–114, 123–126. Quotation on p. 125.

24. Boylan, "Growing Up Female in Young America," 169; Faye E. Dudden, *Serving*

Women: Household Service in Nineteenth-Century America (Middletown, CT: Wesleyan University Press, 1983).

25. Boylan, "Growing Up Female in Young America," 169–170; Christine Stansell, *City of Women: Sex and Class in New York, 1789–1860* (New York: Knopf, 1986), Chapters 6 and 8.

26. Deborah Gray White, *Ar'n't I A Woman? Female Slaves in the Plantation South* (New York: W. W. Norton, 1985), 91–118; and Wilma King, *Stolen Childhood: Slave Youth in Nineteenth-Century America* (Bloomington: Indiana University Press, 1995), 21–41, 109–110. See also Marie Jenkins Schwartz, *Born in Bondage: Growing Up Enslaved in the Antebellum South* (Cambridge, MA: Harvard University Press, 2000).

27. Boylan, "Growing Up Female in Young America," 160. The efforts by the Lowell managers to make education a condition for factory work was supported by the Massachusetts legislature. In 1836, it passed the first compulsory attendance law, requiring children under age 15 employed in manufacturing to have had three months of schooling during the previous year. See Reinier, *From Virtue to Character*, 136.

28. Hunter, *How Young Ladies Became Girls*, 11–37, 169–171; Boylan, "Growing Up Female in Young America," 160–164; Kett, *Rites of Passage*, 138; Barbara Miller Solomon, *In the Company of Educated Women* (New Haven, CT: Yale University Press, 1985), 27–42; Cott, 101–125; Constance A. Nathanson, *Dangerous Passage: The Social Control of Sexuality in Women's Adolescence* (Philadelphia: Temple University Press, 1991), 85–86.

29. Hunter, *How Young Ladies Became Girls*, 169–221; Kelly, *In the New England Fashion*, 70–76; John L. Rury, *Education and Women's Work: Female Schooling and the Division of Labor in Urban America, 1870–1930*, SUNY Series on Women and Work (Albany: State University of New York Press, 1991), 64.

30. Boylan, "Growing Up Female in Young America," 162; Joseph M. Hawes, *The Children's Rights Movement: A History of Advocacy and Protection* (Boston: Twayne, 1991), 15–17; Barbara M. Brenzel, *Daughters of the State: A Social Portrait of The First Reform School for Girls in North America, 1856–1905* (Cambridge, MA: MIT Press, 1983). On age-grading in nineteenth-century public schools, see Chudacoff, *How Old Are You?* 29–38.

31. Kett, *Rites of Passage*, 62–70, 75–79; Juster and Vinovskis, "Nineteenth-Century America, Adolescence in," 699–700; Mary P. Ryan, *Cradle of the Middle Class: The Family in Oneida County, New York, 1790–1865* (Cambridge: Cambridge University Press, 1981), 75–83. For additional sources on girls' and women's religious experiences in this period, see the "Essay on Sources."

32. See the "Essay on Sources" for scholarship that explores the rise of the nineteenth-century domestic family.

33. For a close analysis of the shift in married women's domestic labor, see Jeanne Boydston, *Home and Work: Housework, Wages, and the Ideology of Labor in the Early Republic* (New York: Oxford University Press, 1990). On the rise of "intensive" motherhood, see Julia Grant, *Raising Baby by the Book: The Education of American Mothers* (New Haven, CT: Yale University Press, 1998), 15–18; Ruth H. Bloch, "American Feminine Ideals in Transition: The Rise of the Moral Mother, 1785–1815," *Feminist Studies* 4 (June 1978): 101–126; and Jan Lewis, "Mother's Love: The Construction of an Emotion in Nineteenth-Century America," in *Social History and Issues in Human Consciousness: Some Interdisciplinary Connections*, ed. Andrew E. Barnes and Peter N. Stearns (New York: New York University Press, 1989), 209–229.

On the rise of breadwinning as the primary characteristic of Victorian fatherhood, see Robert L. Griswold, *Fatherhood in America: A History* (New York: Basic Books, 1993), 10–33. On the role of nineteenth-century fathers not only as providers, but as teachers, caregivers, and governors as well, see Rotundo, *American Manhood*, 25–27; and Shawn Johansen, *Family Men: Middle-Class Fatherhood in Early Industrializing America* (New York: Routledge, 2001). On the fall in birth rates, see Ryan, *Cradle of the Middle Class*, 155–157; Robert V. Wells, "Family History and the Demographic Transition," *Journal of Social History* 9 (Fall 1975): 1–19; and Daniel Scott Smith, "Family Limitation, Sexual Control, and Domestic Feminism in Victorian America," *Feminist Studies* 1, nos. 3–4 (1973): 40–57.

34. On the meaning of new family forms and functions for girls' lives and relationships, see especially, Hunter, *How Young Ladies Became Girls*; Boylan, "Growing Up Female in Young America," 162–163; Carroll Smith-Rosenberg, *Disorderly Conduct: Visions of Gender in Victorian America* (New York: Alfred A. Knopf, 1985), 53–76; and Nancy M. Theriot, *The Biosocial Construction of Femininity: Mothers and Daughters in Nineteenth-Century America* (New York: Greenwood Press, 1988). For examination of how changing expectations of family life reflected larger cultural tensions and produced conflict in five Victorian families, see Steven Mintz, *A Prison of Expectations: The Family in Victorian Culture* (New York: New York University Press, 1983), Chapters 4–7.

35. Smith-Rosenberg, *Disorderly Conduct*, 60–67.

36. Hunter, *How Young Ladies Became Girls*, 188–193; Kelly, *In the New England Fashion*, 77–92; Dublin, *Farm to Factory*, 21–24.

37. Hunter, *How Young Ladies Became Girls*, 38–90. For scholarship on girls' literary culture in the nineteenth-century Anglo-American context, see the "Essay on Sources."

38. See the "Essay on Sources" for historical studies of Victorian masculinity that explore changing experiences and meanings of boyhood and male youth.

39. Reinier, *From Virtue to Character*, 125–133; W. J. Rorabaugh, *The Craft Apprentice: From Franklin to the Machine Age in America* (New York: Oxford University Press, 1986). For examples of changes in the artisan-apprentice path in this period, see Graff, *Conflicting Paths*, 80–91.

40. Reinier, *From Virtue to Character*, 126, 136–137; Rorabaugh, *The Craft Apprentice*, 61, 63.

41. Reinier, *From Virtue to Character*, 131–133.

42. Graff, *Conflicting Paths*, 69, 173. Graff argues that much was shared across lines of class, as well. However, even those strategies, beliefs, practices, and activities held in common by different groups of families, children, and youth were marked by "significant divergences" and "yielded diverse results" (p. 71).

43. This did not mean, however, that all working-class parents were forced to choose between work and new educational opportunities for their sons. The flexible quality of antebellum schooling allowed some boys intermittent access to education, which they were able to pursue in conjunction with their wage work. Juster and Vinovskis, "Nineteenth-Century America, Adolescence in," 702.

44. Reinier, *From Virtue to Character*, 126, 138–146; Hawes, *The Children's Rights Movement*, 15–16.

45. Kett, *Rites of Passage*, 29.

46. Ryan, *Cradle of the Middle Class*, 158–165; Rotundo, *American Manhood*, Chapter 2.

47. Ryan, *Cradle of the Middle Class*, 165–179; Demos, *Past, Present, and Personal*, 99–103. For examples of the often discontinuous, bumpy, complicated paths to growing up pursued by middle-class male youth, see Graff, *Conflicting Paths*, 91–109, 173–182.

48. Rotundo, *American Manhood*, 52–53, 56–61; Ryan, *Cradle of the Middle Class*, 173–176.

49. Kett, *Rites of Passage*, 51–59; Demos, *Past, Present, and Personal*, 101–102.

50. Rotundo, *American Manhood*, 62–71; Kett, *Rites of Passage*, 38–40; Ryan, *Cradle of the Middle Class*, 176–177.

51. Rotundo, *American Manhood*, 75–90.

52. Hemphill, *Bowing to Necessities*, 67, 88–89, 93–95. See also Rodney Hessinger, *Seduced, Abandoned, and Reborn: Visions of Youth in Middle-Class America, 1780–1850* (Philadelphia: University of Pennsylvania Press, 2005).

53. In her "genealogy of girlhood," cultural theorist Catherine Driscoll likewise describes late modern adolescence as framed as "a dialectic of dependence and independence," p. 52. It is a formulation, she says, that has been essential to "ideas about modernity and the modern subject": "Adolescence defines the ideal coherence of the modern subject—individuality, agency, and adult (genital) sexuality—while not necessarily ensuring its achievement. When critics such as [Immanuel] Kant or [F. R.] Leavis understand the modern subject as immature —as threatened by possible immaturity, as engaged with self-doubt—they understand it as adolescent. This alignment becomes more overt in late modern critical theory and popular culture after the impact of what Julia Kristeva calls Freud's 'Copernican' realization that the subject was 'split' or not 'self-identical.' Kristeva argues that this modern subjectivity locates no coherent Subject but rather . . . a subject in process . . . ," p. 53. See also Carol Dyhouse, *Girls Growing Up in Late Victorian and Edwardian England* (London: Routledge & Kegan Paul, 1981), 119.

54. See especially, John R. Morss, *The Biologising of Childhood: Developmental Psychology and the Darwinian Myth* (Hove and London: Lawrence Erlbaum Associates, 1990); and Peter J. Bowler, *The Non-Darwinian Revolution: Reinterpreting a Historical Myth* (Baltimore: Johns Hopkins University Press, 1988). See the "Essay on Sources" for more works that explore the history of ideas about development and evolution from ancient times to the present.

55. Richard M. Lerner, *Concepts and Theories of Human Development,* 2nd edition (Mahwah, NJ: Lawrence Erlbaum Associates, 1997), 8–12; Michael Wertheimer, "The Evolution of the Concept of Development in the History of Psychology," in *Contributions to a History of Developmental Psychology*, ed. Georg Eckardt, Wolfgang G. Bringmann, and Lothar Sprung (Berlin: Mouton Publishers, 1985), 13–21; Thomas Laqueur, *Making Sex: Body and Gender from the Greeks to Freud* (Cambridge, MA: Harvard University Press, 1990), 169, 173–175; Stephen Jay Gould, *Ontogeny and Phylogeny* (Cambridge, MA: Belknap Press of Harvard University Press, 1977), 13–28; Arthur O. Lovejoy, *The Great Chain of Being: A Study of the History of an Idea* (Cambridge, MA: Harvard University Press, 1936).

56. Roger A. Dixon and Richard M. Lerner, "A History of Systems in Developmental Psychology," in *Developmental Psychology: An Advanced Textbook,* 3rd edition, ed. Marc H. Bornstein and Michael E. Lamb (Hillsdale, NJ: Lawrence Erlbaum Associates, 1992), 10–11; Sheldon H. White, "The Idea of Development in Developmental Psychology," in *Developmental Psychology: Historical and Philosophical Perspectives,* ed. Richard M. Lerner (Hillsdale, NJ: Lawrence Erlbaum Associates, 1983), 55–77; Morss, *Biologising of Childhood*, 5–6; Laqueur,

Making Sex, 169; Gould, *Ontogeny and Phylogeny*, 17–18, 33–39. While the more extreme positions of preformationism were successfully challenged by the epigenetic view in embryology by the 1830s, elements of both persisted in the organic paradigm of postnatal child development that dominated in the nineteenth century. According to this paradigm, children possessed an inherent nature, the structures and functions of which also changed progressively from the simple to the complex over time. As I explore in this and the next chapter, the relationship between these elements in conceptions of child development was gendered. That is, girls were more often described as possessing an essential female nature that statically unfolded, whereas boys, especially during the period of youth, were seen to develop dynamically, in interaction with the external environment, toward ever higher states of being.

57. Carolyn Steedman, *Strange Dislocations: Childhood and the Idea of Human Interiority, 1780–1930* (Cambridge, MA: Harvard University Press, 1995), 50–62.

58. See the "Essay on Sources" for those scholars' interpretations of Locke and Rousseau's writings on child development that informed my own.

59. Jay Fliegelman, *Prodigals and Pilgrims: The American Revolution against Patriarchal Authority, 1750–1800* (Cambridge: Cambridge University Press, 1982), 13.

60. John Locke, "Some Thoughts Concerning Education," in *John Locke on Education*, ed. Peter Gay (New York: Teachers College, Columbia University, 1964), 65. Although here Locke takes the boy as his subject, his prescriptions for the rearing of young children were intended to apply to both boys and girls. However, as I subsequently argue, his prescriptions for youth dealt specifically with male development. For Rousseau, too, the boy serves as the model for normative youthful development. Hence, I deliberately use the masculine pronouns throughout my discussion of these philosophers' developmental theories.

61. Fliegelman, *Prodigals and Pilgrims*, 12–15; See also Reinier, *From Virtue to Character*, 2–5; Hemphill, *Bowing to Necessities*, 88–89; and Grant, *Raising Baby by the Book*, 18–19.

62. Fliegelman, *Prodigals and Pilgrims*, 14–15; Hemphill, *Bowing to Necessities*, 88–89.

63. Locke, "Some Thoughts Concerning Education," 73, 74–76; Fliegelman, *Prodigals and Pilgrims*, 5, 20, 33.

64. Reinier, *From Virtue to Character*, 18.

65. Jean-Jacques Rousseau, *Emile, or On Education,* introduction, translation, and notes by Allan Bloom (New York: Basic Books, 1979), Books I–III.

66. Rousseau, *Emile*, 211, 212.

67. Rousseau, *Emile*, 215.

68. Rousseau, *Emile*, 216.

69. On the Scottish Enlightenment philosophers, see Fliegelman, *Prodigals and Pilgrims*, 23–26; Reinier, *From Virtue to Character*, 8–9; and Ernest Freeberg, *The Education of Laura Bridgman: First Deaf and Blind Person to Learn Language* (Cambridge, MA: Harvard University Press, 2001), 96–97.

70. Rousseau, *Emile*, 212–214, 221.

71. Rousseau, *Emile*, 220.

72. Rousseau, *Emile*, 211.

73. Rousseau, *Emile*, Book V.

74. Steedman makes a similar claim about the importance of early-nineteenth-century

popular physiology texts in Britain to theories about "the connections between growth, death, and childhood" that would be elaborated later in the century "in the fields of child psychology and emergent psychoanalysis"; *Strange Dislocations*, 63. Her analysis focuses on popular medical guides' treatment of infancy and early childhood and not at all on the phenomenon of puberty or the age of life thought to be associated with it.

75. James C. Whorton, *Crusaders for Fitness: The History of American Health Reformers* (Princeton, NJ: Princeton University Press, 1982), 22–29; Steedman, *Strange Dislocations*, 50–62. See the "Essay on Sources" for additional scholarship on nineteenth-century health reform.

76. Martha H. Verbrugge, *Able-Bodied Womanhood: Personal Health and Social Change in Nineteenth-Century Boston* (New York: Oxford University Press, 1988), 30–31 , 57, 59–62. For a close examination of the relationship between orthodox and unorthodox medicine among advocates of vegetarianism, see Margaret Puskar-Pasewicz, " 'For the Good of the Whole': Vegetarianism in 19th-Century America," Indiana University Dissertation, 2003, 25–32, 42–44.

77. Whorton, *Crusaders for Fitness*, 4, 29–61; Freeberg, *Education of Laura Bridgman*, 72–77; Verbrugge, *Able-Bodied Womanhood*, 29–30; Steven Mintz, *Moralists and Moralizers: America's Pre-Civil War Reformers* (Baltimore: Johns Hopkins University Press, 1995), 20–24.

78. Sylvester Graham, *A Lecture to Young Men* (Providence, RI, 1834; reprint edition, New York: Arno Press, 1974), 11–12. On Graham's role in health reform, see Whorton, *Crusaders for Fitness*, Chapters 2–4; Jayme A. Sokolow, *Eros and Modernization: Sylvester Graham, Health Reform, and the Origins of Victorian Sexuality* (Rutherford, NJ: Farleigh Dickinson University Press, 1983); and Stephen Nissenbaum, *Sex, Diet, and Debility in Jacksonian America: Sylvester Graham and Health Reform* (Westport, CT: Greenwood Press, 1980).

79. Whorton, *Crusaders for Fitness*, 5–8; Regina Markell Morantz-Sanchez, *Sympathy and Science: Women Physicians in American Medicine* (New York: Oxford University Press, 1985), 33–35. For medical texts that explicated these principles of health reform, see, for example, Amariah Brigham, *Remarks on the Influence of Mental Cultivation and Mental Excitement upon Health*, 2nd edition (Boston, 1833); Andrew Combe, *The Principles of Physiology Applied to the Preservation of Health and to the Improvement of Physical and Mental Education*, 7th edition (New York, 1849); William Sweetser, *Mental Hygiene; or, an Examination of the Intellect and Passions Designed to Show How They Affect and Are Affected by the Bodily Functions, and Their Influences on Health and Physiology* (New York, 1850); and O. S. Fowler, *Physiology, Animal and Mental: Applied to the Preservation and Restoration of Health of Body, and Power of Mind*, 6th edition (New York, 1853).

80. Combe, *Principles of Physiology*, 34. The American edition of Combe's *The Principles of Physiology* was edited by American phrenologist and hygiene enthusiast Orson Fowler, who declared that "[n]o family should be without it and no young man or woman should fail to peruse and reperuse every page of it" (p. xi).

81. Brigham, *Mental Cultivation*, vii.

82. William A. Alcott, *The Young Woman's Guide to Excellence*, 16th edition (New York, 1854), 45–46.

83. On Alcott's career as a health reformer, see Whorton, *Crusaders for Fitness*, 49–59.

84. Cynthia Eagle Russett, *Sexual Science: The Victorian Construction of Womanhood* (Cam-

bridge, MA: Harvard University Press, 1989), 16–19; Freeberg, *Education of Laura Bridgman*, 30–32. For a comprehensive study of the science of phrenology, see Stephen Tomlinson, *Head Masters: Phrenology, Secular Education, and Nineteenth-Century Social Thought* (Tuscaloosa: University of Alabama Press, 2005).

85. Orson Fowler, *Self-Culture and Perfection of Character including the Management of Youth* (New York, 1853), 28, 48–49, iii, 32–33. See also Freeberg, *Education of Laura Bridgman*, 67.

86. Brigham, *Mental Cultivation*, 20. See also Freeberg, *Education of Laura Bridgman*, 40–41.

87. For a discussion of the history of the relationship between the "individual" and the "romantic" self in the nineteenth-century United States, see Rotundo, *American Manhood*, 279–281. Michael B. Katz also addresses the meaning of the "state of nature" for mid-nineteenth-century educational reformers in *The Irony of Early School Reform: Educational Innovation in Mid-Nineteenth Century Massachusetts* (Cambridge, MA: Harvard University Press, 1968), 122–123.

88. Elizabeth Blackwell, *The Laws of Life, with Special Reference to the Physical Education of Girls* (New York: G. P. Putnam, 1852), 10, 171.

89. Sweetser, *Mental Hygiene*, 42–43. See also Brigham, *Mental Cultivation*, 72.

90. Blackwell, *The Laws of Life*, 67–68, 23, 43, 109–111.

91. Blackwell, *The Laws of Life*, 38–53, 147, 121–122.

92. Orson Fowler, *Sexual Science* (Chicago, 1870), 418.

93. Russett, *Sexual Science*, 105–116.

94. Blackwell, *The Laws of Life*, 55–58.

95. Fowler, *Physiology*, 249; Orson Fowler, *Memory and Intellectual Improvement Applied to Self-Education and Juvenile Instruction* (New York, 1853), 41–44, 51; Sweetser, *Mental Hygiene*, 79–86; Brigham, *Mental Cultivation*, 70–71. See also Kett, *Rites of Passage*, 124–125; and Grant, *Raising Baby by the Book*, 32–33.

96. Markell Morantz-Sanchez, *Sympathy and Science*, 40–41.

97. Blackwell, *The Laws of Life*, 28; Graham, *A Lecture to Young Men*, 13–17, 25–27.

98. R. P. Neuman, "Masturbation, Madness, and the Modern Concepts of Childhood and Adolescence," *Journal of Social History* 8 (1975): 1–27; Hessinger, *Seduced, Abandoned, and Reborn*, Chapter 6.

99. Fowler, *Self-Culture*, 105.

100. Horace Bushnell, *Christian Nurture* (New Haven, CT: Yale University Press, 1947), 4. See also, Kett, *Rites of Passage*, 114–120.

101. Sweetser, *Mental Hygiene*, 376.

102. William A. Alcott, *Familiar Letters to Young Men on Various Subjects* (Buffalo, NY, 1850), 61, 135.

103. Blackwell, *The Laws of Life*, 72, 111.

104. Blackwell, *The Laws of Life*, 82–83, 28, 123. See also, William A. Alcott, *The Physiology of Marriage* (Boston: John P. Jewett, 1856), 19.

105. Both Hemphill and Kett interpret the cultural contributions of these reformers in this way. Hemphill, *Bowing to Necessities*, 176–178, 219; Kett, *Rites of Passage*, 139–140.

106. Mintz, *Moralists and Moralizers*, especially xix–xx and 157. Sokolow makes the same point in *Eros and Modernization*: "Although the romantic health reformers were often nos-

talgic about the rural society they once knew, they helped modernize urban Americans by encouraging traits that led to success in an increasingly commercial and urban America" (p. 15).

107. Hemphill, *Bowing to Necessities*, 178.

108. Blackwell, *The Laws of Life*, 139.

109. Fowler, *Physiology*, 243–48; Combe, *The Principles of Physiology*, 25–26.

110. On women in health reform, see Markell Morantz-Sanchez, *Sympathy and Science*, 35–46; Verbrugge, *Able-Bodied Womanhood*, 38–48, 49–80; and Whorton, *Crusaders for Fitness*, 105–109.

111. For a helpful review of the trends among women's historians in thinking about the ideologies of gender that were shaped and deployed within the context of antebellum reform, see Verbrugge, *Able-Bodied Womanhood*, 66–69.

112. Quoted in Markell Morantz-Sanchez, *Sympathy and Science*, 35.

113. Markell Morantz-Sanchez, *Sympathy and Science*, 35–46; Verbrugge, *Able-Bodied Womanhood*, 38–48, 49–80; and Whorton, *Crusaders for Fitness*, 105–109.

114. On Elizabeth Blackwell's life and career, see Markell Morantz-Sanchez, *Sympathy and Science*, Chapter 7; and Elizabeth H. Thomson, "Elizabeth Blackwell," in *Notable American Women, 1607–1950*, ed. Edward T. James (Cambridge, MA: Belknap Press of Harvard University Press, 1971), 161–165.

115. Quoted in Markell Morantz-Sanchez, *Sympathy and Science*, 190.

116. On orthodox medical opinion of the debility of the female body, see Russett, *Sexual Science*, Chapters 1 and 4; Markell Morantz-Sanchez, *Sympathy and Science*, 203–255; Carroll Smith-Rosenberg and Charles Rosenberg, "The Female Animal: Medical and Biological Views of Woman and Her Role in Nineteenth Century America," *Journal of American History* 60, no. 2 (Sept. 1973): 332–356; Carroll Smith-Rosenberg, "Puberty to Menopause: The Cycle of Femininity in Nineteenth Century America," in *Clio's Consciousness Raised*, ed. Mary S. Hartman and Lois Banner (New York: Harper & Row, 1974), 23–37; and John S. Haller, Jr., and Robin M. Haller, *The Physician and Sexuality in Victorian America* (Urbana: University of Illinois Press, 1974). On the challenge to this perspective within nineteenth-century popular culture, see Frances B. Cogan, *All-American Girl: The Ideal of Real Womanhood in Mid-Nineteenth-Century America* (Athens: University of Georgia Press, 1989), Chapter 1.

117. Brigham, *Mental Cultivation*, v; Combe, *The Principles of Physiology*, 129; Jno. Stainback Wilson, MD, *Woman's Home Book of Health: A Work for Mothers and for Families* (Philadelphia, 1860), 156; Blackwell, *The Laws of Life*, 122–142. See also, William A. Alcott, *The Young Woman's Book of Health* (New York: Auburn, Miller, Orton & Mulligan, 1855; 1850), 27–28, 37, 43, 112–113.

118. On this point, see Steedman, *Strange Dislocations*, 8; and Ornella Moscucci, *The Science of Woman: Gynaecology and Gender in England, 1800–1929* (Cambridge: Cambridge University Press, 1990), 16.

119. Karin Calvert, *Children in the House: The Material Culture of Early Childhood, 1600–1900* (Boston: Northeastern University Press, 1992), 110.

120. Analyzing the findings from nineteenth-century embryology as to the homologous

nature of the structures that become the reproductive organs during the earliest stages of fetal life, historian Thomas Laqueur writes: "A stranger surveying the landscape of mid-nineteenth-century science might well suspect that incommensurable sexual difference was created despite, not because of, new discoveries" (*Making Sex,* 169). Understanding the importance of the concept of development to nineteenth-century thought, and particularly ideas about the development of sexual difference during childhood and youth, helps to explain the seeming contradiction between the discovery of such homologies and the dominant scientific and cultural endorsement of dichotomous sexual difference. See also Moscucci, *The Science of Woman,* 17–18.

121. M. Lallemand, *A Practical Treatise on the Causes, Symptoms and Treatment of Spermatorrhoea,* 4th American edition, trans. and ed. Henry J. McDougall (Philadelphia, 1861), 147; William Acton, *The Function and Disorders of the Reproductive Organs* (Philadelphia: Lindsay and Blakiston, 1865), 192, 48, 54–55.

122. Acton, *Reproductive Organs,* 47, 53.

123. Graham, *A Lecture to Young Men,* 21–24, quotation on p. 21; Samuel B. Woodward, *Hints for the Young in Relation to the Health of Body and Mind,* 4th edition (Boston: George W. Light, 1840), 18–19.

124. Alcott, *Familiar Letters,* 237–238, 55, 56–59.

125. William A. Alcott, *The Young Man's Guide to Excellence,* 14th edition (Boston: Perkins and Marvin, 1841), 29.

126. See especially, Smith-Rosenberg, "From Puberty to Menopause."

127. Alcott, *The Young Woman's Book of Health,* 109, 130–135. Quotation on p. 310. See also, Alcott, *The Young Woman's Guide,* 108–111; and Blackwell, *The Laws of Life,* 28, 85–86, 123–135, 138–142.

128. Alcott, *The Young Woman's Book of Health,* 125–126.

129. Brigham, *Mental Cultivation,* 81–82. See also Alcott, *The Young Woman's Book of Health,* 28–29.

130. Alcott, *The Young Woman's Guide,* 143; Fowler, *Physiology, Animal and Mental,* 247.

131. Blackwell, *The Laws of Life,* 49, 118, 136, 148–150.

132. Blackwell, *The Laws of Life,* 142–150.

CHAPTER TWO. "PERSISTENCE" VERSUS "PERIODICITY"

1. Mary Putnam Jacobi, "Mental Action and Physical Health," in *The Education of American Girls,* ed. Anna C. Brackett (New York: G. P. Putnam's Sons, 1874), 259.

2. David Tyack and Elisabeth Hansot, *Learning Together: A History of Coeducation in American Public Schools* (New Haven, CT: Yale University Press, 1990), 151.

3. See the "Essay on Sources" for works that discuss Edward H. Clarke's treatise within the context of the history of female education.

4. Edward H. Clarke, *Sex in Education: Or, A Fair Chance for the Girls* (Boston: James R. Osgood, 1873; reprint edition, New York: Arno Press, 1972), 47, 111.

5. Tyack and Hansot, *Learning Together,* Chapter 2; Linda K. Kerber, *Women of the Republic: Intellect and Ideology in Revolutionary America* (Chapel Hill: University of North Caro-

lina Press, 1980); Barbara Miller Solomon, *In the Company of Educated Women: A History of Women and Higher Education in America* (New Haven, CT: Yale University Press, 1985), Chapter 2.

6. Jane H. Hunter, *How Young Ladies Became Girls: The Victorian Origins of American Girlhood* (New Haven, CT: Yale University Press, 2002), 169. See also Miller Solomon, *Educated Women*, Chapter 3.

7. Tyack and Hansot, *Learning Together*, 45.

8. Tyack and Hansot, *Learning Together*, Chapter 3.

9. Tyack and Hansot, *Learning Together*, 78–100, 104.

10. Tyack and Hansot, *Learning Together*, 92–95, 99–113.

11. Lynn D. Gordon, *Gender and Higher Education in the Progressive Era* (New Haven, CT: Yale University Press, 1990), 17–18, 22.

12. Gordon, *Gender and Higher Education*, 21–30. Quotation on p. 24.

13. Robert H. Wiebe, *The Search for Order, 1877–1920* (Westport, CT: Greenwood Press, 1967).

14. For one treatment of the "woman question," see Elaine Showalter, *Sexual Anarchy: Gender and Culture at the Fin de Siècle* (London: Virago, 1992).

15. Louise Michele Newman, *White Women's Rights: The Racial Origins of Feminism in the United States* (New York: Oxford University Press, 1999), 7–17.

16. Gail Bederman, *Manliness & Civilization: A Cultural History of Gender and Race in the United States, 1880–1917* (Chicago: University of Chicago Press, 1995), 1–44. Quotation on p. 42.

17. Clarke, *Sex in Education*, 5–7. See also, Rosalind Rosenberg, *Beyond Separate Spheres: Intellectual Roots of Modern Feminism* (New Haven, CT: Yale University Press, 1982), 5; and Mary Roth Walsh, *"Doctors Wanted: No Women Need Apply": Sexual Barriers in the Medical Profession, 1835–1975* (New Haven, CT: Yale University Press, 1977), 123–124.

18. See the "Essay on Sources" for the body of literature documenting this point.

19. Clarke, *Sex in Education*, 34, 47. For the treatment of girls as "miniature adults" in nineteenth-century conduct and entertainment guides, see Melanie Dawson, "The Miniaturizing of Girlhood: Nineteenth-Century Playtime and Gendered Theories of Development," in *The American Child: A Cultural Studies Reader*, ed. Caroline F. Levander and Carol J. Singley (New Brunswick, NJ: Rutgers University Press, 2003).

20. The classic work that establishes the importance of Darwin and Spencer to American intellectuals and the educated public is Richard Hofstadter, *Social Darwinism in American Thought, 1860–1915* (Philadelphia: University of Pennsylvania Press, 1944). See the "Essay on Sources" for additional works on the history of evolutionary thought. My summary of Darwin's and Spencer's theories of the cause of sexual difference in human evolution is informed by Louise Michele Newman, ed., *Men's Ideas/Women's Realities: Popular Science, 1870–1915* (New York: Pergamon Press, 1985), 2–8; Newman, *White Women's Rights*, 29–34; Cynthia Eagle Russett, *Sexual Science: The Victorian Construction of Womanhood* (Cambridge, MA: Harvard University Press, 1989), 40–42, 78–84; and Rosenberg, *Beyond Separate Spheres*, 7–9.

21. Newman, *White Women's Rights*, 29.

22. Roger A. Dixon and Richard M. Lerner, "A History of Systems in Developmental Psychology," in *Developmental Psychology: An Advanced Textbook*, 3rd edition, ed. Marc H.

Bornstein and Michael E. Lamb (Hillsdale, NJ: Lawrence Erlbaum Associates, 1992), 13–16; William R. Charlesworth, "Charles Darwin and Developmental Psychology: Past and Present," in *A Century of Developmental Psychology*, ed. Ross D. Parke et al. (Washington, DC: American Psychological Association, 1994), 77–102; John R. Morss, *The Biologising of Childhood: Developmental Psychology and the Darwinian Myth* (Hove and London: Lawrence Erlbaum Associates, 1990); William Kessen, *The Rise and Fall of Development* (Worcester, MA: Clarke University Press, 1990).

23. Dixon and Lerner, "Systems in Developmental Psychology," 15–17. Quotation on p. 16.

24. Until the 1840s, the terms "ontogeny" and "phylogeny" were both used to refer to biological change generally. Following the work of comparative embryologist Karl Ernst von Baer, "ontogeny" first became associated with embryology and then with "development"— the process of change in the individual from conception to adulthood. In the 1860s, German zoologist Ernst Haeckel used "phylogeny" to refer to the changes in species over the course of evolution and formulated the biogenetic law that asserted that ontogeny recapitulated phylogeny. See Charlesworth, "Charles Darwin and Developmental Psychology," 90.

25. Charlesworth, "Charles Darwin and Developmental Psychology," 80; Stephen Jay Gould, *Ontogeny and Phylogeny* (Cambridge, MA: Harvard University Press, 1977), 70–74.

26. Dixon and Lerner, "Systems in Developmental Psychology," 12; Morss, *Biologising of Childhood*, 12.

27. Herbert Spencer, *First Principles*, 4th edition (New York: D. Appleton, 1890 [1862]), especially 190–192, 278–286, 307–400,483–559. Quotation on p. 517. Spencer's definition of evolution reads as follows: "Evolution is an integration of matter and concomitant dissipation of motion; during which the matter passes from an indefinite, incoherent homogeneity to a definite, coherent heterogeneity; and during which the retained motion undergoes a parallel transformation" (italicized in original), 396. My understanding of *First Principles* was assisted by Hofstadter, *Social Darwinism in American Thought*, 35–38; Gould, *Ontogeny and Phylogeny*, 112–114; and Andreas M. Kazamias, ed., *Herbert Spencer on Education* (New York: Teachers College Press, 1966), 34–38, 67–75.

28. Spencer, *First Principles*, 545–547.

29. Herbert Spencer, *The Principles of Biology*, 2 vols. (New York: D. Appleton, 1897 [1866]), 1:349–350.

30. On Spencer's Lamarckianism, see Hofstadter, *Social Darwinism in American Thought*, 39; J. D. Y. Peel, ed., *Herbert Spencer on Social Evolution* (Chicago: University of Chicago Press, 1972), xxii–xxiii; and Claudia Castañeda, *Figurations: Child, Bodies, Worlds* (Durham, NC: Duke University Press, 2002), 20–21.

31. Kazamias, *Herbert Spencer on Education*, 21, 52.

32. Herbert Spencer, *Education: Intellectual, Moral, and Physical* (New York: D. Appleton, 1860), 51. Spencer claimed that Rousseau had no direct influence on either his political views or his ideas about education, although Kazamias says that he was surely acquainted with them, pp. 10–11n20.

33. Spencer, *Education*, 103–115, 51, 106–109. Quotations on pp. 112, 113, 51, 106, 108.

34. Spencer, *Education*, 115, 116

35. Spencer, *Education*, 117–119. On Spencer's ideas about evolution and racial difference,

generally, see John S. Haller, Jr., *Outcasts from Evolution: Scientific Attitudes of Racial Inferiority, 1859–1900* (Urbana: University of Illinois Press, 1971), 121–132. On Spencer's use of the pre-Haeckelian recapitulation tradition, see Gould, *Ontogeny and Phylogeny*, 148–149; and Castañeda, *Figurations*, 28–30.

36. Spencer, *Education*, 120–124. Quotation on p. 120. Italics in original.

37. Spencer, *Education*, 213–215.

38. Spencer, *Education*, 175–197. Quotations on pp. 191 and 197.

39. Spencer, *Education*, 166–174. Quotation on pp. 173–74.

40. Spencer, *Education*, 232.

41. Spencer, *Education*, 95, 284.

42. Spencer, *Education*, 287–288. See also Spencer, *The Principles of Biology*, 1:107–152.

43. Spencer, *Education*, 288.

44. Spencer, *The Principles of Biology*, 2:471; Spencer, "Psychology of the Sexes," *The Popular Science Monthly* 4 (November 1873): 32. Spencer's initial formulation of the conflict between individual development and reproductive development, presented in an 1852 article in the *Westminster Review*, addressed the effects of excessive mental work on the production of sperm. In *The Principles of Biology*, however, he determined that a conflict between individuation and genesis was not normative in male development because the cost of reproduction in males was "so much less than it is to females" (2:486–487). See also Russett, *Sexual Science*, 118–119, and Newman, *Men's Ideas/Women's Realities*, 6–7.

45. Spencer, "Psychology of the Sexes," 32.

46. Spencer, *Education*, 106–109, 220–223. Quotations on pp. 109, 221.

47. Herbert Spencer, *Social Statics, or, The Conditions Essential to Human Happiness Specified, and the First of Them Developed* (New York: D. Appleton, 1872), 213.

48. Spencer, *Social Statics*, 209.

49. Clarke, *Sex in Education*, 129, 130, 35.

50. Spencer, *First Principles*, especially 190–196 and 250–271; Clarke, *Sex in Education*, 120–121.

51. Clarke, *Sex in Education*, 60, 38,120, 124.

52. Clarke, *Sex in Education*, 137, 37, 40–42, 101. On Laycock's theory of "vital periodicity," see Ornella Moscucci, *The Science of Woman: Gynaecology and Gender in England, 1800–1929* (Cambridge, England: Cambridge University Press, 1990), 19–20; and Anne E. Walker, *The Menstrual Cycle* (London: Routledge, 1997), 17.

53. Thomas Laqueur, *Making Sex: Body and Gender from the Greeks to Freud* (Cambridge, MA: Harvard University Press, 1990), 181–192, 207–213. See also Helen King, *The Disease of Virgins: Green Sickness, Chlorosis and the Problems of Puberty* (London: Routledge, 2004), 67–73.

54. Laqueur, *Making Sex*, 213–222. See also Ronald O. Valdiserri, "Menstruation and Medical Theory: An Historical Overview," *Journal of the American Medical Women's Association* 38, no. 3 (May/June 1983): 66–70; Moscucci, *The Science of Woman*, 33–34; and Russett, *Sexual Science*, 16–119.

55. Clarke, *Sex in Education*, 113–114. Maudsley, editor of the *Journal of Mental Science*, published his own book-length polemic against coeducation in 1873, which stirred up similar controversy in Britain. An essay version of the book entitled "Sex in Mind and in Education"

was published in *The Popular Science Monthly* 5 (June 1874): 198–215. See Walker, *The Menstrual Cycle*, 39.

56. Clarke, *Sex in Education*, 36–39. See also Mary Putnam Jacobi, *The Question of Rest for Women during Menstruation* (New York: G. P. Putnam's Sons, 1877), 97–101.

57. Clarke, *Sex in Education*, 54, 19.

58. Clarke, *Sex in Education*, 135, 61–117, 140, 128.

59. Clarke, *Sex in Education*, 29–30, 131–33, 178–79.

60. Clarke, *Sex in Education*, 11–30.

61. For birth and marriage rates during the second half of the nineteenth century, see Newman, *White Women's Rights*, 88–89.

62. Clarke, *Sex in Education*, 64, 112, 138–39, 30.

63. Henry Lyman, et al., *The Practical Home Physician* (Houston, 1885), 875. See also, for example, Thomas Addis Emmet, *The Principles and Practice of Gynecology*, 3rd edition (Philadelphia: Henry C. Lea's Son, 1884), 18; George Napheys, *The Physical Life of Woman: Advice to the Maiden, Wife, and Mother* (Philadelphia, 1872), 15–35; and William Capp, *The Daughter: Her Health, Education and Wedlock* (Philadelphia: F. A. Davis, 1891), 52–92.

64. Michale B. Katz, *The Irony of Early School Reform: Educational Innovation in Mid-Nineteenth Century Massachusetts* (Cambridge, MA: Harvard University Press, 1968), 126–128.

65. In addition to its developmental implications, the image of the budding girl also had resonance with the metaphor of the menses as "flowers," which was common from the medieval period onward. As King explains in *The Disease of Virgins*, "the loss of virginity is called deflowering because of the floral appearance of the vaginal entrance" (p. 157 n23).

66. John Harvey Kellogg, *Plain Facts for Old and Young* (Burlington, IA: I. F. Segner, 1882), 472.

67. G. Stanley Hall, *Adolescence: Its Psychology and Its Relations to Physiology, Anthropology, Sociology, Sex, Crime, Religion and Education*, 2 vols. (New York: D. Appleton, 1904), 2:625.

68. George F. Comfort and Mrs. Anna Manning Comfort, MD, *Woman's Education and Woman's Health: Chiefly in Reply to "Sex in Education"* (Syracuse, NY: Thos. W. Durston, 1874), 16–19. See also Tyack and Hansot, *Learning Together*, 153.

69. Carla Jean Bittel, "The Science of Women's Rights: The Medical and Political Worlds of Mary Putnam Jacobi," Cornell University Dissertation, 2003, 28–44. For additional accounts of Jacobi's life and career, see Rhoda Truax, *The Doctors Jacobi* (Boston: Little, Brown, 1952); "Mary Corinna Putnam Jacobi," in *Notable American Women: A Biographical Dictionary, 1607–1950*, ed. Edward T. James (Cambridge, MA: Belknap Press of Harvard University Press, 1971), 2:263–265; Regina Markell Morantz-Sanchez, *Sympathy and Science: Women Physicians in American Medicine* (New York: Oxford University Press, 1985), 55–68, 184–202; and Roth Walsh, *"Doctors Wanted,"* 59–60, 99.

70. Bittel, "The Science of Women's Rights," 45–58.

71. Bittel, "The Science of Women's Rights," 65–82, 87–145.

72. Bittel, "The Science of Women's Rights," 150–191, 256–292.

73. For extensive discussion of the differences between Blackwell and Jacobi's approaches to medicine, see Markell Morantz-Sanchez, *Sympathy and Science*, 184–202.

74. Jacobi, "Mental Action," 258; Jacobi, *The Question of Rest*, 25. *The Question of Rest* won

Harvard University's Boylston Prize in 1876. For analysis of *The Question of Rest* in relation to the history of ideas about menstruation, see Valdiserri, "Menstruation and Medical Theory"; and Laqueur, *Making Sex*, 222–225.

75. Eliza Bisbee Duffey, *No Sex in Education: An Equal Chance for Both Girls and Boys* (Philadelphia: J. M. Stoddart, 1874), 12. For similar sentiments, see Elizabeth Stuart Phelps, "Chapter VII," in *Sex and Education: A Reply to Dr. E. H. Clarke's "Sex in Education,"* ed. Julia Ward Howe (Boston: Roberts Brothers, 1874; reprint edition, New York: Arno Press, 1972), 127–130; and Caroline H. Dall, "The Other Side," in Brackett, *The Education of American Girls*, 167.

76. On the contribution female doctors were to make to knowledge about and treatment of women's health, M. B. Jackson declared, "When women are permitted to add the light of science and art to their personal experiences and similar organizations, we may look for a healthier race of women" (Chapter IX, in Howe, *Sex and Education*, 158). For an extensive examination of the various approaches women doctors took to medical theory and treatment during the late nineteenth century, see Markell Morantz-Sanchez, *Sympathy and Science*, 203–231. "Just as men lacked unanimity on many medical issues," she concludes, "women physicians also differed significantly with each other. As females struggling to strike a balance between science, professionalism, and their own womanhood, they were bound to develop individual solutions to the problems of female health. The historian is hard pressed, therefore, to uncover a uniform approach among these women on how to treat, diagnose, or prevent illness. Women internalized many 'male' values, just as men were sometimes advocates of 'female' positions" (p. 222). King examines this issue as well, specifically in relation to women physicians' treatment of chlorosis, in *The Disease of Virgins*, pp. 134–138.

77. Jacobi, *The Question of Rest*, 62–63, 115.

78. Jacobi, *The Question of Rest*, 168–174, 186.

79. The characterization of nature as such is Duffey's, *No Sex in Education*, 24. In an 1895 letter to the *Boston Medical and Surgical Journal,* Jacobi argued against Spencer's assertion of the antagonism between individuation and genesis: "To suppose that cerebral activity could dwarf sexual activity (which is often alleged) is absurd, or rather, though now noted by ethnologists, that sexual passion is far more highly developed among highly civilized peoples than among savages, shows that normally the two poles of existence develop *pari passu* and not in antagonism to each other." "Modern Female Invalidism," in *Mary Putnam Jacobi, M.D.: A Pathfinder in Medicine*, ed. Women's Medical Association of New York City (New York: G. P. Putnam's Sons, 1925), 478. The correlation between female sexual passion and the rise of civilization was most extensively elaborated in this period by British sexologist Havelock Ellis and is discussed in the next chapter.

80. Jacobi, "Mental Action," 259.

81. Jacobi, *The Question of Rest*, 73–79, 98.

82. Jacobi, *The Question of Rest*, 81–83.

83. Jacobi, *The Question of Rest*, 81–83; Jacobi, "Mental Action," 269–275.

84. Jacobi, *The Question of Rest*, 97–98.

85. Jacobi, "Mental Action," 285–286. See also, *The Question of Rest*, 173–174.

86. See, for example, Duffey, 24, 38, 48–49; M. B. Jackson in Howe, *Sex and Education*, 159–160; and Marion Harland, *Eve's Daughters or Common Sense for Maid, Wife, and Mother* (New York: Charles Scribner's Sons, 1885), 45–52.

87. Jacobi, *The Question of Rest*, 27.

88. Duffey, *No Sex in Education*, 67–68.

89. Brackett, *The Education of American Girls*, 52.

90. Jacobi, "Mental Action," 303.

91. This point is influenced by Newman's argument in *White Women's Rights*, 39–42.

92. Duffey, *No Sex in Education*, 71, 73, 56.

93. Duffey, *No Sex in Education*, 72, 30–35; Mrs. Horace Mann, "Chapter III," in Howe, *Sex and Education*, 58–59; Comfort and Comfort, *Woman's Education and Woman's Health*, 140–142.

94. Elizabeth Stuart Phelps, "Chapter VII," in Howe, *Sex and Education*, 136; Caroline H. Dall, "Chapter V," in Howe, *Sex and Education*, 107.

95. Howe, "Sex and Education," in Howe, *Sex and Education*, 28.

96. Mann, Chapter III, in Howe, *Sex and Education*, 55, 58; Edna D. Cheyney, "A Mother's Thought on the Education of Girls," in Brackett, *The Education of American Girls*, 142; Jacobi, "Modern Female Invalidism," 482.

97. "Boston Daily Advertiser," "Chapter VIII," in Howe, *Sex and Education*, 147.

98. Brackett, *The Education of American Girls*, 15.

99. Duffey, *No Sex in Education*, 52, 73–74; Eliza B. Duffey, *What Women Should Know: A Woman's Book about Women* (Philadelphia: J. M. Stoddart, 1873; reprint edition, New York: Arno Press, 1974), 61.

100. Jacobi, *The Question of Rest*, 172–173; "Mental Action," 285, 287, 298.

101. Jacobi, "Mental Action," 298–302.

102. Duffey, *What Women Should Know*, 56–57. Clarke was more concerned with the problems for girls of being educated in the same way as boys than he was with educating them according to different methods, but in proximity to, or "juxtaposition" with, one another (Clarke, *Sex in Education*, 122, 146–147).

103. Jacobi, "Mental Action," 302–304.

104. Elizabeth Blackwell, *The Human Element in Sex, Being a Medical Inquiry into the Relation of Sexual Physiology to Christian Morality* (London, 1894 [1884]), 17–30.

105. Duffey, *No Sex in Education*, 53. See also, "Boston Daily Advertiser," in Howe, *Sex and Education*, 152.

106. Duffey, *No Sex in Education*, 29; Caroline H. Dall, "The Other Side," in Brackett, *The Education of American Girls*, 160–161. See also, Howe, "Introduction," in Howe, *Sex and Education*, 9; "Boston Daily Advertiser," in Howe, *Sex and Education*, 144; Comfort and Comfort, *Woman's Education and Woman's Health*, 85–88.

107. Duffey, *No Sex in Education*, 30, 39.

108. Kathy Peiss, *Cheap Amusements: Working Women and Leisure in Turn-of-the-Century New York* (Philadelphia: Temple University Press, 1986); Mary E. Odem, *Delinquent Daughters: Protecting and Policing Adolescent Female Sexuality in the United States, 1885–1920* (Chapel Hill: University of North Carolina Press, 1995); and Ruth M. Alexander, *The "Girl Problem": Female Sexual Delinquency in New York, 1900–1930* (Ithaca, NY: Cornell University Press, 1995).

109. Ada Shepard Badger, "Chapter IV," in Howe, *Sex and Education*, 74; Marie A. Elmore, "Chapter XII," in Howe, *Sex and Education*, 175.

110. R. B. Leach, "A Physician's Standpoint," *The Arena* 12 (April 1895): 285; O. Edward Janney, "A Physician's View of These Laws," *The Arena* 11 (January 1895): 207.

111. Emily Blackwell, "Another Physician Speaks," *The Arena* 11 (January 1895): 212–215.

112. A. C. Tompkins, "The Age of Consent from a Physio-Psychological Standpoint," *The Arena* 13 (July 1895): 223.

113. Tyack and Hansot, *Learning Together*, 114; Gordon, *Gender and Higher Education*, 2; Miller Solomon, *Educated Women*, 58, 63.

114. Margaret A. Lowe, *Looking Good: College Women and Body Image, 1875–1930* (Baltimore: Johns Hopkins University Press, 2003), 1, 13–53. See also, Rosenberg, *Beyond Separate Spheres*, 12; Newman, *White Women's Rights*, 93; Markell Morantz-Sanchez, *Sympathy and Science*, 217.

CHAPTER THREE. FROM "BUDDING GIRL" TO "FLAPPER AMERICANA NOVISSIMA"

1. G. Stanley Hall, "The Budding Girl," *Appleton's Magazine* 13 (January 1909): 47. For similar renderings of the adolescent girl's development as baffling problem for science to solve, see G. Stanley Hall, "The Awkward Age," *Appleton's Magazine* 12 (August 1908): 149–156; G. Stanley Hall, "A Medium in the Bud," *American Journal of Psychology* 29 (April 1918): 144–158; and G. Stanley Hall, "Flapper Americana Novissima," *Atlantic Monthly* 129, no. 6 (June 1922), 771–780.

2. Joseph F. Kett and John R. Gillis both deem the first half of the twentieth century to be the "era of adolescence," in the United States and Europe, respectively. Joseph F. Kett, *Rites of Passage: Adolescence in America 1790 to the Present* (New York: Basic Books, 1977), Chapter 8, quotation on p. 213; John R. Gillis, *Youth and History: Tradition and Change in European Age Relations 1770–Present* (New York: Academic Press, 1981), Chapter 4, quotation on p. 133. See also Grace Palladino, *Teenagers: An American History* (New York: Basic Books, 1996), part 1; and Nancy Lesko, *Act Your Age! A Cultural Construction of Adolescence* (New York: Routledge, 2001), Chapters 1–2.

3. See the "Essay on Sources" for works arguing for Hall's influence on the establishment of a professional science of child development, broadly, and on his contributions to the construction of the modern category of adolescence, specifically.

4. Gail Bederman, *Manliness & Civilization: A Cultural History of Gender and Race in the United States, 1880–1917* (Chicago: University of Chicago Press, 1995), Chapter 3; Lesko, *Act Your Age!* 34; Kett, *Rites of Passage*, especially the "Introduction" and Chapters 5–8; Gillis, *Youth and History*, 114–115, and Chapter 4. Gillis's assessment of the "universalizing" of adolescence in this regard is devastating: "The emphasis on the physical and psychological sides of adolescence . . . was ultimately to reduce this phase of life to an object of scientific observation and clinical treatment by adults. What began as an effort to allow the young to live by the rules of nature, ended in chaining them to a new conformity sanctioned by positivist social science. Furthermore, in an attempt to protect the adolescent against the decadent world of adults, the young were separated from those civil and social rights which were their only real protection against the elders" (p. 142).

5. See also Melanie Dawson, "The Miniaturizing of Girlhood: Nineteenth-Century Play-time and Gendered Theories of Development," in *The American Child: A Cultural Studies Reader,* ed. Caroline F. Levander and Carol J. Singley (New Brunswick, NJ: Rutgers University

Press, 2003), 75. Catherine Driscoll notices that Hall "focused on adolescent girls while not centrally discussing them," although she does not provide an in-depth analysis of what that "focus" entailed. Catherine Driscoll, *Girls: Feminine Adolescence in Popular Culture and Cultural Theory* (New York: Columbia University Press, 2002), 7, 53–57. Quotation on p. 57. Carol Dyhouse devotes a chapter to Hall's ideas about the education of adolescent girls in *Girls Growing Up in Late Victorian and Edwardian England* (London: Routledge, 1981), 115–138.

6. G. Stanley Hall, *Adolescence: Its Psychology and Its Relations to Physiology, Anthropology, Sociology, Sex, Crime, Religion and Education*, 2 vols. (New York: D. Appleton, 1904), 2:625. Other scholars have noticed Hall's characterization of adolescence as such. See Dorothy Ross, *G. Stanley Hall: The Psychologist as Prophet* (Chicago: University of Chicago Press, 1972), 338–39; Driscoll, *Girls*, 7, 54–57; T. J. Jackson Lears, *No Place of Grace: Anti-Modernism and the Transformation of American Culture, 1880–1920* (New York: Pantheon, 1981), 247–251; Rosalind Rosenberg, *Beyond Separate Spheres: Intellectual Roots of Modern Feminism* (New Haven, CT: Yale University Press, 1982), 22, 42; Cynthia Eagle Russett, *Sexual Science: The Victorian Construction of Womanhood* (Cambridge, MA: Harvard University Press, 1989), 54–55; and Kenneth B. Kidd, *Making American Boys: Boyology and the Feral Tale* (Minneapolis: University of Minnesota Press, 2004), 59–63.

7. On Hall's sponsoring of Freud's visit, see Ross, *G. Stanley Hall*, 386–394.

8. Ross, *G. Stanley Hall*, 131, 121. See G. Stanley Hall, "The Moral and Religious Training of Children," *Princeton Review* 9 (1882): 26–48.

9. For a summary of Hall's academic accomplishments, see Robert E. Grinder, "The Concept of Adolescence in the Genetic Psychology of G. Stanley Hall," *Child Development* 40, no. 2 (June 1969): 355–356; and Ross, *G. Stanley Hall*, xiii.

10. Kett, *Rites of Passage*, 221–244; Ross, *G. Stanley Hall*, Chapters 15 and 17, especially 345–367; Julia Grant, *Raising Baby by the Book: The Education of American Mothers* (New Haven, CT: Yale University Press, 1998), 36–38.

11. On the origins of the new, scientific psychology and Hall's training in it, see Ross, *G. Stanley Hall*, Chapters 5–6.

12. Ross, *G. Stanley Hall*, 148–168; 106–107, 123–124, 157, 168, 262–265; Grinder, "Genetic Psychology of G. Stanley Hall," 356–358; Robert E. Grinder and Charles Strickland, "G. Stanley Hall and the Social Significance of Adolescence," *Teacher's College Record* 64, no. 5 (February 1963): 393; Kett, *Rites of Passage*, 217–218. The professional disappointments and personal tragedy that moved Hall toward his focus on child study and the creation of a developmental psychology are recounted in Ross, *G. Stanley Hall*, Chapters 12–14.

13. G. Stanley Hall, "Modern Methods in the Study of the Soul," *Christian Register* 75 (February 1886): 131. See also Ross, *G. Stanley Hall*, 263. John R. Morss provides a helpful summation of the relationship between the emergence of experimental psychology and developmental psychology in *The Biologising of Childhood: Developmental Psychology and the Darwinian Myth* (Hove and London: Lawrence Erlbaum Associates, 1990), 6–8.

14. Ross, *G. Stanley Hall*, 12–14, 19–21, 115–119. Lesko emphasizes the importance of the German Youth Movement to Hall's ideas about adolescence (*Act Your Age!* 51–54).

15. Ross, *G. Stanley Hall*, 3–12; Bederman, *Manliness & Civilization*, 79–83; Jeffrey P. Moran, *Teaching Sex: The Shaping of Adolescence in the 20th Century* (Cambridge, MA: Harvard

University Press, 2000), 2–4. See also, G. Stanley Hall, *Life and Confessions of a Psychologist* (New York: D. Appleton, 1923).

16. Kett, *Rites of Passage*, 219–220.

17. Ross, *G. Stanley Hall*, 279, 325, 336. See G. Stanley Hall, *Youth, Its Education, Regimen and Hygiene* (New York: D. Appleton, 1906).

18. For a synthesis of the basic tenets of Hall's genetic psychology, see Ross, *G. Stanley Hall*, Chapter 16; Grinder, "Genetic Psychology of G. Stanley Hall," 358–361; and Sheldon H. White, "G. Stanley Hall: From Philosophy to Developmental Psychology," in *A Century of Developmental Psychology*, ed. Ross D. Parke et al. (Washington, DC: American Psychological Association, 1994), 112–120.

19. On the rise and fall of recapitulation in nineteenth- and early-twentieth-century science, see Stephen Jay Gould, *Ontogeny and Phylogeny* (Cambridge, MA: Belknap Press of Harvard University Press, 1977), 33–206. On Hall's use of recapitulation theory, see Ross, *G. Stanley Hall*, 261–265; Bederman, *Manliness & Civilization*, 92–94, 109–110; Morss, *The Biologising of Childhood*, 32–37; and Russett, *Sexual Science*, 50–63. Morss points out that Hall's reliance on Larmarckianism, as Darwin's, was more accurately a neo-Lamarckianism because for Lamarck the inheritance of acquired characteristics "had been of quite minor significance" in his evolutionary theory (p. 3). He also emphasizes that, while the "stronger, causal versions" of recapitulation theory, such as Hall's, were rejected under the challenge by Mendelian genetics, recapitulatory logic has continued to pervade developmental psychology: "[A]ny set of laws and any sequence of stages which seeks to describe development 'in general' is likely to carry with it the recapitulationary tendency. Such a tendency may certainly have become attenuated, with attention narrowed down to the empirical study of the individual. But the competence remains, even if the performance is suppressed" (p. 229).

20. Hall, *Adolescence*, 1:viii; Hall, *Life and Confessions*, 360.

21. Hall, *Adolescence*, 1:xiii–xv; G. Stanley Hall, "How Far Is the Present High-School and Early College Training Adapted to the Nature and Needs of Adolescents?" *The School Review* 9 (December 1901): 649.

22. Hall, *Adolescence*, 1:49–50. For one critique of Hall's privileging of adolescence as a unique life cycle stage marked by "transcendent power," as well as of the educational implications that Hall expected to follow from such an assumption, see Harvard University President Charles Eliot's response to Hall's 1901 address before the New England Association of Colleges and Secondary Schools, published in *The School Review* 9 (December 1901): 665–667. Here, Eliot expressed "some misgivings with regard to the existence of any well defined period in the ordinary span of human life, of any period which can be given a beginning and an end and be said to have remarkable characteristics of its own." He continued: "My fundamental belief is that love and freedom and the nursing of nature would make human life a progress, a growth, and expansion, a triumph, from beginning to end . . . I believe the methods of teaching should be all one, from the lap of the mother to the lap of the university."

23. This argument is most forcefully made by Bederman, *Manliness & Civilization*, 10–15, and Chapter 3; Lesko, *Act Your Age!* Chapters 1–2; and Kidd, *Making American Boys*, especially the introduction and Chapter 2. Putting less emphasis on Hall's contribution here, Kidd singles out William Byron Forbush, author of *The Boy Problem* (1901), and Henry William

Gibson, author of *Boyology or Boy Analysis* (1916) as "the most visible spokesmen of boyology," which he defines as "the post-1900 spate of organization and publication" aimed at giving meaning to and shaping the experiences of boys and boyhood (pp. 68, 67).

24. Bederman, *Manliness & Civilization*, 1–44.

25. Kett, *Rites of Passage*, 144–162.

26. Kett, *Rites of Passage*, 162–204; Lears, *No Place of Grace*, 220–221.

27. George M. Beard, *American Nervousness: Its Causes and Consequences* (New York: G. P. Putnam's Sons, 1881). For analysis of the construction of the problem of neurasthenia in late-nineteenth-century American culture, see Bederman, *Manliness & Civilization*, 14, 84–88; Lears, *No Place of Grace*, 49–57; F. G. Gosling, *Before Freud: Neurasthenia and the American Medical Community* (Urbana: University of Illinois Press, 1987); E. Anthony Rotundo, *American Manhood: Transformations in Masculinity from the Revolution to the Modern Era* (New York: Basic Books, 1993), 185–193.

28. Kett, *Rites of Passage*, 162–168; Bederman, *Manliness & Civilization*, 15–23; Rotundo, *American Manhood*, 222–232.

29. For a discussion of antimodern constructions of masculinity that functioned both as protest and accommodation to corporate capitalism, see Lears, *No Place of Grace*, Chapters 1 and 3. For the same point, see also Kett, *Rites of Passage*, 164.

30. Kett, *Rites of Passage*, 173–211.

31. Bederman, *Manliness & Civilization*, 91–92.

32. Bederman, *Manliness & Civilization*, 93; Rotundo, *American Manhood*, 31–36; Kidd, *Making American Boys*, 14–15; Julia Grant, "A 'Real Boy' and Not a Sissy: Gender, Childhood and Masculinity, 1890–1940, *Journal of Social History* 37, no. 4 (Summer 2004): 832–833.

33. Morss, *The Biologising of Childhood*, 23–26; Anne McClintock, *Imperial Leather: Race, Gender and Sexuality in the Colonial Conquest* (New York: Routledge, 1995), 50–51; Claudia Castañeda, *Figurations: Child, Bodies, Worlds* (Durham, NC: Duke University Press, 2002), 13–15; John S. Haller Jr., *Outcasts from Evolution: Scientific Attitudes of Racial Inferiority, 1859–1900* (Urbana: University of Illinois Press, 1971).

34. G. Stanley Hall, "Psychic Arrest in Adolescence," *Journal of Proceedings and Addresses of the National Educational Association* (July 1903): 811–813. For Spencer and Fiske on this point, see Haller, *Outcasts from Evolution*, 121–138.

35. Kett, *Rites of Passage*, 224; See H. W. Gibson, *Boyology, or Boy Analysis* (New York, 1916).

36. Hall, *Adolescence*, 1:202.

37. A. Caswell Ellis and G. Stanley Hall, "A Study of Dolls," *Pedagogical Seminary* 4 (December 1896): 129–175. Quotations on pp. 159, 129, 162. See also Hall, *Adolescence*, 1:208–209.

38. Hall, *Adolescence*, 1:209.

39. Ellis and Hall, "A Study of Dolls," 161.

40. Ellis and Hall, "A Study of Dolls," 161.

41. Hall, *Adolescence*, 1:203, 205.

42. Hall, *Adolescence*, 1:207–210, 217–223. Quotation on p. 217.

43. Hall, *Adolescence*, 1:206.

44. G. Stanley Hall, "The Needs and Methods of Educating Young People in the Hygiene of Sex," *Pedagogical Seminary* 15 (March 1908): 82–91. Quotation on p. 84; G. Stanley Hall,

"Education in Sex Hygiene," *Eugenics Review* 1 (January 1910): 242–253. See also, Ross, *G. Stanley Hall*, 383–386; Bederman, *Manliness & Civilization*, 80–84; and Moran, *Teaching Sex*, 1–4.

45. Hall, *Adolescence*, 1:223–224.

46. Hall, *Adolescence*, 1:205. On Hall's use of the concept of sublimation, see Ross, *G. Stanley Hall*, 372 n8.

47. Hall, *Adolescence*, 2:620, 452.

48. Hall, *Adolescence*, 2:452–53, 1:308.

49. Hall, *Adolescence*, 1:308–309.

50. For Hall's indictment of civilization's "effeminizing" effects on the boy, see G. Stanley Hall, "Feminization in School and Home," *World's Work* 16 (May 1908): 10237–10244.

51. Hall, *Adolescence*, 2:561. See also Hall, *Youth*, 158.

52. Ross, *G. Stanley Hall*, 338–339; Lears, *No Place of Grace*, 247–251.

53. Hall, "The Awkward Age," 149–156.

54. Hall, *Adolescence*, 1:14; Hall, "The Awkward Age," 149.

55. Hall, *Adolescence*, 1:44–45, xiii, 35, 47, 36.

56. Hall, "The Awkward Age," 149, 151.

57. Hall, *Adolescence*, 1:42–43.

58. Hall, "The Awkward Age," 150; Hall, *Adolescence*, 1:42, 1:36, 1:416.

59. Hall, *Adolescence*, 1:128. See also Hall, *Adolescence*, 2:58.

60. For Hall's treatment of the "diseases of body and mind" of adolescence and their relationship to normal adolescent psychology, see *Adolescence*, 1:237–324. Quotations on pp. 1:241, 2:71, 2:68, 1:304, 1:266.

61. Hall, *Adolescence*, 1:267. Hall explores the "antithetic impulses" vying for expression in the adolescent psyche in *Adolescence*, 2:75–88. See also Hall, "The Awkward Age," 152.

62. Hall, *Adolescence*, 1:321, 2:89, 2:72.

63. Hall, "The Awkward Age," 155; Hall, *Adolescence*, 1:294.

64. G. Stanley Hall, "Coeducation," *Journal of Proceedings and Addresses of the National Educational Association* (June–July 1904), 538; Hall, *Adolescence*, 1:501. See also *Adolescence*, 2:76. For a broader discussion of the role of rhythm in human development, see Hall, "Modern Methods in the Study of the Soul," 132.

65. Hall, *Adolescence*, 1:494, 2:89.

66. Hall, *Adolescence*, 1:547, 2:59. See also Hall, "The Needs and Methods of Educating Young People in the Hygiene of Sex," 90.

67. Hall, "Modern Methods in the Study of the Soul," 131–132; F. H. Saunders and G. Stanley Hall, "Pity," *American Journal of Psychology* 11 (July 1900): 574; Hall, "The Budding Girl," 49; G. Stanley Hall, "Coeducation in the High School," *Journal of Proceedings and Addresses of the National Educational Association* (July 1903): 451.

68. Hall, *Adolescence*, 2:81. See also Hall, *Life and Confessions*, 460–463; and Hall, *Youth*, 314, 321, 363, 365.

69. Hall, *Life and Confessions*, 460–461. For a discussion of Hall's social vision, see Grinder and Strickland, "Social Significance of Adolescence," 391–392.

70. Saunders and Hall, "Pity," 572; Hall, *Adolescence*, 2:390–392.

71. Hall, *Adolescence*, 2:132–136. For the sexual ideologies of Ellis and Key, see Mari Jo

Buhle, *Feminism and Its Discontents: A Century of Struggle with Psychoanalysis* (Cambridge, MA: Harvard University Press, 1998), 35–42.

72. Hall, *Adolescence*, 1:123–124.

73. Hall, "The Awkward Age," 154; Hall, *Adolescence*, 2:304, 2:125; 2:374, 2:303. See also, Hall, *Adolescence*, 2:281–362, 1:225.

74. Hall, *Adolescence*, 2:303; Saunders and Hall, "Pity," 590–591. See also Grinder and Strickland, "Social Significance of Adolescence," 392.

75. Hall, *Adolescence*, 2:624.

76. Hall, *Adolescence*, 2:90, 66, 70.

77. Hall, *Adolescence*, 2:72, 1:315–316, 2:375–377; Hall, *Youth*, 176.

78. G. Stanley Hall, *Recreations of a Psychologist* (New York: D. Appleton, 1920), 128–146. Hall treats the theme of the bisexual nature of the psyche in another story in the same collection, entitled "Preestablished Harmony," 175–183. See also Ross, *G. Stanley Hall*, 257–259; and Lears, *No Place of Grace*, 248–249.

79. Hall, *Recreations*, 128–131.

80. Hall, *Recreations*, 133–135.

81. Hall, *Recreations*, 136–145.

82. Hall, *Recreations*, 135.

83. Hall, *Adolescence*, 2:391, 2:369.

84. Saunders and Hall, "Pity," 567. See also, G. Stanley Hall, "The Dangerous Age," *Pedagogical Seminary* 28 (September 1921): 275.

85. Hall, *Adolescence*, 2:621; Hall, *Life and Confessions*, 37, 44, 41.

86. Jane F. Gerhard provides similar analysis of Freud's theory of psychosexual development in *Desiring Revolution: Second-Wave Feminism and the Rewriting of American Sexual Thought 1920–1982* (New York: Columbia University Press, 2001), 31.

87. Hall, *Adolescence*, 2:117.

88. Hall, *Youth*, 281, 285; Hall, "The Awkward Age," 151.

89. Hall, *Life and Confessions*, 45; Hall, *Adolescence*, 2:392, 2:382.

90. Hall, *Adolescence*, 2:392.

91. On Freud's development of the theory of the Oedipus complex, see Buhle, *Feminism and Its Discontents*, 66–74.

92. Hall, *Adolescence*, 2:610.

93. Early-twentieth-century feminist psychologists' conceptions of female development are examined in the next chapter.

94. Scholarly works informing my analysis of Hall's treatment of adolescent sexuality are listed in the "Essay on Sources."

95. Nathan G. Hale Jr., *Freud and the Americans: The Beginning of Psychoanalysis in the United States, 1876–1917* (New York: Oxford University Press, 1971), 25–35.

96. John D'Emilio and Estelle B. Freedman, *Intimate Matters: A History of Sexuality in America* (New York: Harper & Row, 1988), 172–173.

97. Gerhard, *Desiring Revolution*, 14; Buhle, *Feminism and Its Discontents*, 24–29; Kathy Peiss, *Cheap Amusements: Working Women and Leisure in Turn-of-the-Century New York* (Philadelphia: Temple University Press, 1986); Mary E. Odem, *Delinquent Daughters: Protecting and Policing Adolescent Female Sexuality in the United States, 1885–1920* (Chapel Hill: University of

North Carolina Press, 1995); Joan Jacobs Brumberg, *The Body Project: An Intimate History of American Girls* (New York: Random House, 1997), especially pp. 99–107, 153–157.

98. On the influences of Freud on Hall's developmental psychology, see Ross, *G. Stanley Hall*, 381–412. As Nathan G. Hall Jr. asserts, Hall's most significant engagement with Freudian theory did not occur until after 1904. At the time of the publication of *Adolescence*, Hall was acquainted with Freud's essays up to 1896, although he had no grasp of the Freudian system (Hale, *Freud and the Americans*, 209). Moreover, as Mari Jo Buhle explains, Freud himself came comparatively late to a fully drawn theory of female psychosexual development, elaborating it only in the 1920s as a response to demands by feminists enamored with psychoanalysis that he do so and to the debate engaged by his critics over the concepts of penis envy and genital transference referred to in his earlier work. Buhle, *Feminism and Its Discontents*, Chapters 1 and 2.

99. Paul Robinson, *The Modernization of Sex: Havelock Ellis, Alfred Kinsey, William Masters, and Virginia Johnson* (Ithaca, NY: Cornell University Press, 1989), 3. See also D'Emilio and Freedman, *Intimate Matters*, 224–226.

100. The final volume of the *Studies, Eonism and Other Supplementary Studies* was published in 1928.

101. For the tension between Ellis's hierarchical and egalitarian assertions, see, for example, Havelock Ellis, *The Evolution of Modesty*, 3rd edition (1910), in *Studies in the Psychology of Sex*, vol. 1, part 1 (New York: Random House, 1942), 1–84. For Ellis on the sexual instinct among civilized and primitive peoples, see especially, Ellis, *Appendix A: The Sexual Instinct in Savages*, 2nd edition (1913), in *Studies in the Psychology of Sex*, vol. 1, part 2, 259–276.

102. On this point, see Janice M. Irvine, *Disorders of Desire: Sex and Gender in Modern American Sexology* (Philadelphia: Temple University Press, 1990).

103. Ellis, *Analysis of the Sexual Impulse*, 2nd edition (1913), in *Studies in the Psychology of Sex*, vol. 1, part 2, 13. Ellipses added; first set of brackets added; second set of brackets in the original. See also, Ellis, *Sex in Relation to Society* (1910) in *Studies in the Psychology of Sex*, vol. 2, part 3, 34–35, 85–86.

104. Sigmund Freud, *Three Essays on the Theory of Sexuality* (1905), in *Standard Edition of the Complete Psychological Works of Sigmund Freud*, ed. and trans. James Strachey, vol. 7 (London: Hogarth Press, 1953–1974), 207–208, 219–230. Freud elaborated his theory of the distinct nature of male and female psychosexual development in *Beyond the Pleasure Principle* (1920), *The Ego and the Id* (1923), "The Dissolution of the Oedipus Complex" (1924), "Some Psychical Consequences of the Anatomical Differences between the Sexes" (1925), "Female Sexuality" (1931), and "Femininity" (1933). See Buhle, *Feminism and Its Discontents*, 66–67, 71–74, 78–84.

105. Ellis, *The Sexual Impulse in Women*, 2nd edition (1913), in *Studies in the Psychology of Sex*, vol. 1, part 2, 219, 198.

106. Ellis, *The Evolution of Modesty*, 1–84. Quotations on pp. 1 and 4.

107. Ellis, *Sex in Relation to Society*, 62.

108. Ellis, *The Sexual Impulse in Women*, 251.

109. Ellis, *Auto-Eroticism*, 3rd edition (1910), in *Studies in the Psychology of Sex*, vol. 1, part 1, 187–201. Quotation on p. 197.

110. Ellis, *Auto-Eroticism*, 184–187. Quotation on p. 185.

111. Ellis, *The Evolution of Modesty*, 44.

112. Ellis, *Sex in Relation to Society*, 517–518, 530.

113. Ellis, *The Sexual Impulse in Women*, 226.

114. The "Cumulative Index of Authors" for Ellis's *Studies* lists forty-five citations of Hall's work.

115. Hall, *Adolescence*, 1:280, 2:38, 1:480–481.

116. Hall, *Adolescence*, 2:372–373. See also G. Stanley Hall and Theodate L. Smith, "Showing Off and Bashfulness as Phases of Self-Consciousness," *Pedagogical Seminary* 10 (June 1903): 194; and Hall, "A Medium in the Bud."

117. Hall, *Adolescence*, 2:116–117; Hall, "Budding Girl," 48.

118. Hall, "The Needs and Methods of Educating Young People in the Hygiene of Sex," 86–88. For more on Hall's influence on the sex education movement of the early twentieth century, see Moran, *Teaching Sex*, 40–49.

119. Jane H. Hunter, *How Young Ladies Became Girls: The Victorian Origins of American Girlhood* (New Haven, CT: Yale University Press, 2002), 261–311; Carolyn Kitch, *The Girl on the Magazine Cover: The Origins of Visual Stereotypes in American Mass Media* (Chapel Hill: University of North Carolina Press, 2001). Hunter puts Hall in the tradition of those Victorian architects of girlhood who condemned the adolescent girl's precocity. In my reading, Hall legitimizes the girl's efforts to assert and express certain aspects of a "modern" self but also establishes these sensibilities and behaviors as the limit of the possibilities for her development and offers them up as justifications for new forms of adult oversight and intervention in the life of the adolescent girl.

120. Hall, "Flapper," 772.

121. Hall, "Budding Girl," 53, 48–49.

122. Hall, "Flapper," 776.

123. Hall, "Budding Girl," 53, 51.

124. Hall, "Flapper," 774.

125. Hall, "Budding Girl," 48; Hall, "Flapper," 779.

126. Hall, "Budding Girl," 48–50; Hall, "Flapper," 777.

127. Hall, "Budding Girl," 49–50, 54; Hall, "Flapper," 776.

128. Hall, "Flapper," 776, 771, 777, 780.

129. Hall, "Medium in the Bud," 156. Sigmund Freud, "Femininity" (1933), in *Standard Edition of the Complete Psychological Works of Sigmund Freud*, 22:135.

CHAPTER FOUR. "NEW GIRLS FOR OLD"

1. Mary E. Moxcey, *Girlhood and Character* (New York: Abington Press, 1916), 73.

2. Moxcey, *Girlhood and Character*, 17.

3. Moxcey, *Girlhood and Character*, 11, 17–26. Quotation on p. 23. Italics in original. Moxcey was the author of several other popular books meant to educate adults about the adolescent stage, including *Leadership of Girls' Activities* (New York: Methodist Book Concern, 1919); and *The Psychology of Middle Adolescence* (New York: Claxton Press, 1925).

4. On female juvenile delinquency, see Ruth M. Alexander, *The "Girl Problem": Female Sexual Delinquency in New York, 1900–1930* (Ithaca, NY: Cornell University Press, 1995);

Mary E. Odem, *Delinquent Daughters: Protecting and Policing Adolescent Female Sexuality in the United States, 1885–1920* (Chapel Hill: University of North Carolina Press, 1995); Constance A. Nathanson, *Dangerous Passage: The Social Control of Sexuality in Women's Adolescence* (Philadelphia: Temple University Press, 1991), part 3; Joan Jacobs Brumberg, "'Ruined' Girls: Community Responses to Illegitimacy in Upstate New York, 1890–1920," *Journal of Social History* 18 (Winter 1984): 247–272; and Steven Schlossman and Stephanie Wallach, "The Crime of Precocious Sexuality: Female Juvenile Delinquency in the Progressive Era," *Harvard Educational Review* 48 (February 1978): 65–94. On the flapper, see Paula S. Fass, *The Damned and the Beautiful: American Youth in the 1920s* (Oxford: Oxford University Press, 1977); Carolyn Kitch, *The Girl on the Magazine Cover: The Origins of Visual Stereotypes in American Mass Media* (Chapel Hill: University of North Carolina Press, 2001), Chapter 6; and Kelly Schrum, *Some Wore Bobby Sox: The Emergence of Teenage Girl Culture, 1920–1945* (New York: Palgrave Macmillan, 2004).

5. Elizabeth Lunbeck, *The Psychiatric Persuasion: Knowledge, Gender and Power in Modern America* (Princeton, NJ: Princeton University Press, 1994), 3.

6. Odem, *Delinquent Daughters*, 8–37; Kathy Peiss, *Cheap Amusements: Working Women and Leisure in Turn-of-the-Century New York* (Philadelphia: Temple University Press), 163–178.

7. On progressivism, see John D. Buenker, John C. Burnham, and Robert M. Crunden, *Progressivism* (Cambridge, MA: Schenkman, 1977); Robert H. Wiebe, *The Search for Order* (New York: Hill and Wang, 1967); and Richard Hofstadter, *The Age of Reform: From Bryan to F.D.R.* (New York: Vintage Books, 1955). For helpful overviews of progressivism and child saving, see David I. Macleod, *The Age of the Child: Children in America, 1890–1920* (New York: Twayne, 1998), 26–31; and Joseph M. Hawes, *The Children's Rights Movement: A History of Advocacy and Protection* (Boston: Twayne, 1991), 26–53.

8. On the history of the eugenics movement up through the 1920s, see Daniel J. Kevles, *In the Name of Eugenics: Genetics and the Uses of Human Heredity* (Cambridge, MA: Harvard University Press, 1995 [1985]), Chapters 1–9. On the centrality of women and gender to the movement in this period, see Wendy Kline, *Building a Better Race: Gender, Sexuality, and Eugenics from the Turn of the Century to the Baby Boom* (Berkeley: University of California Press, 2001), Chapters 1–3.

9. Henry H. Goddard, *Juvenile Delinquency* (New York: Dodd, Mead & Company, 1923), 3.

10. On the condition of "feeble-mindedess" and Goddard's role in the eugenics movement, see Kevles, *In the Name of Eugenics*, 77–79, 92–95, 107–111; Carl N. Degler, *In Search of Human Nature: The Decline and Revival of Darwinism in American Social Thought* (New York: Oxford University Press, 1991), 36–55; Leila Zenderland, *Measuring Minds: Henry Herbert Goddard and the Origins of American Intelligence Testing* (New York: Cambridge University Press, 1997). Wendy Kline traces the practices of segregation and sterilization of feeble-minded girls and women at the Sonoma State Home for the Feebleminded in California. "Segregation and sterilization gained worldwide legitimacy as a result of their use at Sonoma," she explains (Kline, *Building a Better Race*, Chapters 1 and 2. Quotation on p. 34).

11. Lunbeck, *The Psychiatric Persuasion*, 61–70, 185–208.

12. Dorothy Ross, "Gendered Social Knowledge: Domestic Discourse, Jane Addams, and the Possibilities of Social Science," in *Gender and American Social Science: The Formative Years*, ed. Helene Silverberg (Princeton, NJ: Princeton University Press, 1998), 240. On the network

of Progressive era female reformers that Addams helped to constitute, see Karen J. Blair, *The Clubwoman as Feminist: True Womanhood Redefined, 1868–1914* (New York: Holmes and Meier, 1980); Ellen Fitzpatrick, *Endless Crusade: Women Social Scientists and Progressive Reform* (New York: Oxford University Press, 1990); Robyn Muncy, *Creating a Female Dominion in Progressive Reform* (New York: Oxford University Press, 1991); and Molly Ladd-Taylor, *Mother-Work: Women, Child Welfare, and the State, 1890–1930* (Urbana: University of Illinois Press, 1994).

13. Ross, "Gendered Social Knowledge," 236. See also Mina Carson, *Settlement Folk: Social Thought and the American Settlement Movement, 1885–1930* (Chicago: University of Chicago Press, 1990); and Mary Jo Deegan, *Jane Addams and the Men of the Chicago School, 1892–1918* (New Brunswick, NJ: Transaction Books, 1990).

14. Jane Addams, *The Spirit of Youth and the City Streets* (New York: Macmillan Company, 1926 [1909]), 15, 96–97.

15. Addams, *Spirit of Youth,* 51, 53, 11.

16. Addams, *Spirit of Youth,* 8–9.

17. Addams, *Spirit of Youth,* 51, 143, 15–16, 161, 20–21. For Addams' statement on the similar "desire for action" and "wish to right wrong and alleviate suffering" on the part of middle-class girls, see *Twenty Years at Hull-House* (New York: Penguin Books, 1998 [1910]), 82–83.

18. Sophonsiba Breckinridge and Edith Abbott, *The Delinquent Child and the Home: A Study of the Delinquent Wards of the Juvenile Court of Chicago* (New York: Arno Press, 1912; reprint edition, 1970); Ruth S. True, *The Neglected Girl* (New York: Survey Associates, 1914), 57, 75, 68. See also, Odem, *Delinquent Daughters,* 100–105; Kathleen W. Jones, *Taming the Troublesome Child: American Families, Child Guidance, and the Limits of Psychiatric Authority* (Cambridge, MA: Harvard University Press, 1999), 47–48; Muncy, *Creating a Female Dominion,* 66–91; and Lela B. Costin, *Two Sisters for Social Justice: A Biography of Grace and Edith Abbott* (Urbana: University of Illinois Press, 1983).

19. Jones, *Taming the Troublesome Child,* 38–43.

20. Jones, *Taming the Troublesome Child,* 49–50, 52–54. On the history of the new dynamic psychiatry, see Gerald N. Grob, *Mental Illness and American Society, 1875–1940* (Princeton, NJ: Princeton University Press, 1983); and Lunbeck, *The Psychiatric Persuasion,* 118. On Meyer's psychobiological approach, see "The Role of the Mental Factors in Psychiatry," *American Journal of Insanity* 65 (1908): 39–56; and Alfred Lief, ed., *The Commonsense Psychiatry of Dr. Adolf Meyer: Fifty-Two Selected Papers* (New York: McGraw-Hill, 1948).

21. William Healy, *The Individual Delinquent: A Text-Book of Diagnosis and Prognosis for All Concerned in Understanding Offenders* (Boston: Little, Brown, 1924 [1915]), 14. See also, William Healy, *Mental Conflicts and Misconduct* (Boston: Little, Brown, 1917); and William Healy and Augusta F. Bronner, *Delinquents and Criminals: Their Making and Unmaking, Studies in Two American Cities* (New York: Macmillan, 1926).

22. Healy, *The Individual Delinquent,* 4, 181, 31, 26.

23. Healy, *The Individual Delinquent,* 234, 590, 710, 234–261, 628, 711.

24. See Odem, *Delinquent Daughters;* and Alexander, *The "Girl Problem."*

25. Alexander, *The "Girl Problem,"* 1, 20–21. See also James R. McGovern, "The American Woman's Pre–World War I Freedom in Manners and Morals," *Journal of American History* 55, no. 2 (September 1968): 315–333, for accounts of nontraditional behaviors on the part of the

middle-class young woman during the first two decades of the twentieth century. McGovern argues that it was these women who pioneered in the adoption of modern behaviors, which then had a "trickle up" effect on the manners and morals of young women of the working class.

26. John R. Gillis, *Youth and History: Tradition and Change in European Age Relations, 1770–Present* (New York: Academic Press, 1981), 137–138.

27. On the metric relationship between the normal and the abnormal, see, for example, William A. White, "Introduction" to Winifred Richmond, *The Adolescent Girl: A Book for Parents and Teachers* (New York: Macmillan, 1925), xi–xii; H. L. Hollingworth, *Mental Growth and Decline: A Survey of Developmental Psychology* (New York: D. Appleton, 1927), 4; Ernest R. Groves and Phyllis Blanchard, *Introduction to Mental Hygiene* (New York: Henry Holt, 1930), 43; Phyllis Blanchard, "The Child with Difficulties of Adjustment," in *A Handbook of Child Psychology*, 2nd edition, revised, ed. Carl Murchison (Worcester, MA: Clark University Press, 1933), 858–859. For works that brought this relationship to bear on the female juvenile delinquent in the 1920s, see, for example, Groves and Blanchard, *Introduction to Mental Hygiene*, 56–89; William I. Thomas, *The Unadjusted Girl* (Boston: Little, Brown, 1923); and Miriam Van Waters, *Youth in Conflict* (New York: Republic Publishing Company, 1925).

28. For a synthesis of the social and cultural changes of the 1920s, see Lynn Dumenil, *The Modern Temper: American Culture and Society in the 1920s* (New York: Hill and Wang, 1995).

29. Joseph M. Hawes, *Children between the Wars: American Childhood, 1920–1940* (New York: Twayne Publishers, 1997), 1–12. Quotation on p. 11; Lawrence A. Cremin, *American Education: The Metropolitan Experience* (New York: Harper & Row, 1988); Joseph F. Kett, *Rites of Passage: Adolescence in America, 1790 to the Present* (New York: Basic Books, 1977), 245–264; Howard P. Chudacoff, *How Old Are You? Age Consciousness in American Culture* (Princeton, NJ: Princeton University Press, 1989), 98–106; Heather Munro Prescott, *A Doctor of Their Own: The History of Adolescent Medicine* (Cambridge, MA: Harvard University Press, 1998), 20–21; Jones, *Taming the Troublesome Child*, 120–136; and Fass, *The Damned and the Beautiful*, 8–9, 14, 21–22. For the widely cited study of one community's attempts to adjust to and make sense of the social and cultural changes of the 1920s, see Robert S. Lynd and Helen Merrell Lynd, *Middletown: A Study in American Culture* (New York: Harcourt Brace, 1929).

30. Dorothy Ross, *G. Stanley Hall: The Psychologist as Prophet* (Chicago: University of Chicago Press, 1972), 341–344; Julia Grant, *Raising Baby by the Book: The Education of American Mothers* (New Haven, CT: Yale University Press, 1998), 36–38. For one prominent contemporary critique of Hall's ideas by a fellow psychologist, see Edward L. Thorndike, "The Newest Psychology," *Educational Review* 28 (October 1904): 217–227.

31. Prescott, *A Doctor of Their Own*, 23; Margo Horn, *Before It's Too Late: The Child Guidance Movement in the United States* (Philadelphia: Temple University Press, 1989), 18–19; and Jones, *Taming the Troublesome Child*, 53–54.

32. The first such clinic, the Chicago Juvenile Psychopathic Institute, was founded in 1909 in conjunction with the juvenile court in Chicago and is discussed briefly in this chapter. The clinic that has received the most attention by historians is the Judge Baker Guidance Center in Boston, founded in 1917. For works examining the history of child guidance, see the "Essay on Sources."

33. Jones, *Taming the Troublesome Child*, 92. Horn makes a similar argument in *Before It's Too Late.*

34. Phyllis Blanchard, *The Child and Society: An Introduction to the Social Psychology of the Child* (New York: Longmans, Green, 1928), 325.

35. Jessie Taft, "Mental Hygiene Problems of Normal Adolescence," *Mental Hygiene* 5 (1921): 741.

36. Hawes, *Children between the Wars*, 71–78; Hamilton Cravens, *Before Head Start: The Iowa Station and America's Children* (Chapel Hill: University of North Carolina Press, 1993).

37. Hawes, *Children between the Wars*, 79–85; Grant, *Raising Baby by the Book*, especially Chapters 4 and 5. Grant argues that mothers did not always interpret or use information about the child's development in the way that the experts intended. Her analysis is focused mostly on mothers' relationship to expert information about the development of the young child, rather than the adolescent.

38. For an interesting discussion of the mixed motives of scientific experts in the 1920s, see Hawes, *Children between the Wars*, 8–9. For a helpful synthesis of all of the trends in the study of child development discussed here, see Ann Hulbert, *Raising America: Experts, Parents, and a Century of Advice about Children* (New York: Alfred A. Knopf, 2003), 102–110.

39. The literature on the "new woman," her various incarnations, and her broad range of activities is extensive. See the "Essay on Sources" for some of the helpful forays into her history.

40. On the movement of women into the social sciences during the early twentieth century, see Degler, *In Search of Human Nature*, Chapters 3–8; Fitzpatrick, *Endless Crusade;* Rosalind Rosenberg, *Beyond Separate Spheres: Intellectual Roots of Modern Feminism* (New Haven, CT: Yale University Press, 1982); and Helene Silverberg, ed., *Gender and American Social Science*. Women made significant gains in the graduate programs that prepared them for work in the social sciences during the early twentieth century. In 1920, they received more than 15 percent of Ph.D.'s awarded, up from about 9 percent in 1900. By the end of the 1920s, that number was on the decline and did not recover again until the 1960s and 1970s. Nancy F. Cott, *The Grounding of Modern Feminism* (New Haven, CT: Yale University Press, 1987), 218.

41. Cott, *The Grounding of Modern Feminism*, 215–239. Quotation on p. 234.

42. Hawes, *Children between the Wars*, 2, 66.

43. Carl Murchison ed., *A Handbook of Child Psychology* (Worcester, MA: Clark University Press, 1931), ix.

44. On the problems faced by men in the psychological sciences in claiming authority over domestic knowledge, see Lunbeck, *The Psychiatric Persuasion*, 34–38.

45. Cott, *The Grounding of Modern Feminism*, 138. See also Mari Jo Buhle, *Feminism and Its Discontents: A Century of Struggle with Psychoanalysis* (Cambridge, MA: Harvard University Press, 1998), Chapters 1 and 2.

46. Cott, *The Grounding of Modern Feminism*, 152–156, 225–239.

47. For information on Hollingworth's life and career, see Rosenberg, *Beyond Separate Spheres*, 84–113; Degler, *In Search of Human Nature*, 124–130; Linda Kreger Silverman, "Leta Stetter Hollingworth: Champion of the Psychology of Women and Gifted Children," *Journal of Educational Psychology* 84, no. 1 (1992): 20–27; and Harry L. Hollingworth, *Leta Stetter Hollingworth: A Biography* (Lincoln: University of Nebraska Press, 1943).

48. Rosenberg, *Beyond Separate Spheres*, 91–95; Ross, *G. Stanley Hall*, 346–348.

49. See Leta Hollingworth, "The Frequency of Amentia as Related to Sex," *Medical Record* 84 (1913): 753–756; Leta Hollingworth, *Functional Periodicity: An Experimental Study of the Mental and Motor Abilities of Women during Menstruation* (New York: Teachers College, Columbia University, 1914); Leta Hollingworth, "Variability as Related to Sex Differences in Achievement: A Critique," *The American Journal of Sociology* 19 (1914): 510–530; Leta Hollingworth and Helen Montague, "The Comparative Variability of the Sexes at Birth," *American Journal of Sociology* 20 (1914): 335–370; Leta Hollingworth and Max Schlapp, "An Economic and Social Study of Feeble-minded Women," *Medical Record* 85 (1914): 1025–1228; Leta Hollingworth, "Sex Differences in Mental Traits," *The Psychological Bulletin* 13 (1916): 377–383; Leta Hollingworth, "Comparison of the Sexes in Mental Traits," *The Psychological Bulletin* 15 (1918): 427–432; Leta Hollingworth, "Comparison of the Sexes in Mental Traits," *The Psychological Bulletin* 16 (1919): 371–373; Leta Hollingworth, "Differential Action upon the Sexes of Forces Which Tend to Segregate the Feeble-Minded," *Journal of Abnormal and Social Psychology* 17 (1922): 35–57. The only other woman since Mary Putnam Jacobi to study systematically the effects of menstruation on women's mental capacity was Clelia Duel Mosher, whose findings reinforced those of Hollingworth. See Mosher, "Functional Periodicity in Women and Some Modifying Factors," *California Journal of Medicine* (January–February 1911): 1–21; and Rosenberg, *Beyond Separate Spheres*, 98–99.

50. Some of Hollingworth's most important publications in this area include Leta Hollingworth, *The Psychology of Subnormal Children* (New York: Macmillan, 1920); Leta Hollingworth, *Special Talents and Defects* (New York: Macmillan, 1923); Leta Hollingworth, *Gifted Children: Their Nature and Nurture* (New York: Macmillan, 1926); and Leta Hollingworth, *Children above 180 IQ Stanford-Binet: Origin and Development* (Yonkers-on-Hudson, New York: World Book, 1942).

51. Leta Hollingworth, *The Psychology of the Adolescent* (New York: D. Appleton, 1928), ix. Kreger Silverman calls Hollingworth's *The Psychology of the Adolescent* "the standard text in the field of adolescent psychology for two decades" (p. 23). See also, Leta Hollingworth, "The Adolescent in the Family," *Child Study* 37 (1926): 5–6, 13; Leta Hollingworth, "Getting Away from the Family: The Adolescent and His Life Plans," in *Concerning Parents: A Symposium on Modern Parenthood*, reprint edition (New York: New Republic, 1926), 71–82; Leta Hollingworth, "After High School-What?" *Parents* 4 (June 1929): 21, 60; Leta Hollingworth, "Developmental Problems of Middle Adolescence," *Westminster Leader* (1931): 5, 20–21; Leta Hollingworth, "Late Adolescence," *Westminster Leader* (1931): 5, 24–25; and Leta Hollingworth, "The Adolescent Child," in *A Handbook of Child Psychology*, 2nd edition, 882–908.

52. Leta Hollingworth, *The Psychology of the Adolescent*, ix.

53. H. L. Hollingworth, *Mental Growth and Decline*, 206–213, 367–368, 6. See also Douglas A. Thom, *Everyday Problems of the Everyday Child* (New York: D. Appleton, 1928), 1–11.

54. On the rise of environmentalism in the American social sciences during the first three decades of the twentieth century, see Degler, *In Search of Human Nature*, Chapters 3–8; Rosenberg, *Beyond Separate Spheres;* and Hamilton Cravens, *The Triumph of Evolution: American Scientists and the Heredity-Environment Controversy* (Philadelphia: University of Pennsylvania Press, 1978), Chapters 3–4.

55. Arnold Gesell, *The Mental Growth of the Pre-School Child* (New York: Macmillan, 1925).

On the normative orientation of Gesell's ideas, see Heather Munro Prescott, "'I was a Teenage Dwarf': The Social Construction of 'Normal' Adolescent Growth and Development in the United States," in *Formative Years: Children's Health in the United States, 1880–2000*, ed. Alexandra Minna Stern and Howard Markel (Ann Arbor: University of Michigan Press, 2005), 160, 169–170; and Hulbert, *Raising America*, 154–187. On his reception by parents, see Grant, 215–218.

56. Kevles, *In the Name of Eugenics*, 118–128; and Kline, *Building a Better Race*, 94–123.

57. Leta Hollingworth, *The Psychology of the Adolescent*, 19–20; Leta Hollingworth, "The Adolescent Child," 883. See also H. L. Hollingworth, *Mental Growth and Decline*, 241–243.

58. In an early statement, Thorndike recognized that adolescence was a time of rapid and abrupt change and a period of "heightened and unstable emotional condition" due to the onset of the sex instinct. However, he also stressed that emotional upheaval could be avoided if adolescents simply were "taught that these feelings of theirs are of no consequence, that they in no wise reveal anything concerning their welfare, but are mere accidental accompaniments of certain physiological conditions." Nonetheless, he also anticipated the treatment of adolescence by social scientists in the 1920s in his suggestion that some adolescents might turn these feelings to a certain kind of distinctly modern advantage in fashioning and experiencing the self: "The educated boy, especially the one who has some notions about things in general, some sort of a world view, is likely to refer his moods to the constitution of the universe, and so to evolve doubt and even despair, while the untutored, objective youth, takes them as mere matters of fact." Edward L. Thorndike, *Notes on Child Study*, 2nd edition, vol. 8 (New York: Macmillan, 1903), 150–153. Quotations on pp. 151 and 153. In a later analysis of the period of adolescence, Thorndike went on to more unequivocally challenge virtually all of the characteristics that Hall associated with this stage of life: "We must conclude then that the intellectual and moral picture of the high school boy as breaking loose from home allegiance, full of vast enthusiasms, perplexed and tender in conscience, and the like, is likely to prove truer of the college boy. The picture of these changes as occurring so suddenly that the youth is a mystery to himself, seems true of no age." Edward L. Thorndike, "Magnitude and Rate of Alleged Changes at Adolescence," *Educational Review* 54 (September 1917): 140–147. Quotation on p. 147.

59. Leta Hollingworth, *The Psychology of the Adolescent*, Chapters 3–7.

60. See, for example, Blanchard, *The Child and Society*, 228–249; Groves and Blanchard, *Introduction to Mental Hygiene*, 132–149; Grace Loucks Elliott, *Understanding the Adolescent Girl* (New York: Woman's Press, 1930), 40; and H. L. Hollingworth, *Mental Growth and Decline*, 229–267.

61. Groves and Blanchard, *Introduction to Mental Hygiene*, 148–149.

62. David Harvey, *The Condition of Post-Modernity: An Enquiry into the Origins of Cultural Change* (Cambridge: Basil Blackwell, 1990); T. J. Jackson Lears, *No Place of Grace: Antimodernism and the Transformation of American Culture* (New York: Pantheon Books, 1981).

63. Cott, *The Grounding of Modern Feminism*, Chapter 5. For a theoretical exploration of the history of women and individualism, see Elizabeth Fox-Genovese, *Feminism without Illusions: A Critique of Individualism* (Chapel Hill: University of North Carolina Press, 1991), especially Chapter 5.

64. Hawes, *Children between the Wars*, 4.

65. Thomas, *The Unadjusted Girl*, 243–244. See also, Blanchard, *The Child and Society*, vii–ix.

66. V. F. Calverton and S. D. Schmalhausen, eds., Preface in *Sex in Civilization* (New York: Macaulay Company, 1929), 11.

67. V. F. Calverton, Introduction in Blanchard and Manasses, *New Girls for Old* (New York: Macaulay Company, 1930), ix–x.

68. Calverton, in Blanchard and Manasses, *New Girls for Old*, x, xii; Calverton and Schmalhausen, in *Sex in Civilization,* 10. On Calverton and Schmalhausen, see Buhle, *Feminism and Its Discontents*, 93–99; and Leonard Wilcox, *V. F. Calverton: Radical in the American Grain* (Philadelphia: Temple University Press, 1992). The three anthologies edited by Calverton and Schmalhausen are *Sex in Civilization*; *The New Generation: The Intimate Problems of Modern Parents and Children* (New York: Macaulay Company, 1930); and *Woman's Coming of Age: A Symposium* (New York: Horace Liveright, 1931).

69. G. Stanley Hall, "Flapper Americana Novissima," *Atlantic Monthly* (June 1922): 776.

70. Phyllis Blanchard, "The Longest Journey," in *These Modern Women: Autobiographical Essays from the Twenties*, revised edition, ed. Elaine Showalter (New York: Feminist Press, 1989), 105–106; Gwendolyn Stevens and Sheldon Gardner, *The Women of Psychology*. Vol. 2, *Expansion and Refinement* (Cambridge, MA: Schenkman Publishing Company, 1982), 69–70.

71. Phyllis Blanchard, *The Adolescent Girl: A Study from the Psychoanalytic Viewpoint* (New York: Morrat, Yard, 1920), xi. Hall contributed the Preface to *The Adolescent Girl*, where he attested to the importance of Blanchard's contribution: "It is because a true knowledge of woman, as of man, must begin if it does not end in the study of the teens, when nature is trying to add a new and higher story to our being, that I am glad of an opportunity to very heartily commend this book to the attention of all who at this crisis in her history, when woman has so suddenly attained so much, are now asking what is the next step" (pp. vii–viii).

72. Blanchard, "The Longest Journey," 106–109.

73. Blanchard, *The Adolescent Girl*, 26–41, 87–114. Quotations on pp. 47, 16, 104, 37, 38.

74. Blanchard, *The Adolescent Girl*, 37, 87–91, 40.

75. Blanchard, *The Adolescent Girl*, 91, 113, 110, 112, 40, 233.

76. Fass, *The Damned and the Beautiful*, Chapter 2; Cott, *The Grounding of Modern Feminism*, 156–170; Hulbert, *Raising America*, 97–102, 110–115.

77. The papers given at the 1925 Child Study Association Conference on Modern Parenthood, published as *Concerning Parents*, addressed many of these themes. See Beatrice M. Hinkle, "New Relations of Men and Women as Family Members"; Ethel Puffer Howes, "The Mother in the Present-Day Home"; Elton G. Mayo, "The Father in the Present-Day Home"; Ernest R. Groves, "The Family as Coordinator of Community Forces"; Anna Garlin Spencer, "Parents, The Constant and Inevitable Educators of Their Children"; and Dorothy Canfield Fisher, "Freedom for the Child—What Does It Mean?" See also the contributions to V. F. Calverton and S. D. Schmalhausen's *The New Generation*, especially John B. Watson, "After the Family-What?"; B. Liber, "The Pathos of Parenthood"; T. Swann Harding, "What Price Parenthood?"; Agnes de Lima "The Dilemma of Modern Parenthood"; and Sidonie Matsner Gruenberg, "New Parents for Old."

78. Leta Hollingworth, "The Adolescent in the Family," 5. See also Leta Hollingworth,

"Getting Away from the Family"; Leta Hollingworth, "The Adolescent Child"; and Frankwood E. Williams, "Confronting the World: The Adjustments of Later Adolescence," in *Concerning Parents*, 137–159.

79. On these points, see, for example, Sidonie Matner Gruenberg, "New Parents for Old"; Agnes De Lima, "The Dilemma of Modern Parenthood."

80. S. D. Schmalhausen, "Family Life: A Study in Pathology," in *The New Generation*, 275, 281, 296.

81. Leta Hollingworth, *The Psychology of the Adolescent*, 42. See also, Leta Hollingworth, "The Adolescent Child"; Leta Hollingworth, "The Adolescent in the Family"; and Leta Hollingworth, "Getting Away from the Family."

82. John B. Watson, *Psychological Care of Infant and Child* (New York: W. W. Norton, 1928), especially 69–87. On Watson and Behaviorism, see Hulbert, *Raising America*, 122–153. On the ideas and practices of mother-blaming in the child guidance movement of the 1920s and 1930s and for mothers' responses to it; see Jones, *Taming the Troublesome Child*, 174–204.

83. For the classic psychoanalytic statements about the deficiencies in female development in this regard, see Sigmund Freud, "The Transformations of Puberty," in *Three Essays on the Theory of Sexuality* (1905), *Standard Edition of the Complete Psychological Works of Sigmund Freud*, ed. and trans. James Strachey (London: Hogarth Press, 1953–1974), 7:207–230, especially 225, 227–228; and Freud, "Some Psychical Consequences of the Anatomical Distinction between the Sexes" (1925), *Standard Edition*, 19:242–258.

84. Jones, *Taming the Troublesome Child*, 185.

85. Leta Hollingworth, *The Psychology of the Adolescent*, 42–43, 49–51; Leta Hollingworth, "Social Devices Impelling Women to Bear and Rear Children," *American Journal of Sociology* 22 (July 1916): 19–29. Quotation on p. 24.

86. Lorine Pruette, *Women and Leisure: A Study of Social Waste* (New York: E. P. Dutton & Company, 1924), 207. See also Elizabeth Goldsmith, "Emotional Problems in Children," in *The New Generation*, 485–487.

87. Blanchard and Manasses, *New Girls for Old*, 140, 127–134, 124. See also, Blanchard, *The Child and Society*, 228–233.

88. Blanchard, *The Child and Society*, 97; Blanchard and Manasses, *New Girls for Old*, 44–45; Richmond, *The Adolescent Girl*, 163. See also, Blanchard, *The Adolescent Girl*, x; Groves and Blanchard, *Introduction to Mental Hygiene*, 28; Elliott, *Understanding the Adolescent Girl*, 25–26; and Joseph Jastrow, "Introduction," in H. L. Hollingworth, *Vocational Psychology: Its Problems and Methods* (New York: D. Appleton, 1916), xvi–xvii.

89. Willystine Goodsell, *The Education of Women: Its Social Background and Its Problems* (New York: Macmillan, 1923), 78, 82.

90. Blanchard and Manasses, *New Girls for Old*, 235–236, 145–161; Blanchard, *The Child and Society*, 209, 215–227; Leta Hollingworth, *The Psychology of the Adolescent*, 60–61, 77–82; Richmond, *The Adolescent Girl*, 136–157, 163–185.

91. Goodsell, *The Education of Women*, 347, 343, 342.

92. Degler, *In Search of Human Nature*, parts 1 and 2; and Rosenberg, *Beyond Separate Spheres*.

93. See the review of the extensive attention by psychologists of the 1920s to sex differences in intelligence and personality development in Beth L. Wellman, "Sex Differences," in

A Handbook of Child Psychology, 2nd edition, 626–649. Wellman is both critical of the unexamined assumptions her fellow psychologists make about sex differences and willing to entertain the possibility that "[s]mall differences may be more crucial than is sometimes believed" (p. 627).

94. Cott, *The Grounding of Modern Feminism*, 148–152; Fass, *The Damned and the Beautiful*, Chapter 6.

95. Girls' pursuit of such activities is most thoroughly explored by Schrum, in *Some Wore Bobby Sox*.

96. For historical works providing analysis of the liberalization and domestication of female sexuality during the first three decades of the twentieth century, see the "Essay on Sources."

97. Hall, "Flapper Americana Novissima," 780.

98. Leta Hollingworth, *The Psychology of the Adolescent*, 123–126.

99. Williams, "Confronting the World," 150–159. Quotation on p. 151. See also, Groves and Blanchard, *Introduction to Mental Hygiene*, 139–143; Leta Hollingworth, *The Psychology of the Adolescent*, 2–18, 100–147.

100. Richmond, *The Adolescent Girl*, 40–47; Elliott, *Understanding the Adolescent Girl*, 65–74; Blanchard and Manasses, *New Girls for Old*, 24–25, 77–95, 98–109; Thom, *Everyday Problems of the Everyday Child*, 262–288. On developmental psychologists' deeming of the homosexual stage to be normal in adolescent girls, see John C. Spurlock, "From Reassurance to Irrelevance: Adolescent Psychology and Homosexuality in America," *History of Psychology* 5, no. 1 (February 2002): 38–51.

101. Nelly Oudshoorn, *Beyond the Natural Body: An Archeology of Sex Hormones* (London: Routledge, 1994), 1–64, 22–23. See also Margaret Marsh and Wanda Ronner, *The Empty Cradle: Infertility in America from Colonial Times to the Present* (Baltimore: Johns Hopkins University Press, 1996), 134–142.

102. William P. Graves, *Gynecology*, 4th edition (Philadelphia: W. B. Saunders Company, 1929), 34, 58–59, 65.

103. Blanchard, *The Adolescent Girl*, 44. Blanchard borrowed the term "sex complex" from Blair W. Bell, *The Sex Complex* (London: Bailliere, Tinkall & Cox, 1916). See also Robert T. Frank, *The Female Sex Hormone* (Springfield, IL: Charles C. Thomas, 1929), 7–19.

104. See the many pieces on these themes by psychologists and other social scientists in Calverton and Schmalhausen's *Sex in Civilization*.

105. Blanchard, "Sex in the Adolescent Girl," in *Sex in Civilization*, 93; Blanchard and Manasses, *New Girls for Old*, 61.

106. Blanchard, *The Adolescent Girl*, 45–46.

107. Ethel Dummer, "Foreward" in Thomas, *The Unadjusted Girl*, viii–ix.

108. See Blanchard, "Sex in the Adolescent Girl," 538–556; Groves and Blanchard, *Introduction to Mental Hygiene*, 139–143; Blanchard and Manasses, *New Girls for Old*, Chapter 4; Thomas, *The Unadjusted Girl*, 109, 110–116.

109. Blanchard, *The Adolescent Girl*, 82–84; Pruette, *Women and Leisure*, 134; Leta Hollingworth, *The Psychology of the Adolescent*, 109; Leta Hollingworth, "The Vocational Aptitudes of Women," in *Vocational Psychology*, 238–240.

110. Margaret Sanger, *What Every Boy and Girl Should Know*, reprint edition (Elmsford, NY: Maxwell Reprint Company, 1969 [1927]), 73.

111. Christina Simmons, "Modern Sexuality and the Myth of Victorian Repression," in *Passion and Power: Sexuality in History*, ed. Kathy Peiss and Christina Simmons (Philadelphia: Temple University Press, 1989), 167–169.

112. Lorine Pruette, "The Flapper," in *The New Generation*, 580.

113. Blanchard, "Sex in the Adolescent Girl," 547; Blanchard, *The Child and Society*, 244.

114. Blanchard and Manasses, *New Girls for Old*, 58. See also, Blanchard, *The Child and Society*, 258–259.

115. Lorine Livingston Pruette, "The Evolution of Disenchantment," in *These Modern Women*, 68–73.

116. Pruette, "The Flapper," 588–589.

117. Pruette, "The Flapper," 589–590; Beatrice Forbes-Robertson Hale, "Women in Transition," in *Sex and Civilization*, 80.

118. Blanchard and Manasses, *New Girls for Old*, 5; Moxcey, *Girlhood and Character*, 311.

119. Cott, *The Grounding of Modern Feminism*, 147–165; Fass, *The Damned and the Beautiful*, 66–68; Dumenil, *The Modern Temper*, 31–40, 68–71.

120. Cott, *The Ground of Modern Feminism*, 117–142, 179–211.

121. Ruth Shonle Cavan and Jordan True Cavan, *Building a Girl's Personality: A Social Psychology of Later Girlhood* (New York: The Abingdon Press, 1932), 63, 64; Leta Hollingworth, "The Adolescent Child," 892–894; Leta Hollingworth, *The Psychology of the Adolescent*, 59–99. See also, for example, Blanchard and Manasses, *New Girls for Old*, 5, 7, 40–41, 145–161; Richmond, *The Adolescent Girl*, 163–185, Goodsell, *The Education of Women*, 108–121.

122. Blanchard and Manasses, *New Girls for Old*, 237; Leta Hollingworth, *The Psychology of the Adolescent*, 42–43; Cavan and Cavan, *Building a Girl's Personality*, Chapter 4; Goodsell, *The Education of Women*, 337–339.

123. Leta Hollingworth, "The Adolescent Child," 893–94; Leta Hollingworth, "After High School, What?" 60; Leta Hollingworth, *The Psychology of the Adolescent*, 81.

124. Blanchard and Manasses, *New Girls for Old*, 20–21, 236–237.

125. Pruette, *Women and Leisure*, 45, 117–118. The typical girl surveyed by Pruette was a white middle-class 16-year-old high school student from a southern state. For the demographics of the survey population, see pp. 118–122.

126. Pruette, *Women and Leisure*, 187, 189, 122, 124, 131, 199.

127. Pruette, *Women and Leisure*, 186–187. See also Pruette's rumination on this theme in "Why Women Fail," in *Woman's Coming of Age*, 240–259.

128. Blanchard and Manasses, *New Girls for Old*, 20–21.

129. Pruette, *Women and Leisure*, 188.

130. Blanchard and Manasses, *New Girls for Old*, 174–175, 182. For similar findings from survey data, see Cavan and Cavan, *Building a Girl's Personality*, 63.

131. Blanchard and Manasses, *New Girls for Old*, 240–249. Quotations on pp. 245 and 246.

132. Leta Hollingworth, *The Psychology of the Adolescent*, 148–150.

133. Degler, *In Search of Human Nature*, Chapters 2 and 7; Kevles, *In the Name of Eugenics*, 77–84, 129–147. See also Stephen J. Gould, *Mismeasure of Man* (New York: W. W. Norton, 1981).

134. Blanchard, *The Child and Society*, 91–92.

135. Leta Hollingworth, *The Psychology of the Adolescent*, 157, 148; H. L. Hollingworth, *Mental Growth and Decline*, 242–243.

136. Rosenberg, *Beyond Separate Spheres*, Chapters 3 and 4. See also, Degler, *In Search of Human Nature*, Chapter 5; Cynthia Eagle Russett, *Sexual Science: The Victorian Construction of Womanhood* (Cambridge, MA: Harvard University Press, 1989), 164–174.

137. Helen Thompson Woolley, "A Review of the Recent Literature on the Psychology of Sex," *The Psychological Bulletin* 7 (October 1910): 335–352; Woolley, "The Psychology of Sex," *The Psychological Bulletin* 11 (October 15, 1919): 353–379; Leta Hollingworth, "Sex Differences in Mental Traits"; Leta Hollingworth, "Comparison of the Sexes in Mental Traits" (1918); Leta Hollingworth, "Comparison of the Sexes in Mental Traits" (1919).

138. Leta Hollingworth, "Comparison of the Sexes in Mental Traits," 1918, 427.

139. Leta Hollingworth, "Comparison of the Sexes in Mental Traits," 1919, 373.

140. Lewis Madison Terman, "Were We Born That Way?" *World's Work* 44 (October 1922): 660. Quoted in Degler, *In Search of Human Nature*, 130.

141. Blanchard, *The Child and Society*, 88; Elliott, *Understanding the Adolescent Girl*, 61–63; Thom, *Everyday Problems of the Everyday Child*, 303–327.

142. Blanchard, *The Child and Society*, 93.

143. Blanchard, *The Child and Society*, 93 and Chapter 3; Leta Hollingworth, *The Psychology of the Adolescent*, 169; Helen Thompson Woolley, "A New Scale of Mental and Physical Measurements for Adolescents and Some of Its Uses," *The Journal of Educational Psychology* 6 (November 1915): 532, 534, 536; Thomas, *The Unadjusted Girl*, 220–221; Cavan and Cavan, *Building a Girl's Personality*, 112.

144. Elliott, *Understanding the Adolescent Girl*, 62.

145. Hall, "Flapper Americana Novissima," 780; Wellman, "Sex Differences," 626.

146. Joseph Jastrow, "The Implications of Sex," in *Sex in Civilization*, 130.

147. Jastrow, "The Implications of Sex," 135.

148. Jastrow, "The Implications of Sex," 137–142.

149. Miriam Lewin, " 'Rather Worse Than Folly?': Psychology Measures Masculinity and Femininity, 1," in *In the Shadow of the Past: Psychology Portrays the Sexes*, ed. Miriam Lewin (New York: Columbia University Press, 1984), 153–178; and Cott, 153–154, 334n17.

CHAPTER FIVE. ADOLESCENT GIRLHOOD COMES OF AGE?

1. Havelock Ellis, "The Sexual Impulse in Women," in *Studies in the Psychology of Sex,* 2nd edition, vol. 1 (New York: Random House, 1942 [1913]), 250n2.

2. Miriam Van Waters, "The Adolescent Girl Among Primitive Peoples," *Journal of Religious Psychology* 6, no. 4 (1913): 376.

3. Margaret Mead was 23 years old when she left for Samoa in the summer of 1925. Jane Howard, *Margaret Mead: A Life* (New York: Simon and Schuster, 1984), 76. Margaret Mead, *Coming of Age in Samoa: A Psychological Study of Primitive Youth for Western Civilisation,* with a foreword by Franz Boas (New York: Morrow, 1928), 5; Susan Hegeman, *Patterns for America: Modernism and the Concept of Culture* (Princeton, NJ: Princeton University Press, 1999), 30.

4. For examples of uses made of Mead's conclusions in *Coming of Age*, see Gerald H. J.

Pearson, "What the Adolescent Girl Needs in Her Home," *Mental Hygiene* (January 1930): 42–43; and Leta Hollingworth, "The Adolescent Child," in *A Handbook of Child Psychology*, 2nd edition, ed. Carl Murchison (Worcester, MA: Clark University Press, 1933), 883.

5. Mead, *Coming of Age*, 195–248.

6. George W. Stocking, Jr., "The Ethnographic Sensibility of the 1920s and the Dualism of the Anthropological Tradition," in *Romantic Motives: Essays on Anthropological Sensibility*, ed. George W. Stocking Jr. (Madison: University of Wisconsin Press, 1989), 267 n18. See also, Carl N. Degler, *In Search of Human Nature: The Decline and Revival of Darwinism in American Social Thought* (New York: Oxford University Press, 1991), 135–136; and Rosalind Rosenberg, *Beyond Separate Spheres: Intellectual Roots of Modern Feminism* (New Haven, CT: Yale University Press, 1982), 226–237.

7. Louise Michele Newman, *White Women's Rights: The Racial Origins of Feminism in the United States* (New York: Oxford University Press, 1999), Chapter 7.

8. Margaret Mead, "Adolescence in Primitive and Modern Society," in *The New Generation: The Intimate Problems of Modern Parents and Children*, ed. V. F. Calverton and Samuel D. Schmalhausen (New York: Macaulay Company, 1930), 169–188.

9. Mead, *Coming of Age*, 9. Sociologist and anthropologist Elsie Clews Parsons maintained that "a woman student [of ethnology] would have many opportunities for observing the life of women and children that male ethnographers lacked." Quoted in Rosenberg, *Beyond Separate Spheres*, 166. Parsons accurately anticipated the value of such access to modern anthropology, as the customs and beliefs surrounding child-rearing practices became increasingly recognized as providing important insight into the formation of cultural patterns. See also Kamala Visweswaran, " 'Wild West' Anthropology and the Disciplining of Gender," in *Gender and American Social Science: The Formative Years*, ed. Helene Silverberg (Princeton, NJ: Princeton University Press, 1998), 87; Degler, *In Search of Human Nature*, 136; Robert H. Lowie, *The History of Ethnological Theory* (New York: Farrar & Rinehart, 1937), 134; and Margaret Mead, "Theoretical Setting—1954," in *Childhood in Contemporary Cultures*, ed. Margaret Mead and Martha Wolfenstein (Chicago: University of Chicago Press, 1955), 4–5.

10. Mead, *Coming of Age*, 3.

11. Margaret Mead, *Blackberry Winter: My Earlier Years* (New York: William Morrow & Company, 1972), 139; Mead, *Coming of Age*, 9.

12. For an in-depth analysis of the emergence, assumptions, and complexities of cultural evolutionism, see George W. Stocking Jr., *Victorian Anthropology* (New York: Free Press, 1987), 144–185. See also Stocking, *Race, Culture, and Evolution: Essays in the History of Anthropology* (New York: Free Press, 1968); Newman, *White Women's Rights*, 11, 29–30; Gail Bederman, *Manliness & Civilization: A Cultural History of Gender and Race in the United States, 1880–1917* (Chicago: University of Chicago Press, 1995), 28–29; and Degler, *In Search of Human Nature*, 14–22.

13. Desley Deacon, *Elsie Clews Parsons: Inventing Modern Life* (Chicago: University of Chicago Press, 1997); and Estelle B. Freedman, *Maternal Justice: Miriam Van Waters and the Female Reform Tradition* (Chicago: University of Chicago Press, 1996).

14. On Boas's contributions to the formulation of the culture concept, see Stocking, *Race, Culture, and Evolution*, 133–233; and Degler, *In Search of Human Nature*, 59–83. Hegeman qualifies the "mythical-heroic reading" of Boas's role by Degler, in particular. She admits that

"in a more complex way [than Degler's reading ascertains] Boas was central to the creation of both the culture concept and the professional discipline of anthropology in America . . . " However, she also asserts that "it is something of a misreading to see Boas's contribution as being fundamentally antihierarchical or evaluatively relativist in nature; rather his crucial intervention might be more properly described as a *spatial* reorganization of human differences" (pp. 16, 32).

15. Indeed, Boas did not provide a written definition of culture until 1930, after his students Edward Sapir and Alfred L. Kroeber "had begun to use the word as a technical term." Hegeman, *Patterns for America*, 39.

16. Mead sets up Hall as her foil in the introduction to *Coming of Age*, 2. See also, Mead, *Blackberry Winter*, 139.

17. Margaret Mead, *Growing up in New Guinea: A Comparative Study of Primitive Education*, with a new preface by Mead (New York: Morrow Quill Paperbacks, 1975; reprint, 1930), iv–v. See also Mead, *Blackberry Winter*, 166; and Margaret Mead, "The Primitive Child," in *A Handbook of Child Psychology*, ed. Carl Murchison (Worcester, MA: Clark University Press, 1931), 671–672.

18. Stocking, *Victorian Anthropology*, 15, 167–169, 228–229; and Stocking, *Race, Culture, and Evolution*, 114.

19. Quoted in Sheldon H. White, foreword to *Cultural Psychology: A Once and Future Discipline*, by Michael Cole (Cambridge, MA: Belknap Press of Harvard University Press, 1996), x.

20. Stocking, *Victorian Anthropology*, 228–230; Cole, *Cultural Psychology*, 12–14.

21. Francis Galton, "Hereditary Talent and Character," *Macmillan's Magazine* 12 (1865): 325–26, quoted in Stocking, *Victorian Anthropology*, 95.

22. Dudley Kidd, *Savage Childhood: A Study of Karfir Children* (London: A. and C. Black, 1906), viii–ix, quoted in Cole, *Cultural Psychology*, 18.

23. Stocking, *Victorian Anthropology*, 229–30.

24. On nineteenth-century Lamarckianism and early twentieth-century neo-Lamarkianism, see Degler, *In Search of Human Nature*, 20–25; and Stocking, *Race, Culture, and Evolution*, 238–269. On Hall's use of the comparative method and the theory of acquired characteristics, see Dorothy Ross, *G. Stanley Hall: The Psychologist as Prophet* (Chicago: University of Chicago Press, 1972), 371–372; and Stocking, *Race, Culture, and Evolution*, 125–126, 242, 254–255.

25. Bederman, *Manliness & Civilization*, 88–110.

26. G. Stanley Hall, *Adolescence: Its Psychology and Its Relations to Physiology, Anthropology, Sociology, Sex, Crime, Religion, and Education*, 2 vols. (New York: D. Appleton, 1904), 2:650.

27. Bederman, *Manliness & Civilization*, 110–117; Ross, *G. Stanley Hall*, 412–416.

28. On the assumptions about the degradation of women and children in primitive societies and the superiority of the civilized family held by ethnographers, reformers, and missionaries during the nineteenth and early twentieth centuries, see Stocking, *Victorian Anthropology*, 84, 90; Newman, *White Women's Rights*, 34, 42, 116–131; Elizabeth Fee, "The Sexual Politics of Victorian Social Anthropology," in *Clio's Consciousness Raised: New Perspectives on the History of Women*, eds. Mary S. Hart and Lois Banner (New York: Harper & Row,

1974), 86–102; and Margaret D. Jacobs, *Engendered Encounters: Feminism and Pueblo Cultures 1879–1934* (Lincoln: University of Nebraska Press, 1999), 34–37.

29. Joan Jacobs Brumberg, "Zenanas and Girlless Villages: The Ethnology of American Evangelical Women, 1870–1910," *Journal of American History* 69, no. 2 (1982): 347–371. Quotation on p. 365.

30. Hall, *Adolescence,* 2:232.

31. Hall, *Adolescence,* 2:232–239; 2:245–246; 1:479–480. On the idea of "survivals" in anthropological thought, see Lowie, *The History of Ethnological Theory,* 25–26; and Stocking, *Victorian Anthropology,* 127–128, 162–163.

32. Hall, *Adolescence,* 1:479–480.

33. Hall, *Adolescence,* 1:472–512. Quotations on pp. 478, 472, and 511.

34. On the direction Boas provided for Mead's first field trip, see Mead, *Blackberry Winter,* 124–142; and Howard, *Margaret Mead,* 67–70, 76–77.

35. On Boas's liberal upbringing, see Stocking, *Race, Culture and Evolution,* 149; and Degler, *In Search of Human Nature,* 73.

36. Stocking, *Race, Culture and Evolution,* 156–159, 169–70; Hegeman, *Patterns for America,* 37, 47–51.

37. Stocking, *Race, Culture and Evolution,* 165; Ross, *G. Stanley Hall,* 196–197, 293. Quotation on p. 196.

38. Stocking, *Race, Culture and Evolution,* 161–194. Boas resigned from Clark University in 1892, along with several other faculty members, following a struggle with Hall over issues of administrative control, financial management, and academic freedom. For the details of this struggle, see Ross, *G. Stanley Hall,* 207–230.

39. See Franz Boas, "Anthropological Investigations in Schools," *Science* 17 (June 1891): 351–352; Boas, "The Growth of Children," *Science* 19 (May 1892): 256–257; Boas, "The Growth of Children—II," *Science* 19 (May 1892): 281–282; "The Growth of Children," *Science* 20 (December 1892): 351–352; Boas, "On Dr. William Townsend Porter's Investigation of the Growth of the School Children of St. Louis," *Science,* n.s., 1 (March 1895): 225–230; Boas, "The Growth of First-Born Children," *Science,* n.s., 1 (March 1895): 402–404; Boas, "The Growth of Children," *Science,* n.s., 5 (April 1897): 570–573. Boas's growth studies consisted of analysis and critique of the physical measurements of children previously obtained by other scientists and analysis of data he himself collected in a study of the growth of children in the Worcester, Massachusetts, public schools begun in 1891, which was aborted because of his resignation from Clark. He also subsequently initiated growth studies in Oakland, California, and Toronto, Canada, which were conducted in conjunction with the anthropological work of the 1893 World's Columbian Exposition in Chicago. See Stocking, *Race, Culture and Evolution,* 170–172; and J. M. Tanner, "Boas' Contributions to Knowledge of Human Growth and Form," in *The Anthropology of Franz Boas, Essays on the Centennial of His Birth,* ed. Walter Goldschmidt (Menasha, WI: American Anthropological Association, 1959), 76–87.

40. Boas, "Anthropological Investigations in Schools," 351.

41. On this point, see Boas, "Anthropological Investigations in Schools," 351; and Boas, "The Growth of Children," *Science* 20 (December 1892), 351. See also Stocking, *Race, Culture and Evolution,* 171; and Tanner, "Boas' Contributions," 81–82.

42. Stocking, *Race, Culture and Evolution*, 171; and Tanner, "Boas' Contributions," 77–81.

43. Boas, "On Dr. William Townsend Porter's Investigation of the Growth of the School Children of St. Louis," 227–229. See also Boas, "The Growth of Children," *Science*, n.s., 5 (April 1897): 571–573.

44. Stocking, *Race, Culture and Evolution*, 190, 172.

45. Boas published two additional papers during what J. M. Tanner refers to as his "first period of the study of growth": Franz Boas and C. Wissler, "Statistics of Growth," in *Report of the U.S. Commissioner of Education, 1904, 1906* (Washington, DC: U.S. Commissioner of Education), 25–32; and Boas, "The Growth of Children," *Science*, n.s., 36 (December 1912): 815–818. In these papers, Boas drew connections between his own earlier findings about growth and the concept of developmental age then being formulated by other scientists. See Tanner, "Boas' Contributions," 85–86.

46. Stocking, *Race, Culture and Evolution*, 175–180; Degler, *In Search of Human Nature*, 63–66; Tanner, "Boas' Contributions," 99–103; Hegeman, *Patterns for America*, 47–49.

47. On Boas's "second growth period," which lasted from 1930 to 1942, see Tanner, "Boas' Contributions," 87–93.

48. Margaret Mead and Frances Cooke Macgregor, *Growth and Culture: A Photographic Study of Balinese Childhood* (New York: G. P. Putnam's Sons, 1951).

49. Deacon, *Elsie Clews Parsons*, 1–7, 11–12, 16–19; Rosenberg, *Beyond Separate Spheres*, 147–148. For additional biographical information on Parsons, see the "Essay on Sources."

50. Deacon, *Elsie Clews Parsons*, 28–37, 45; Rosenberg, *Beyond Separate Spheres*, 148–156.

51. Deacon, *Elsie Clews Parsons*, 39–60; Rosenberg, *Beyond Separate Spheres*, 154–156.

52. Elsie Clews Parsons, *The Family: An Ethnographical and Historical Outline with Descriptive Notes, Planned as a Text-book for the Use of College Lecturers and of Directors of Home-reading Clubs* (New York: G. P. Putnam's Sons, 1906), vi; Deacon, *Elsie Clews Parsons*, 46–47, 61.

53. Parsons, *The Family*, vii, ix, 340. Italics in original.

54. Parsons, *The Family*, 31, 339–340, 60–61, 115.

55. Parsons, *The Family*, 91, 90–93, 340–341. Italics in original.

56. Parsons, *The Family*, 341, xi, 346, 347–355.

57. Deacon, *Elsie Clews Parsons*, 68–71; Rosenberg, *Beyond Separate Spheres*, 162–163.

58. Deacon, *Elsie Clews Parsons*, 97–121; Rosenberg, *Beyond Separate Spheres*, 167–168.

59. Rosenberg, *Beyond Separate Spheres*, 169–177.

60. Elsie Clews Parsons, *The Old-Fashioned Woman: Primitive Fancies about the Sex* (New York: G. P. Putnam's Sons, 1913), 11–23. Quotation on p. 14.

61. Parsons, *The Old-Fashioned Woman*, 15.

62. Parsons, *The Old-Fashioned Woman*, 24–30. Quotations on p. 24.

63. Parsons, *The Old-Fashioned Woman*, 30.

64. Parsons, *The Old-Fashioned Woman*, 24–30.

65. Elsie Clews Parsons, *Fear and Conventionality* (New York: G. P. Putnam's Sons, 1914), 119–135, 176–196. Quotations on pp. 119 and 134. Elsie Clews Parsons, *Social Rule: A Study of the Will to Power* (New York: G. P. Putnam's Sons, 1916), 12–57.

66. Elsie Clews Parsons, "The Ceremonial of Growing Up," *School and Society* 2, no. 38 (1915): 408–411; Parsons, *Social Rule*, 24–29.

67. Parsons, *Fear and Conventionality*, xxxvi–xxxvii, 205–218.

68. Elsie Clews Parsons, *Social Freedom: A Study of the Conflicts between Social Classification and Personality* (New York: G. P. Putnam's Sons, 1915), 105–106.

69. Miriam Van Waters, "The Adolescent Girl Among Primitive Peoples," *Journal of Religious Psychology* 7.1 (1914): 113; Van Waters, "The Adolescent Girl" (1913): 383.

70. Van Waters published two widely read books on juvenile delinquency during the 1920s, *Youth in Conflict* (New York: Republic Publishing, 1925) and *Parents on Probation* (New York: New Republic, 1927). On Van Waters's career as a juvenile justice reformer, see Freedman, *Maternal Justice*, Chapters 4–7; and Mary E. Odem, *Delinquent Daughters: Protecting and Policing Adolescent Female Sexuality in the United States, 1885–1920* (Chapel Hill: University of North Carolina Press, 1995), Chapter 5.

71. Quoted in Freedman, *Maternal Justice*, 54.

72. Freedman, *Maternal Justice*, 3–21.

73. Freedman, *Maternal Justice*, 22–32.

74. Freedman, *Maternal Justice*, 35–56.

75. Van Waters, "The Adolescent Girl" (1913): 376–77.

76. Van Waters, "The Adolescent Girl" (1913): 382–404. Quotation on p. 404.

77. Van Waters, "The Adolescent Girl" (1913): 382–421; Van Waters, "The Adolescent Girl" (1914): 95–105. Quotations on pp. 95 and 96–97.

78. Van Waters, "The Adolescent Girl" (1913): 375; Van Waters, "The Adolescent Girl" (1914): 105–116. Quotations on pp. 112 and 114.

79. Van Waters, "The Adolescent Girl" (1914): 89–90, 104–105; Van Waters, "The Adolescent Girl" (1913): 389–390.

80. Van Waters, "The Adolescent Girl" (1914): 115.

81. For the subsequent unfolding of the careers of Parsons and Van Waters, see Rosenberg, *Beyond Separate Spheres*, 176–177; Jacobs, *Engendered Encounters*, 72–105; Deacon, *Elsie Clews Parsons*, Chapters 6–16; and Freedman, *Maternal Justice*, Chapters 4–17.

82. Lois W. Banner, *Intertwined Lives: Margaret Mead, Ruth Benedict, and Their Circle* (New York: Alfred A. Knopf, 2003), especially Chapter 3.

83. Banner, *Intertwined Lives*, 68–73, 77–79; Rosenberg, *Beyond Separate Spheres*, 208–211; Howard, *Margaret Mead*, 21–36; Mead, *Blackberry Winter*, 9–29.

84. Banner, *Intertwined Lives*, 68–69, 72–77, 79–81; Rosenberg, *Beyond Separate Spheres*, 210–211; Howard, *Margaret Mead*, 21–36; Mead, *Blackberry Winter*, 30–56.

85. Banner, *Intertwined Lives*, 155–168; Rosenberg, *Beyond Separate Spheres*, 212–215; Howard, *Margaret Mead*, 37–50; Mead, *Blackberry Winter*, 88–111.

86. Banner, *Intertwined Lives*, 172–175, 183–184; Rosenberg, *Beyond Separate Spheres*, 215–219; Mead, *Blackberry Winter*, 111–115, 122.

87. Rosenberg, *Beyond Separate Spheres*, 220–227; Howard, *Margaret Mead*, 64–67; Mead, *Blackberry Winter*, 124–125.

88. Banner, *Intertwined Lives*, 229–230; Rosenberg, *Beyond Separate Spheres*, 226–229; Howard, *Margaret Mead*, 67–70, 76–77; Mead, *Blackberry Winter*, 124–142. Stephen O. Murray sees Mead's trip to Samoa less as the result of compromise between Mead and Boas than Mead's "manipulation" of Boas to get to do fieldwork in Polynesia in "On Boasians and Margaret Mead: Reply to Freeman," *Current Anthropology* 32, no. 4 (August–October 1991): 448–452.

89. Mead's research methods are detailed in Appendices II and V to *Coming of Age in Samoa*.

90. Quoted in Mead, *Blackberry Winter*, 138.

91. Mead, *Coming of Age*, 196.

92. Mead, *Coming of Age*, 197.

93. For a helpful synthesis of American life in the 1920s, see Lynn Dumenil, *The Modern Temper: American Culture and Society in the 1920s* (New York: Hill and Wang, 1995).

94. Mead, *Coming of Age*, 259-260.

95. Mead, *Coming of Age*, 39-58, 208-216. Quotations on pp. 40 and 212. Boas notes that Mead's study posed a challenge to the essentialism of psychoanalytic theories of child development in his Foreward to *Coming of Age*, xiv-xv.

96. See, for example, Robert H. Lowie, "Review of *Coming of Age in Samoa*," *American Anthropologist* (July 1929): 532-534; and Freda Kirchwey, "Sex in the South Seas," *The Nation* (October 24, 1928): 427. Malinowski revealed comparable findings and made similar arguments about the sexuality of South Sea Islanders as those of Mead in his study, *Sex and Repression in Savage Society*, published in 1927.

97. Mead, *Coming of Age*, 131-138. Quotation on p. 138.

98. Mead, *Coming of Age*, 28-33, 38, 144-146.

99. Mead, *Coming of Age*, 86-109, 221-223. Quotations on pp. 92, 103-104, 221, 147, 138, 222.

100. Mead, *Coming of Age*, Chapters 13 and 14.

101. Mead, *Coming of Age*, Chapters 3-13. Quotation on p. 206. See also Mead, "The Primitive Child," 669-670. In Appendix III of *Coming of Age*, entitled "Samoan Civilisation as it is To-Day," Mead cautions her readers to "not mistake the conditions [of Samoan life and child rearing] which have been described for the aboriginal ones, nor for typical primitive ones" and makes some attempt to account for the role of historical change in shaping the lives of contemporary Samoan adolescents. Before the advent of "white influence," she explains, Samoan children were tyrannically controlled by adult heads of households, unchaste girls were severely punished, boys suffered great anxiety in competition for social rank, and puberty was experienced as a traumatic period for boys and girls alike. According to Mead, it was the imposition of the American legal system and of Christianity that expunged the worst brutalities of aboriginal Samoan culture, without replacing these with new forms of social or economic misery. Mead's recognition here of the "flexibility" of Samoan culture does less to complicate her depiction of it as static in the body of the text than to bolster her argument that ontogeny is, indeed, variously shaped by cultural conditions. Significantly, the stresses that she describes were experienced by aboriginal Samoan adolescents were of the merely torturing sort long documented by cultural evolutionists and not of the individually or culturally regenerative kind experienced by their modern American counterparts (pp. 266-277; quotations on pp. 272, 273).

102. Mead, *Coming of Age*, Chapter 8. Quotations on pp. 118, 121.

103. Mead, "The Primitive Child," 669-670.

104. For the late-twentieth-century scholarly debate over the methods and interpretations of Mead's cultural anthropology, see the "Essay on Sources."

105. On Mead's reflections on the relationship between biology and culture in her early

work, see, for example, Margaret Mead, *From the South Seas: Studies of Adolescence and Sex in Primitive Societies* (New York: William Morrow, 1939), v–xxxi; Mead and Macgregor, *Growth and Culture*, 9–16, 22–23, 185–186; Mead, "Theoretical Setting—1954," 5–12.

106. Mead, *Coming of Age*, 247–248.

107. Mead, *Coming of Age*, 207. See also, Mead, "The Primitive Child," 669–670.

108. Mead, "Adolescence in Primitive and Modern Society," 186–187. Rosenberg and Newman come to similar conclusions in their analyses of Mead's feminism. See Rosenberg, *Beyond Separate Spheres*, Chapter 8; and Newman, *White Women's Rights*, Chapter 7.

109. Mead, "Adolescence in Primitive and Modern Society," 179; Mead, *Coming of Age*, 33, 38.

110. Mead, *Coming of Age*, Chapters 13 and 14; Mead, *From the South Seas*, xxvi–xxxi.

111. Mead, *Coming of Age*, 26–27, 59–73, 215.

112. Mead, *Coming of Age*, 138.

113. Mead, *Coming of Age*, 212.. Louise Michele Newman argues that Charlotte Perkins Gilman (1860–1935) was the first feminist intellectual who "effectively dismantled her society's hegemonic association of (the civilized white) 'woman' with the supposedly natural and ineradicable sexual differences produced by woman's relegation to the home" (p. 135).

114. Phyllis Blanchard, *The Adolescent Girl: A Study from the Psychoanalytic Viewpoint* (New York: Morrat, Yard and Company, 1920), 233.

115. Mead, *Coming of Age*, 200–206. Quotations on p. 202.

116. Mead, *Coming of Age*, 208–216. Quotations on pp. 216, 212, 213.

117. The concept of psychological weaning is discussed in the previous chapter.

118. Mead, *Coming of Age*, 236–238. See also Mead, "Adolescence in Primitive and Modern Society," 183.

119. Mead, "Adolescence in Primitive and Modern Society," 179.

120. Mead, *Growing Up in New Guinea*, part 1.

121. Mead, "Adolescence in Primitive and Modern Society," 185.

EPILOGUE

1. Kelly Schrum, *Some Wore Bobby Sox: The Emergence of Teenage Girls' Culture, 1920–1945* (New York: Palgrave Macmillan). See also Grace Palladino, *Teenagers: An American History* (New York: Basic Books, 1996).

2. Juliet B. Schor, *Born To Buy: The Commercialized Child and the New Commercial Culture* (New York: Scribner, 2004), 43–47, 202–203. See also Daniel Thomas Cook, *The Commodification of Childhood: The Children's Clothing Industry and the Rise of the Child Consumer, 1917–1962* (Durham, NC: Duke University Press, 2004).

3. Arnold Gesell, Frances L. Ilg, and Louise Bates Ames, *Youth: The Years from Ten to Sixteen* (New York: Harper & Row, 1956), 4. *Youth* focused on the first 7 years of the adolescent stage of life, which Gesell and his colleagues deemed to extend up to the age of 21.

4. Gesell et al., *Youth*, 27, 23, 21, 253, 13.

5. On Gesell, see Heather Munro Prescott, " 'I was a Teenage Dwarf': The Social Construction of 'Normal' Adolescent Growth and Development in the United States" in *Formative Years: Children's Health in the United States, 1880–2000*, ed. Alexandra Minna Stern and

Howard Markel (Ann Arbor: The University of Michigan Press, 2005), 160, 169–170; Ann Hulbert, *Raising America: Experts, Parents, and a Century of Advice about Children* (New York: Alfred A. Knopf, 2003), 154–187; and Julia Grant, *Raising Baby by the Book: The Education of American Mothers* (New Haven, CT: Yale University Press, 1998), 215–218.

6. Gesell et al., *Youth*, 12, 503–507, 6, 22, 27, 28.

7. Gesell et al., *Youth*, 27, 23, 21, 253, 13.

8. Heather Munro Prescott, *A Doctor of Their Own: The History of Adolescent Medicine* (Cambridge, MA: Harvard University Press, 1998), 24–29; Munro Prescott, " 'I was a Teenage Dwarf,' 165–172; Hulbert, 104–106. For examples of findings on adolescence from the longitudinal studies, see Jerome Kagan and Howard A. Moss, *Birth to Maturity: The Fels Study of Psychological Development* (New York: Wiley, 1962); and Mary Cover Jones et al., eds., *The Course of Human Development: Selected Papers from the Longitudinal Studies, Institute of Human Development, The University of California at Berkeley* (Waltham, MA: Xerox College Publishing, 1971).

9. Prescott, *A Doctor of Their Own*, 25–29. For Frank's writings on these themes, see Lawrence K. Frank, "Society as the Patient," *American Journal of Sociology* 42, no. 3 (1936): 335–344; Lawrence K. Frank, "Certain Problems of Puberty and Adolescence," *Journal of Pediatrics* 19, no. 3 (September 1941): 294–301; Lawrence K. Frank, "Introduction: Adolescence as a Period of Transition" in *Forty-Third Yearbook of the National Society for the Study of Education, part 1: Adolescence*, ed. Nelson B. Henry, (Chicago: Department of Education, University of Chicago, 1944), 1–7; Lawrence K. Frank, "The Adolescent and the Family" in *Forty-Third Yearbook*, 240–254; Lawrence K. Frank, *Society as the Patient: Essays on Culture and Personality* (New Brunswick, NJ: Rutgers University Press, 1948).

10. Prescott, *A Doctor of Their Own*, 29–33.

11. Lawrence J. Friedman, *Identity's Architect: A Biography of Erik H. Erikson* (New York: Scribner, 1999), 27–81. Quotations on pp. 42 and 47. See also Kit Welchman, *Erik Erikson: His Life, Work, and Significance* (Buckingham, England: Open University Press, 2000); and Robert Coles, *Erik H. Erikson: The Growth of His Work* (Boston: Little, Brown, 1970).

12. Friedman, *Identity's Architect*, 49–56, 60, 87–101, 109–139.

13. Erik H. Erikson, *Childhood and Society* (New York: W. W. Norton, 1950).

14. Erikson, *Childhood and Society*, 228; Friedman, *Identity's Architect*, 149–241.

15. Erikson, *Childhood and Society*; Erik H. Erikson, *Insight and Responsibility: Lectures on the Ethical Implications of Psychoanalytic Insight* (New York: W. W. Norton, 1964); Erik H. Erikson, *Identity: Youth and Crisis* (New York: W. W. Norton, 1968); Erik H. Erikson, *Young Man Luther: A Study in Psychoanalysis and History* (New York: W. W. Norton, 1958); Erik H. Erikson, *Gandhi's Truth: On the Origins of Militant Nonviolence* (New York: W. W. Norton, 1969).

16. Erikson, "Inner and Outer Space: Reflections on Womanhood," in *The Woman in America*, ed. Robert Jay Lifton (Boston: Beacon Press, 1964), 1–26. Quotations on pp. 14, 25, 6, 19, 21, 20. Italics in original. "Inner and Outer Space" was first published in *Daedalus* in 1964 and was also reprinted in Erikson, *Identity: Youth and Crisis* and slightly revised and reprinted in Erik H. Erikson, *Life History and the Historical Moment* (New York: W. W. Norton, 1975).

17. Erikson, "Inner and Outer Space," 26, 21.

18. Friedman, 220–227, 423–426. Liberal feminism, which was the orientation of those who founded the National Organization for Women in 1966, emphasized woman's common

humanity with man, sought to secure women's individual political and legal rights, and campaigned for women's equal access to and opportunity within existing political, economic, social institutions. Cultural feminism emerged in full force by around 1975 out of earlier radical strains in the women's liberation movement. Cultural feminists emphasized and affirmed women's differences from men, celebrated the contributions of a distinctive "women's culture," and sought to create separate social institutions based on women's supposedly superior values of compassion, nurturance, and cooperation. See Sara M. Evans, *Tidal Wave: How Women Changed America at Century's End* (New York: Free Press, 2003), 24–26, 143–144.

19. For Piaget's work, see, for example, Jean Piaget, *The Psychology of Intelligence* (London: Routledge & Kegan Paul, 1950); Jean Piaget, *The Origins of Intelligence in Children* (New York: International Universities Press, 1952); Jean Piaget, *The Moral Judgment of the Child* (New York: Free Press, 1965); Jean Piaget, "The Intellectual Development of the Adolescent," in *Adolescence: Psychosocial Perspectives*, ed. Gerald Caplan and Serge Lebovici (New York: Basic Books, 1969), 22–26. For Kohlberg, see Lawrence Kohlberg, "The Development of Children's Orientations toward a Moral Order. I. Sequence in the Development of Moral Thought," *Vita Humana* 6 (1963): 11–33; Lawrence Kohlberg, "Moral Development and Identification," in *Child Psychology, 62nd Yearbook of the National Society for the Study of Education*, ed. Harold W. Stevenson (Chicago: University of Chicago Press, 1963).

20. Anna Freud, "Adolescence," in *The Writings of Anna Freud. Vol. V: Research at the Hampstead Child-Therapy Clinic and Other Papers, 1956–1965* (New York: International Universities Press, 1969): 136–166. Quotations on pp. 145, 149, 155, 164, 150. Anna Freud read "Adolescence" at the thirty-fifth anniversary of the Worcester Youth Guidance Center on September 18, 1957. It was first published in *The Psychoanalytic Study of the Child* 13 (1955): 255–278. For midcentury works that challenged the notion of biological storm and stress at adolescence, see Albert Bandura, "The Stormy Decade: Fact or Fiction?" *Psychology in the School* 1 (1964): 224–231; Elizabeth Douvan and Joseph Adelson, *The Adolescent Experience* (New York: Wiley, 1966); and Daniel Offer, *The Psychological World of the Teen-Ager: A Study of Normal Adolescent Boys* (New York: Basic Books, 1969).

21. Prescott, *A Doctor of Their Own*, 37–73.

22. Prescott, *A Doctor of Their Own*, 74–117. Quotations on pp. 99 and 109.

23. For a full treatment of these feminist challenges, see Mari Jo Buhle, *Feminism and Its Discontents: A Century of Struggle with Psychoanalysis* (Cambridge, MA: Harvard University Press, 1998), Chapters 6 and 7; and Jane F. Gerhard, *Desiring Revolution: Second-Wave Feminism and the Rewriting of American Sexual Thought, 1920–1982* (New York: Columbia University Press, 2001), Chapters 2–5.

24. "Listening to a Different Voice: Celia Kitzinger Interviews Carol Gilligan," *Feminism & Psychology* 4, no. 3 (1994), 417. For an overview of Gillman's early career, see Francine Prose, "Confident at 11, Confused at 16," in *New York Times Magazine* (January 7, 1990): 22–25, 37–38, 40, 45–46.

25. Carol Gilligan, *In a Different Voice: Psychological Theory and Women's Development* (Cambridge, MA: Harvard University Press, 1982), 1, 12, 8–9.

26. Gilligan, *In a Different Voice*, 8–9; 156–157, 2, 63. For Nancy Chodorow's ideas about development, see *The Reproduction of Mothering: Psychoanalysis and the Sociology of Gender*

(Berkeley: University of California Press, 1978). On the intellectual relationship between Gilligan and Chodorow's contributions to cultural feminism, see Buhle, *Feminism and Its Discontents*, 263–265.

27. See the "Essay on Sources" for the publications generated by Gilligan and her colleagues. The quotation is from Jill McLean Taylor, Carol Gilligan, and Amy M. Sullivan, *Between Voice and Silence: Women and Girls, Race and Relationship* (Cambridge, MA: Harvard University Press, 1995), 14. On Gilligan's *Time* citation, see Rosalind Barnett and Caryl Rivers, *Same Difference: How Gender Myths are Hurting Our Relationships, Our Children, and Our Jobs* (New York: Basic Books, 2004), 31.

28. McLean Taylor, Gilligan, and Sullivan, *Between Voice and Silence*, 23–24, 18; Carol Gilligan, "Prologue," in *Making Connections: The Relational Worlds of Adolescent Girls at Emma Willard School*, ed. Carol Gilligan et al. (Cambridge, MA: Harvard University Press, 1990), 4.

29. For an overview of the various arguments of Gilligan's critics and of the broad cultural impact of her ideas, see Barnett and Rivers, *Same Difference*, 25–43. Examples of critiques of Gilligan's work include: Debra Nails, "Gilligan's Mismeasure of Man," *Social Research* 50 (1983): 642–666; Lawrence J. Walker, "Sex Differences in the Development of Moral Reasoning: A Critical Review," *Child Development* 55 (1984): 677–691; Linda K. Kerber et al., "On *In a Different Voice:* An Interdisciplinary Forum," *Signs* 11, no. 2 (Winter 1986): 304–333; Ann Colby and William Damon, "Listening to a Different Voice: A Review of Gilligan's *In a Different Voice*," in *The Psychology of Women: Ongoing Debates*, ed. Mary Roth Walsh (New Haven, CT: Yale University Press, 1987), 321–329; Faye J. Crosby, *Juggling: The Unexpected Advantages of Balancing Career and Home for Women and Their Families* (New York: Free Press, 1993), 119–132; Sheila Greene, *The Psychological Development of Girls and Women: Rethinking Change in Time* (London: Routledge, 2003), 61–66. The entire issue of *Feminism & Psychology* 4, no. 3 (1994) is devoted to criticism of the work of the Harvard Project on female development. It also contains replies to these criticisms by Gilligan's colleagues and an interview with Gilligan herself.

30. Carol Tavris, "Reply to Brown and Gilligan," *Feminism & Psychology* 4, no. 3 (1994): 360.

31. For recent studies of girls' anger and aggression, see Lyn Mikel Brown, *Raising Their Voices: The Politics of Girls' Anger* (Cambridge, MA: Harvard University Press, 1998); Lyn Mikel Brown, *Girlfighting: Betrayal and Rejection among Girls* (New York: New York University Press, 2003); and Sharon Lamb, *The Secret Lives of Girls: What Good Girls Really Do—Sex Play, Aggression, and Their Guilt* (New York: Free Press, 2001).

32. Kenneth B. Kidd, *Making American Boys: Boyology and the Feral Tale* (Minneapolis: University of Minnesota Press, 2004), 167–188; Dan Kindlon and Michael Thompson, *Raising Cain: Protecting the Emotional Life of Boys* (New York: Ballantine Books, 1999); Eli H. Newberger, *The Men They Will Become: The Nature and Nurture of Male Character* (Reading, MA: Perseus Books, 1999); William S. Pollack, *Real Boys: Rescuing Our Sons from the Myths of Boyhood* (New York: Random House, 1998); William S. Pollack, *Real Boys' Voices: Boys Speak Out about Drugs, Sex, Violence, Bullying, Sports, School, Parent, and So Much More* (New York: Random House, 2000).

33. Michael Gurian, *The Wonder of Boys: What Parents, Mentors, and Educators Can Do to Shape Boys into Exceptional Men* (New York: Tarcher-Penguin Putnam, 1998); Christina Hoff

Sommers, *The War against Boys: How Misguided Feminism Is Harming Our Young Men* (New York: Simon & Schuster, 2000); Kidd, *Making American Boys*, 180–186.

34. For recent popular coverage of "The Boy Crisis," see the *Newsweek* cover story by Peg Tyre, "The Trouble with Boys," *Newsweek* (January 30, 2006): 44–52. Carol Gilligan contributes a sidebar to the article, " 'Mommy, I Know You': A Feminist Scholar Explains How the Study of Girls Can Teach Us about Boys," 53. Gilligan rejects the approaches of those who would solve the contemporary boy crisis by "reinstituting traditional codes of manhood, including a return to the patriarchal family." Instead, she contends that what psychologists have learned about girls' development ought to be brought to bear on the problems boys are now experiencing. Praising the extent to which the "[e]motions and relationships" characteristic of girls and women "have become desirable attributes of manhood," she also holds girls up as exemplars of some of the qualities of the fully developed human being. "With a clearer understanding of both boys' and girls' development," she writes, "we now have an opportunity to redress a system of gender relationships that endangers both sexes. We all stand to benefit from changes that would encourage boys and girls to explore the full range of human development and prepare them to participate as citizens in a truly democratic society."

35. Erica Burman, *Deconstructing Developmental Psychology* (London: Routledge, 1994), 188. Valerie Walkerdine, "Beyond Developmentalism?" *Theory & Psychology* 3.4 (1993): 451–469. Quotations on pp. 456, 461, 463.

36. Erica Burman, "Feminism and Discourse in Developmental Psychology: Power, Subjectivity and Interpretation" *Feminism and Psychology* 2 (1992): 45–60. Quotation on p. 50. See also Greene, *The Psychological Development of Girls and Women*, 8.

Essay on Sources

PRIMARY SOURCES

The primary sources for this project consist of published writings by scientists and scientifically minded intellectuals working in the fields of medicine, biology, psychology, and anthropology from 1830 to 1930.

For John Locke and Jean-Jacques Rousseau's ideas about development and youth, see John Locke, "Some Thoughts Concerning Education," in *John Locke on Education,* ed. Peter Gay (New York: Teachers College, Columbia University, 1964); and Jean-Jacques Rousseau, *Emile, or On Education,* introduction, translation, and notes by Allan Bloom (New York: Basic Books, 1979).

Health reformers' works discussed in the first chapter are William A. Alcott, *The Young Man's Guide to Excellence,* 14th edition (Boston, 1841); Alcott, *Familiar Letters to Young Men on Various Subjects* (Buffalo, NY, 1850); Alcott, *The Young Woman's Guide to Excellence,* 16th edition (New York, 1854); Alcott, *The Young Woman's Book of Health* (New York, 1855); Alcott, *The Physiology of Marriage* (Boston, 1856); Elizabeth Blackwell, *The Laws of Life, with Special Reference to the Physical Education of Girls* (New York, 1852); Amariah Brigham, *Remarks on the Influence of Mental Cultivation and Mental Excitement upon Health,* 2nd edition (Boston, 1833); Andrew Combe, *The Principles of Physiology Applied to the Preservation of Health and to the Improvement of Physical and Mental Education,* 7th edition (New York, 1849); O. S. Fowler, *Physiology, Animal and Mental: Applied to the Preservation and Restoration of Health of Body, and Power of Mind,* 6th edition (New York, 1853); Fowler, *Self-Culture and Perfection of Character including the Management of Youth* (New York, 1853); Fowler, *Sexual Science* (Chicago, 1870); Sylvester Graham, *A Lecture to Young Men* (Providence, RI, 1834); William Sweetser, *Mental Hygiene; or, an Examination of the Intellect and Passions Designed to Show How They Affect and Are Affected by the Bodily Functions, and Their Influences on Health and Physiology* (New York, 1850); Jno. Stainback Wilson, *Woman's Home Book of Health: A Work for Mothers and for Families* (Philadelphia, 1860); Samuel B. Woodward, *Hints for the Young in Relation to the Health of Body and Mind,* 4th edition (Boston, 1840).

The two works by Edward H. Clarke in which he lays out his protests against coeducation and expounds on his ideas about development are Clarke, *Sex in Education: Or, a Fair Chance for the Girls* (Boston: James R. Osgood, 1873; reprint edition, Arno Press, 1972); and Clarke, *Building of a Brain* (Boston: Houghton Mifflin, 1874). The sources for Herbert Spencer's theories of evolution and development analyzed here are Spencer, *Education: Intellectual, Moral, and*

Physical (New York: D. Appleton, 1860); Spencer, *First Principles*, 4th edition (New York: D. Appleton, 1890 [1862]); Spencer, *The Principles of Biology* 2 vols. (New York: D. Appleton, 1897 [1866]); Herbert Spencer, *Social Statics, or, The Conditions Essential to Human Happiness Specified, and the First of Them Developed* (New York: D. Appleton, 1872); Spencer, "Psychology of the Sexes," *The Popular Science Monthly* 4 (November 1873): 30–38. See also Henry Maudsley, "Sex in Mind and in Education," *The Popular Science Monthly* 5 (June 1874): 198–215.

Mary Putnam Jacobi's counterpoint to Clarke's and Spencer's conceptions of female development are presented in: Jacobi, "Mental Action and Physical Health," in *The Education of American Girls*, ed. Anna C. Brackett (New York: G. P. Putnam's Sons, 1874); Jacobi, *The Question of Rest for Women during Menstruation* (New York: G. P. Putnam's Sons, 1877). See also Jacobi, "Modern Female Invalidism," in *Mary Putnam Jacobi, M.D.: A Pathfinder in Medicine*, ed. Women's Medical Association of New York City (New York: G. P. Putnam's Sons, 1925), 478–482. Direct protests against Clarke's work were also voiced in George F. Comfort and Mrs. Anna Manning Comfort, MD, *Woman's Education and Woman's Health: Chiefly in Reply to "Sex in Education"* (Syracuse, NY: Thos. W. Durston, 1874); Eliza Bisbee Duffey, *No Sex in Education: An Equal Chance for Both Girls and Boys* (Philadelphia: J. M. Stoddart, 1874); Julia Ward Howe, ed. *Sex and Education: A Reply to Dr. E.H. Clarke's "Sex in Education"* (Boston: Roberts Brothers, 1874; reprint edition, New York: Arno Press, 1972); and Bracket, *The Education of American Girls*. Also see Duffey's *What Women Should Know: A Woman's Book about Women* (Philadelphia: J. M. Stoddart, 1873; reprint edition, New York: Arno Press, 1974). Marion Harland incorporates conclusions from both Clarke and Jacobi in *Eve's Daughters or Common Sense for Maid, Wife, and Mother* (New York: Charles Scribner's Sons, 1885). Some of the many other late-nineteenth-century doctors who weighed in on the coeducation debate and contributed to competing conceptualizations of female development include George Napheys, *The Physical Life of Woman: Advice to the Maiden, Wife, and Mother* (Philadelphia, 1872); John Harvey Kellogg, *Plain Facts for Old and Young* (Burlington, IA: I. F. Segner, 1882); Kellogg, *Ladies' Guide in Health and Disease* (Des Moines, IA: W. D. Condit, 1883); Kellogg, *The Home Handbook of Domestic Hygiene and Rational Medicine*, revised and enlarged edition (1896); Elizabeth Blackwell, *The Human Element in Sex, Being a Medical Inquiry into the Relation of Sexual Physiology to Christian Morality* (London, 1894 [1884]); Henry Lyman et al., *The Practical Home Physician* (Houston, 1885); Thomas Addis Emmet, *The Principles and Practice of Gynecology*, 3rd edition (Philadelphia: Henry C. Lea's Son, 1884); William Capp, *The Daughter: Her Health, Education and Wedlock* (Philadelphia: F. A. Davis, 1891); and Anna M. Galbraith, *The Four Epochs of Woman's Life: A Study in Hygiene* (Philadelphia: W. B. Saunders, 1903).

The works of G. Stanley Hall discussed in chapter three and throughout the book are Hall, "The Moral and Religious Training of Children," *Princeton Review* 9 (1882): 26–48; Hall, "Modern Methods in the Study of the Soul," *Christian Register* 75 (February 1886): 131–133; Hall and Respondents, "How Far Is the Present High-School and Early College Training Adapted to the Nature and Needs of Adolescents?" *The School Review* 9 (December 1901): 649–681; Hall, "Psychic Arrest in Adolescence," *Journal of Proceedings and Addresses of the National Educational Association* (July 1903): 811–813; Hall, "Coeducation in the High School," *Journal of Proceedings and Addresses of the National Educational Association* (July 1903): 446–451; Hall, "Coeducation," *Journal of Proceedings and Addresses of the National Educational Association*

(June–July 1904): 538–542; Hall, *Adolescence: Its Psychology and Its Relations to Physiology, Anthropology, Sociology, Sex, Crime, Religion and Education*, 2 vols. (New York: D. Appleton, 1904); Hall, *Youth, Its Education, Regimen and Hygiene* (New York: D. Appleton, 1906); Hall, "The Needs and Methods of Educating Young People in the Hygiene of Sex," *Pedagogical Seminary* 15 (March 1908): 82–91; Hall, "Feminization in School and Home," *World's Work* 16 (May 1908): 10237–10244; Hall, "From Generation to Generation: With Some Plain Language about Race Suicide and the Instruction of Children during Adolescence," *American Magazine* (July 1908): 249–254; Hall, "The Awkward Age," *Appleton's Magazine* 12 (August 1908): 149–156; Hall, "The Budding Girl," *Appleton's Magazine* 13 (January 1909): 47–54; Hall, "Education in Sex Hygiene," *Eugenics Review* 1 (January 1910): 242–253; Hall, "A Medium in the Bud," *American Journal of Psychology* 29 (April 1918): 144–158; Hall, *Recreations of a Psychologist* (New York: D. Appleton, 1920); Hall, "The Dangerous Age," *Pedagogical Seminary* 28 (September 1921): 275–294; Hall, "Flapper Americana Novissima," *Atlantic Monthly* 126, no. 6 (June 1922), 771–780; Hall, *Life and Confessions of a Psychologist* (New York: D. Appleton, 1923); A. Caswell Ellis and G. Stanley Hall, "A Study of Dolls," *Pedagogical Seminary* 4 (December 1896): 129–175; F. H. Saunders and G. Stanley Hall, "Pity," *American Journal of Psychology* 11 (July 1900): 534–591; and G. Stanley Hall and Theodate L. Smith, "Showing Off and Bashfulness as Phases of Self-Consciousness," *Pedagogical Seminary* 10 (June 1903): 159–199.

Six volumes of Ellis's *Studies in the Psychology* were published between 1890 and 1910, with the last published in 1928. Those sections particularly helpful for discerning Ellis's conception of female adolescence are "The Evolution of Modesty," "The Phenomenon of Sexual Periodicity," "Auto-Eroticism," "Analysis of the Sexual Impulse," "The Sexual Impulse in Women," and "Sex in Relation to Society." These are collected in Ellis, *Studies in the Psychology of Sex*, 2 vols. (New York: Random House, 1942).

The writings by Sigmund Freud about female psychosexual development are *Three Essays on the Theory of Sexuality* (1905); *Beyond the Pleasure Principle* (1920); *The Ego and the Id* (1923); "The Dissolution of the Oedipus Complex" (1924); "Some Psychical Consequences of the Anatomical Differences between the Sexes (1925); "Female Sexuality" (1931); and "Femininity" (1933). They are collected in the *Standard Edition of the Complete Psychological Works of Sigmund Freud*, ed. and trans. James Strachey (London: Hogarth Press, 1953–1974).

For Phyllis Blanchard's writings on child development, child guidance, and female adolescence, see Phyllis Blanchard, *The Adolescent Girl: A Study from the Psychoanalytic Viewpoint* (New York: Morrat, Yard, 1920); Blanchard, *The Child and Society: An Introduction to the Social Psychology of the Child* (New York: Longmans, Green, 1928); Blanchard, "The Child with Difficulties of Adjustment," in *A Handbook of Child Psychology*, 2nd edition, revised, ed. Carl Murchison (Worcester, MA: Clark University Press, 1933), 858–881; Phyllis Blanchard and Carlyn Manasses, *New Girls for Old* (New York: Macaulay, 1930); Ernest R. Groves and Phyllis Blanchard, *Introduction to Mental Hygiene* (New York: Henry Holt, 1930). See also, Blanchard, "The Longest Journey," in *These Modern Women: Autobiographical Essays from the Twenties*, revised edition, ed. Elaine Showalter (New York: Feminist Press, 1989), 105–109.

Leta Hollingworth wrote extensively about the psychology of women and children. For her writings about sexual difference, see Leta Hollingworth, "The Frequency of Amentia as Related to Sex," *Medical Record* 84 (1913): 753–756; Hollingworth, *Functional Periodicity: An Experimental Study of the Mental and Motor Abilities of Women during Menstruation* (New York:

Teachers College, Columbia University, 1914); Hollingworth, "Variability as Related to Sex Differences in Achievement: A Critique," *American Journal of Sociology* 19 (1914): 510–530; Hollingworth, "Sex Differences in Mental Traits," *Psychological Bulletin* 13 (1916): 377–383; Hollingworth, "Social Devices Impelling Women to Bear and Rear Children," *American Journal of Sociology* 22 (July 1916): 19–29; Hollingworth, "Comparison of the Sexes in Mental Traits," *Psychological Bulletin* 15 (1918): 427–432; Hollingworth, "Comparison of the Sexes in Mental Traits," *Psychological Bulletin,* 16 (1919): 371–373; Hollingworth, "Differential Action upon the Sexes of Forces which Tend to Segregate the Feeble-Minded," *Journal of Abnormal and Social Psychology* 17 (1922): 35–57; Hollingworth and Helen Montague, "The Comparative Variability of the Sexes at Birth," *American Journal of Sociology* 20 (1914): 335–370; Hollingworth and Max Schlapp, "An Economic and Social Study of Feeble-Minded Women," *Medical Record* 85 (1914): 1025–1228.

For some of Hollingworth's publications in the area of children's intelligence, see Hollingworth, *The Psychology of Subnormal Children* (New York: Macmillan, 1920); Hollingworth, *Special Talents and Defects* (New York: Macmillan, 1923); Hollingworth, *Gifted Children: Their Nature and Nurture* (New York: Macmillan, 1926); and Hollingworth, *Children above 180 IQ Stanford-Binet: Origin and Development* (Yonkers-on-Hudson, New York: World Book, 1942). On adolescence, see Hollingworth, "The Adolescent in the Family," *Child Study* 37 (1926): 5–6, 13; Hollingworth, "Getting Away from the Family: The Adolescent and His Life Plans," in *Concerning Parents: A Symposium on Modern Parenthood*, reprint edition (New York: New Republic, 1926); Hollingworth, *The Psychology of the Adolescent* (New York: D. Appleton, 1928), 71–82; Hollingworth, "After High School-What?" *Parents* 4 (June 1929): 21, 60; Hollingworth, "Developmental Problems of Middle Adolescence," *Westminster Leader* (1931): 5, 20–21; Hollingworth, "Late Adolescence," *Westminster Leader* (1931): 5, 24–25; and Hollingworth, "The Adolescent Child," in *A Handbook of Child Psychology*, 2nd edition, revised, ed. Carl Murchison (Worcester, MA: Clark University Press, 1933), 882–908.

For Lorine Pruette's writings about girls and women, see *Women and Leisure: A Study of Social Waste* (New York: E. P. Dutton, 1924); Pruette, "The Flapper," in *The New Generation: The Intimate Problems of Modern Parents and Children*, ed. V. F. Calverton and S. D. Schmalhausen (New York: Macaulay, 1930), 572–590; Pruette, "Why Women Fail," in *Woman's Coming of Age: A Symposium*, ed. V. F. Calverton and S. D. Schmalhausen (New York: Horace Liveright, 1931), 240–259. See also Pruette, "The Evolution of Disenchantment," in *These Modern Women*, 68–73.

For other key writings about child development, girls, and adolescence by reformers, social scientists, and child guidance professionals from the Progressive era through the early 1930s, see Edward L. Thorndike, *Notes on Child Study*, 2nd edition, vol. 8 (New York: Macmillan, 1903), 150–153; Thorndike, "The Newest Psychology," *Educational Review* 28 (October 1904): 217–227; Thorndike, "Magnitude and Rate of Alleged Changes at Adolescence," *Educational Review* 54 (September 1917): 140–147; Jane Addams, *The Spirit of Youth and the City Streets* (New York: Macmillan, 1926 [1909]); Sophonsiba Breckinridge and Edith Abbott, *The Delinquent Child and the Home: A Study of the Delinquent Wards of the Juvenile Court of Chicago* (New York: Arno Press, 1970; reprint edition, 1912); Ruth S. True, *The Neglected Girl* (New York: Survey Associates, 1914); Helen Thompson Woolley, "A New Scale of Mental and Physical Measurements for Adolescents and some of Its Uses," *Journal of Educational Psychology* 6

(November 1915): 521–550; William Healy, *The Individual Delinquent: A Text-Book of Diagnosis and Prognosis for All Concerned in Understanding Offenders* (Boston: Little, Brown, 1924 [1915]); Healy, *Mental Conflicts and Misconduct* (Boston: Little, Brown, 1917); Healy and Augusta F. Bronner, *Delinquents and Criminals: Their Making and Unmaking, Studies in Two American Cities* (New York: Macmillan, 1926); Mary E. Moxcey, *Girlhood and Character* (New York: Abington Press, 1916); Moxcey, *Leadership of Girls' Activities* (New York: Methodist Book Concern, 1919); Moxcey, *The Psychology of Middle Adolescence* (New York: Claxton Press, 1925); H. L. Hollingworth, *Vocational Psychology: Its Problems and Methods* (New York: D. Appleton, 1916); Hollingworth, *Mental Growth and Decline: A Survey of Developmental Psychology* (New York: D. Appleton, 1927); Jessie Taft, "Mental Hygiene Problems of Normal Adolescence," *Mental Hygiene* 5 (1921): 741–751; Henry H. Goddard, *Juvenile Delinquency* (New York: Dodd, Mead, 1923); William I. Thomas, *The Unadjusted Girl* (Boston: Little, Brown, 1923); Willystine Goodsell, *The Education of Women: Its Social Background and Its Problems* (New York: Macmillan, 1923); Winifred Richmond, *The Adolescent Girl: A Book for Parents and Teachers* (New York: Macmillan, 1925); *Concerning Parents: A Symposium on Modern Parenthood*, reprint edition (New York: New Republic, 1926); Douglas A. Thom, *Everyday Problems of the Everyday Child* (New York: D. Appleton, 1928); John B. Watson, *Psychological Care of Infant and Child* (New York: W. W. Norton, 1928); V. F. Calverton and S. D. Schmalhausen, eds. *Sex in Civilization* (New York: Macaulay, 1929); Calverton and Schmalhausen, eds., *The New Generation: The Intimate Problems of Modern Parents and Children* (New York: Macaulay, 1930); Calverton and Schmalhausen, eds., *Woman's Coming of Age: A Symposium* (New York: Horace Liveright, 1931); Grace Loucks Elliott, *Understanding the Adolescent Girl* (New York: Woman's Press, 1930); Gerald H. Pearson, "What the Adolescent Girl Needs in Her Home," *Mental Hygiene* (January 1930): 40–53; Carl Murchison, ed., *A Handbook of Child Psychology* (Worcester, MA: Clark University Press, 1931 and 1933 editions); Ruth Shonle Cavan and Jordan True Cavan, *Building a Girl's Personality: A Social Psychology of Later Girlhood* (New York: Abingdon Press, 1932).

For Franz Boas's writings about children's growth, see "Anthropological Investigations in Schools," *Science* 17 (June 1891): 351–352; Boas, "The Growth of Children," *Science* 14 (May 1892): 256–257; Boas, "The Growth of Children—II," *Science* 14 (May 1892): 281–282; "The Growth of Children," *Science* 20 (December 1892): 351–352; Boas, "On Dr. William Townsend Porter's Investigation of the Growth of the School Children of St. Louis," *Science*, n.s., 1 (March 1895): 225–230; Boas, "The Growth of First-Born Children," *Science*, n.s., 1 (March 1895): 402–404; Boas, "The Growth of Children," *Science*, n.s., 5 (April 1897): 570–573; and Boas, "The Growth of Children," *Science*, n.s., 36 (December 1912): 815–818; Franz Boas and C. Wissler, "Statistics of Growth" in *Report of the U.S. Commissioner of Education—for 1904, 1906* (Washington, DC: U.S. Commissioner of Education 1904, 1906), 25–32.

Margaret Mead's writings about childhood and adolescence in cross-cultural perspective include Mead, *Coming of Age in Samoa: A Psychological Study of Primitive Youth for Western Civilization*, with a foreword by Franz Boas (New York: Morrow, 1928); Mead, *Growing Up in New Guinea: A Comparative Study of Primitive Education*, reprint, with a new preface by Mead (New York: Morrow Quill Paperbacks, 1975 [1930]); Mead, "Adolescence in Primitive and Modern Society," in *The New Generation: The Intimate Problems of Modern Parents and Children*, ed. V. F. Calverton and S. D. Schmalhausen (New York: Macaulay, 1930); Mead, "The Primitive Child," in *A Handbook of Child Psychology,* ed. Carl Murchison (Worcester, MA: Clark University Press, 1931); Mead, *From the South Seas: Studies of Adolescence and Sex in Primitive Societies* (New York: William Morrow, 1939); Margaret Mead and Frances Cooke Macgregor, *Growth and Culture: A Photographic Study of Balinese Childhood* (New York: G. P. Putnam's Sons, 1951); Margaret Mead and Martha Wolfenstein, eds., *Childhood in Contemporary Cultures* (Chicago: University of Chicago Press, 1955). See also, Mead, *Blackberry Winter, My Earlier Years* (New York: William Morrow, 1972).

The works of Elsie Clews Parsons examined here are Parsons, *The Family: An Ethnographical and Historical Outline with Descriptive Notes, Planned as a Text-book for the Use of College Lecturers and Directors of Home-Reading Clubs* (New York: G. P. Putnam's Sons, 1906); Parsons, *The Old-Fashioned Woman: Primitive Fancies about the Sex* (New York: G. P. Putnam's Sons, 1913); Parsons, *Fear and Conventionality* (New York: G. P. Putnam's Sons, 1914); Parsons, "The Ceremonial of Growing Up," *School and Society* 2, no. 38 (1915): 408–411; Parsons, *Social Freedom: A Study of the Conflicts between Social Classification and Personality* (New York: G. P. Putnam's Sons, 1915); Parsons, *Social Rule: A Study of the Will to Power* (New York: G. P. Putnam's Sons, 1916).

For Miriam Van Waters writings on adolescent girlhood, see Van Waters, "The Adolescent Girl among Primitive Peoples," *Journal of Religious Psychology* 6, no. 4 (1913): 375–421; and 7, no. 1 (1914): 75–120; Van Waters, *Youth in Conflict* (New York: Republic Publishing Company, 1925); Van Waters, *Parents on Probation* (New York: New Republic, 1927).

The writings about child development, adolescence, and girlhood in the biological and social sciences from the mid-twentieth century briefly discussed in the "Epilogue" are Arnold Gesell, Frances L. Ilg and Louise Bates, *Youth: The Years from Ten to Sixteen* (New York: Harper & Row, 1954); Lawrence K. Frank, "Society as the Patient," *American Journal of Sociology* 42 (1936): 335–344; Frank, "Certain Problems of Puberty and Adolescence," *Journal of Pediatrics* 19, no. 3 (September 1941): 294–301; Frank, "Introduction: Adolescence as a Period of Transition," in *Forty-Third Yearbook of the National Society for the Study of Education. Part 1. Adolescence,* ed. Nelson B. Henry (Chicago: Department of Education, University of Chicago, 1944), 1–7; Frank, "The Adolescent and the Family," in *Forty-Third Yearbook,* 240–254; Frank, *Society as the Patient: Essays on Culture and Personality* (New Brunswick, NJ: Rutgers University Press, 1948); Erik H. Erikson, *Childhood and Society* (New York: W. W. Norton, 1950); Erikson, *Young Man Luther: A Study in Psychoanalysis and History* (New York: W. W. Norton, 1958); Erikson, *Insight and Responsibility: Lectures on the Ethical Implications of Psychoanalytic Insight* (New York: W. W. Norton, 1964); Erikson, "Inner and Outer Space: Reflections on Womanhood," in *The Woman in America,* ed. Robert Jay Lifton (Boston: Beacon Press, 1964), 1–26; Erikson, *Identity: Youth and Crisis* (New York: W. W. Norton, 1968); Erikson, *Gandhi's Truth: On the Origins of Militant Nonviolence* (New York: W. W. Norton, 1969); Jean Piaget, *The Psychology of Intelligence* (London: Routledge & Kegan Paul, 1950); Piaget, *The Origins of Intelligence in Children* (New York: International Universities Press, 1952); Piaget, *The Moral Judgment of the Child* (New York: Free Press, 1965); Piaget, "The Intellectual Development of the Adolescent," in *Adolescence: Psychosocial Perspectives,* ed. Gerald Caplan and Serge Lebovici (New York: Basic Books, 1969), 22–26; Anna Freud, "Adolescence," in *The Writings of Anna Freud. Vol. 5: Research at the Hampstead Child-Therapy Clinic and Other Papers, 1956–1965* (New York: International Universities Press, 1969 [1955]):

136–166; Lawrence Kohlberg, "The Development of Children's Orientations toward a Moral Order. I. Sequence in the Development of Moral Thought," *Vita Humana* 6 (1963): 11–33; Kohlberg, "Moral Development and Identification," in *Child Psychology: 62nd Yearbook of the National Society for the Study of Education*, ed. Harold W. Stevenson (Chicago: University of Chicago Press, 1963); Albert Bandura, "The Stormy Decade: Fact or Fiction?" *Psychology in the School* 1 (1964): 224–231; Elizabeth Douvan and Joseph Adelson, *The Adolescent Experience* (New York: Wiley, 1966); and Daniel Offer, *The Psychological World of the Teen-Ager: A Study of Normal Adolescent Boys* (New York: Basic Books, 1969).

For the publications by Carol Gilligan and her colleagues on female development, see Carol Gilligan, *In a Different Voice: Psychological Theory and Women's Development* (Cambridge, MA: Harvard University Press, 1982); Carol Gilligan et al., eds., *Mapping the Moral Domain: A Contribution of Women's Thinking to Psychological Theory and Education* (Cambridge, MA: Center for the Study of Gender, Education and Human Development, 1988); Carol Gilligan et al., eds., *Making Connections: The Relational World of Adolescent Girls at Emma Willard School* (Cambridge, MA: Harvard University Press, 1990); Carol Gilligan et al., eds., *Women, Girls & Psychotherapy: Reframing Resistance* (New York: Harrington Park Press, 1991); Lyn Mikel Brown and Carol Gilligan, *Meeting at the Crossroads: Women's Psychology and Girls' Development* (Cambridge, MA: Harvard University Press, 1992); Jill McLean Taylor, Carol Gilligan, and Amy M. Sullivan, *Between Voice and Silence: Women and Girls, Race and Relationship* (Cambridge, MA: Harvard University Press, 1995); Lyn Mikel Brown, *Raising Their Voices: The Politics of Girls' Anger* (Cambridge, MA: Harvard University Press, 1998); Mikel Brown, *Girlfighting: Betrayal and Rejection Among Girls* (New York: New York University Press, 2003); Sharon Lamb, *The Secret Lives of Girls: What Good Girls Really Do—Sex Play, Aggression, and Their Guilt* (New York: Free Press, 2001). For works making use of Gilligan's paradigm aimed at a popular audience, see Elizabeth Debold et al., *Mother Daughter Revolution: From Betrayal to Power* (Reading, MA: Addison-Wesley, 1993); Judy Mann, *The Difference: Growing Up Female in America* (New York: Warner Books, 1994); Peggy Orenstein, *School Girls: Young Women, Self-Esteem, and the Confidence Gap* (New York: Doubleday, 1994).

For contemporary works on male development, see Michael Gurian, *The Wonder of Boys: What Parents, Mentors, and Educators Can Do to Shape Boys into Exceptional Men* (New York: Tarcher-Penguin Putnam, 1998); Dan Kindlon and Michael Thompson, *Raising Cain: Protecting the Emotional Life of Boys* (New York: Ballantine Books, 1999); Eli H. Newberger, *The Men They Will Become: The Nature and Nurture of Male Character* (Reading, MA: Perseus Books, 1999); William S. Pollack, *Real Boys: Rescuing Our Sons from the Myths of Boyhood* (New York: Random House, 1998); William S. Pollack, *Real Boys' Voices: Boys Speak Out about Drugs, Sex, Violence, Bullying, Sports, School, Parents, and So Much More* (New York: Random House, 2000); Christina Hoff Sommers, *The War against Boys: How Misguided Feminism is Harming Our Young Men* (New York: Simon & Schuster, 2000).

SECONDARY SOURCES
Introduction

The most important works on the history of youth and adolescence from the Colonial period through the third decade of the twentieth century in the United States that informed my analysis throughout this study include John and Virginia Demos, "Adolescence in Historical Perspective," in *The American Family in Social-Historical Perspective*, ed. Michael Gordon (New York: St. Martin's Press, 1973), 209–221; Joseph F. Kett, *Rites of Passage: Adolescence in America, 1790–the Present* (New York: Basic Books, 1977); Paula S. Fass, *The Damned and the Beautiful: American Youth in the 1920s* (Oxford: Oxford University Press, 1977); John Demos, *Past, Present and Personal: The Family and the Life Course in American History* (New York: Oxford University Press, 1986); Roger Thompson, "Adolescent Culture in Colonial America," *Journal of Family History* 9 (Summer 1984): 131–141; Ross W. Beales Jr., "In Search of the Historical Child: Miniature Adulthood and Youth in Colonial New England," in *Growing Up in America: Children in Historical Perspective*, ed. N. Ray Hiner and Joseph M. Hawes (Urbana: University of Illinois Press, 1985), 17–24; John Modell, *Into One's Own: From Youth to Adulthood in the United States, 1920–1975* (Berkeley: University of California Press, 1989); Judith Graham, *Puritan Family Life: The Diary of Samuel Sewall* (Boston: Northeastern University Press, 2000); and Rodney Hessinger, *Seduced, Abandoned, and Reborn: Visions of Youth in Middle-Class America, 1780–1850* (Philadelphia: University of Pennsylvania Press, 2005).

On girlhood and female adolescence, see also James R. McGovern, "The American Woman's Pre–World War I Freedom in Manners and Morals," *Journal of American History* 55, no. 2 (September 1968): 315–333; Carroll Smith-Rosenberg, "From Puberty to Menopause: The Cycle of Femininity in Nineteenth Century America," in *Clio's Consciousness Raised: New Perspectives on the History of Women*, ed. Mary S. Hartman and Lois Banner (New York: Harper & Row, 1974), 23–37; Smith-Rosenberg, *Disorderly Conduct: Visions of Gender in Victorian America* (New York: Alfred A. Knopf, 1985), 53–76; Joseph M. Hawes, "The Strange History of Female Adolescence in the United States," *Journal of Psychohistory* 13, no. 1 (Summer 1985): 51–63; Joan Jacobs Brumberg, "Chlorotic Girls, 1870–1920: A Historical Perspective on Female Adolescence," in *Women and Health in America*, ed. Judith Walzer Leavitt (Madison: University of Wisconsin Press, 1984), 186–195; Brumberg, " 'Ruined' Girls: Community Responses to Illegitimacy in Upstate New York, 1890–1920," *Journal of Social History* 18 (Winter 1984): 247–272; Brumberg, *Fasting Girls: The Emergence of Anorexia Nervosa as a Modern Disease* (Cambridge, MA: Harvard University Press, 1988); Brumberg, *The Body Project: An Intimate History of American Girls* (New York: Random House, 1997); Anne M. Boylan, "Growing Up Female in Young America, 1800–1860," in *American Childhood: A Research Guide and Historical Handbook*, ed. Joseph M. Hawes and N. Ray Hiner (Westport, CT: Greenwood Press, 1985), 153–184; Kathy Peiss, *Cheap Amusements: Working Women and Leisure in Turn-of-the-Century New York* (Philadelphia: Temple University Press, 1986); Frances B. Cogan, *All-American Girl: The Ideal of Real Womanhood in Mid-Nineteenth Century America* (Athens: University of Georgia Press, 1989); Mary E. Odem, *Delinquent Daughters: Protecting and Policing Adolescent Female Sexuality in the United States, 1885–1920* (Chapel Hill: University of North Carolina Press, 1995); Ruth M. Alexander, *The Girl Problem: Female Sexual Delinquency in New York, 1900–1930* (Ithaca, NY: Cornell University Press, 1995); Caroline Kitch, *The Girl on the Magazine Cover: The Origins of Visual Stereotypes in American Mass Media* (Chapel Hill: University of North Carolina Press, 2001); Jane H. Hunter, *How Young Ladies Became Girls: The Victorian Origins of American Girlhood* (New Haven, CT: Yale University Press, 2002); Melanie Dawson, "The Miniaturizing of Girlhood: Nineteenth-Century Playtime and Gendered Theories of Development," in *The American Child: A Cultural Studies Reader*, ed.

Caroline F. Levander and Carol J. Singley (New Brunswick, NJ: Rutgers University Press, 2003); Kelly Schrum, *Some Wore Bobby Sox: The Emergence of Teenage Girl Culture, 1920–1945* (New York: Palgrave Macmillan, 2004).

For studies of youth and adolescence in medieval and early modern Europe, see Steven R. Smith, "The London Apprentices as Seventeenth-Century Adolescents," *Past and Present* 61 (1973); Smith, "Religion and Conception of Youth in Seventeenth-Century England," *History of Childhood Quarterly* 2 (1974): 493–516; Lawrence Stone, *The Family, Sex and Marriage in England 1500–1800* (New York: Harper & Row, 1977); Barbara A. Hanawalt, *The Ties That Bound: Peasant Families in Medieval England* (New York: Oxford University Press, 1986); Hanawalt, " 'The Child of Bristowe' and the Making of Middle-Class Adolescence," in *Bodies and Disciplines: Intersections of Literature and History in Fifteenth-Century England,* ed. Barbara A. Hanawalt and David Wallace (Minneapolis: University of Minnesota Press, 1996), 155–178; Ilana Krausman Ben-Amos, *Adolescence and Youth in Early Modern England* (New Haven, CT: Yale University Press, 1994); Paul Griffiths, *Youth and Authority: Formative Experiences in England 1560–1640* (Oxford: Clarendon Press, 1996); Konrad Eisenbichler, ed., *The Premodern Teenager: Youth in Society, 1150–1650* (Toronto: Centre for Reformation and Renaissance Studies, 2002); Helen King, *The Disease of Virgins: Green Sickness, Chlorosis and the Problems of Puberty* (London: Routledge, 2004).

Works about girls and boys growing up and the concept of adolescence in Europe in the nineteenth and twentieth centuries include John R. Gillis, *Youth and History: Tradition and Change in European Age Relations, 1770–Present* (New York: Academic Press, 1981); Carol Dyhouse, *Girls Growing Up in Late Victorian and Edwardian England* (London: Routledge & Kegan Paul, 1981); John Neubauer, *The Fin-de-Siècle Culture of Adolescence* (New Haven, CT: Yale University Press, 1992); Sally Mitchell, *The New Girl: Girls' Culture in England, 1880–1915* (New York: Columbia University Press, 1995).

For other general works and surveys in the history of children and childhood in the United States, see Harvey J. Graff, *Conflicting Paths: Growing Up in America* (Cambridge, MA: Harvard University Press, 1995); Jacqueline S. Reinier, *From Virtue to Character: American Childhood, 1775–1850* (New York: Twayne Publishers, 1996); Priscilla Ferguson Clement, *Growing Pains: Children in the Industrial Age, 1850–1890* (New York: Twayne Publishers, 1997); David I. Macleod, *The Age of the Child: Children in America, 1890–1920* (New York: Twayne Publishers, 1998); Ann Hulbert, *Raising America: Experts, Parents, and a Century of Advice about Children* (New York: Alfred A. Knopf, 2003); and Steven Mintz, *Huck's Raft: A History of American Childhood* (Cambridge, MA: Belknap Press of Harvard University Press, 2004).

Useful reference works on the range of topics covered in this study are Richard M. Lerner, Anne C. Petersen, and Jeanne Brooks-Gunn, eds. *Encyclopedia of Adolescence,* 2 vols. (New York: Garland Publishing, 1991); Jacqueline V. Lerner, Richard M. Lerner, and Jordan Finkelstein, eds. *Adolescence in America: An Encyclopedia,* 2 vols. (Santa Barbara, CA: ABC CLIO, 2001); Pricilla Ferguson Clement and Jacqueline S. Reinier, eds., *Boyhood in America: An Encyclopedia,* 2 vols. (Santa Barbara, CA: ABC CLIO, 2001); Miriam Forman-Brunell, ed., *Girlhood in America: An Encyclopedia,* 2 vols. (Santa Barbara, CA: ABC CLIO, 2001); Joseph M. Hawes and Elizabeth F. Shores, eds., *The Family in America: An Encyclopedia,* 2 vols. (Santa Barbara, CA: ABC CLIO, 2001); and Paula S. Fass, ed., *Encyclopedia of Children and Childhood: In History and Society,* 3 vols. (New York: Macmillan Reference, 2004).

Sources by contemporary developmental psychologists that were particularly helpful for this study are William Kessen, "The American Child and Other Cultural Inventions," *American Psychologist* 34, no. 10 (October 1979): 815–820; Kessen, *The Rise and Fall of Development* (Worcester, MA: Clark University Press, 1990); Richard M. Lerner, ed. *Developmental Psychology: Historical and Philosophical Perspectives* (Hillsdale, NJ: Lawrence Erlbaum Associates, 1983); Lerner, *Concepts and Theories of Human Development*, 2nd edition (Mahwah, NJ: Lawrence Erlbaum Associates, 1997); Frank S. Kessel and Alexander W. Siegel, eds., *The Child and Other Cultural Inventions* (New York: Praeger, 1983); Julian Henriques et al., *Changing the Subject: Psychology, Social Regulation, and Subjectivity* (London: Methuen, 1984); Georg Eckardt, Wolfgang G. Bringmann, and Lothar Sprung, eds., *Contributions to a History of Developmental Psychology* (Berlin: Moulton Publishers, 1985); Urie Brofenbrenner, Frank Kessel, William Kessen, and Sheldon White, "Toward a Critical Social History of Developmental Psychology: A Propaedeutic Discussion, *American Psychologist* 41, no. 11 (November 1986): 1218–1230; John M. Broughton, ed., *Critical Theories of Psychological Development* (New York: Plenum Press, 1987); John R. Morss, *The Biologising of Childhood: Developmental Psychology and the Darwinian Myth* (Hove and London: Lawrence Erlbaum Associates, 1990); Morss, "Making Waves: Deconstruction and Developmental Psychology," *Theory & Psychology* 2, no. 4 (1992): 445–465; Marc H. Bornstein and Michael E. Lamb, eds., *Developmental Psychology: An Advanced Textbook*, 3rd edition (Hillsdale, NJ: Lawrence Erlbaum Associates, 1992); Valerie Walkerdine, "Beyond Developmentalism?" *Theory & Psychology* 3, no. 4 (1993): 451–469; Ross D. Parke, Peter A. Ornstein, John J. Rieser, and Carolyn Zahn-Waxler, eds., *A Century of Developmental Psychology* (Washington DC: American Psychological Association, 1994); Erica Burman, "Feminism and Discourse in Developmental Psychology: Power, Subjectivity and Interpretation," *Feminism and Psychology* 2 (1992): 45–60; Burman, *Deconstructing Developmental Psychology* (London: Routledge, 1994); Sheila Greene, *The Psychological Development of Girls and Women: Rethinking Change in Time* (London: Routledge, 2003).

Scholarship in the area of cultural studies that shaped my thinking about the construction of the concept of adolescence includes Angela McRobbie and Mica Nava, eds., *Gender and Generation* (London: Macmillan, 1984); Sue Lees, *Sugar and Spice: Sexuality and Adolescent Girls* (London: Penguin Books, 1993); Nancy Lesko, *Act Your Age: A Cultural Construction of Adolescence* (New York: Routledge, 2001); Claudia Castañeda, *Figurations: Child, Bodies, Worlds* (Durham, NC: Duke University Press, 2002); Catherine Driscoll, *Girls: Feminine Adolescence in Popular Culture and Cultural Theory* (New York: Columbia University Press, 2002); Carol F. Levander and Carol J. Singley, eds., *The American Child: A Cultural Studies Reader* (New Brunswick, NJ: Rutgers University Press, 2003).

Chapter One. "Laws of Life"

For an exploration of age as a salient category of social meaning and experience in the American context, see Howard P. Chudacoff, *How Old Are You? Age Consciousness in American Culture* (Princeton, NJ: Princeton University Press, 1989). In addition to the sources on youth and adolescence, C. Dallett Hemphill's *Bowing to Necessities: A History of Manners in America, 1620–1860* (New York: Oxford University Press, 1999) is particularly helpful for thinking about the changing status and meaning of youth from the Colonial era through the antebellum period.

For sources on girls' and young women's experiences at work and in school up to 1860, in addition to the studies of girlhood and female adolescence listed above, see Nancy F. Cott, *The Bonds of Womanhood: "Woman's Sphere" in New England, 1780–1835* (New Haven, CT: Yale University Press, 1977); Thomas Dublin, *Women at Work: The Transformation of Work and Community in Lowell, Massachusetts, 1826–1860* (New York: Columbia University Press, 1979); Dublin, *Farm to Factory: Women's Letters, 1830–1860*, 2nd edition (New York: Columbia University Press, 1993); Dublin, *Transforming Women's Work: New England Lives in the Industrial Revolution* (Ithaca, NY: Cornell University Press, 1994); Faye E. Dudden, *Serving Women: Household Service in Nineteenth-Century America* (Middletown, CT: Wesleyan University Press, 1983); Christine Stansell, *City of Women: Sex and Class in New York, 1789–1860* (New York: Knopf, 1986); and Catherine E. Kelly, *In the New England Fashion: Reshaping Women's Lives in the Nineteenth Century* (Ithaca, NY: Cornell University Press, 1999).

Nancy F. Cott explains the emergence of separate spheres in *The Bonds of Womanhood*. Barbara Welter describes the characteristics of the ideal woman in this ideology in "The Cult of True Womanhood, 1820–1860," *American Quarterly* 18 (1966): 151–174. Linda K. Kerber offers a useful look at the way the trope of separate spheres has been used by women's historians in "Separate Spheres, Female Worlds, Woman's Place: The Rhetoric of Women's History," *Journal of American History* 75 (1988): 9–39. For two excellent studies that complicate the relationship between home and work and public and private, and the shape these relationships took in women's lives during the early nineteenth century, see Jeanne Boydston, *Home and Work: Housework, Wages, and the Ideology of Labor in the Early Republic* (New York: Oxford University Press, 1990); and Kelly, *In the New England Fashion*.

For studies focusing on social reform efforts directed at neglected and troublesome children and youth, see Barbara M. Brenzel, *Daughters of the State: A Social Portrait of the First Reform School for Girls in North America, 1856–1905* (Cambridge, MA: MIT Press, 1983); and Joseph M. Hawes, *The Children's Rights Movement: A History of Advocacy and Protection* (Boston: Twayne, 1991).

The experiences of girls and boys growing up in slavery in the nineteenth century are explored in Deborah Gray White, *Ar'n't I a Woman? Female Slaves in the Plantation South* (New York: W. W. Norton, 1985); Wilma King, *Stolen Childhood: Slave Youth in Nineteenth Century America* (Bloomington: Indiana University Press, 1995); and Marie Jenkins Schwartz, *Born in Bondage: Growing Up Enslaved in the Antebellum South* (Cambridge, MA: Harvard University Press, 2000).

Studies that focus on girls' and young women's religious experiences in this period include Nancy F. Cott, "Young Women in the Second Great Awakening," *American Quarterly* 3 (1975): 15–29; Joan Jacobs Brumberg, *Mission for Life: The Story of the Family of Adoniram Judson, the Dramatic Events of the First American Foreign Mission, and the Course of Evangelical Religion in the Nineteenth Century* (New York: Free Press, 1980); Susan M. Juster, " 'In a Different Voice': Male and Female Narratives of Religious Conversion in Post-Revolutionary America," *American Quarterly* 41 (1989): 34–62; Nathan O. Hatch, *The Democratization of American Christianity* (New Haven, CT: Yale University Press, 1989); and Catherine A. Brekus, *Strangers and Pilgrims: Female Preaching in America 1740–1845* (Chapel Hill: University of North Carolina Press, 1998).

For middle-class girls' literary culture in the nineteenth-century Anglo American context, see Sheila Rowbotham, *Good Girls Make Good Wives: Guidance for Girls in Victorian Fiction* (Oxford: Basil Blackwell, 1989); Claudia Nelson and Lynne Vallone, eds., *The Girl's Own, Cultural Histories of the Anglo-American Girl, 1830–1915* (Athens: University of Georgia Press, 1994); and Lynne Vallone, *Disciplines of Virtue: Girls' Culture in the Eighteenth and Nineteenth Centuries* (New Haven, CT: Yale University Press, 1995); Sarah Bilston, *The Awkward Age in Women's Popular Fiction, 1850–1900: Girls and the Transition to Womanhood* (Oxford: Clarendon Press, 2004); Hunter, *How Young Ladies became Girls;* and Mitchell, *The New Girl.*

For changes in family life in this period, see Daniel Scott Smith, "Family Limitation, Sexual Control, and Domestic Feminism in Victorian America," *Feminist Studies* 1, nos. 3–4 (1973): 40–57; Robert V. Wells, "Family History and Demographic Transition," *Journal of Social History* 9 (Fall 1975): 1–19; Ruth H. Bloch, "American Feminine Ideals in Transition: The Rise of the Moral Mother, 1785–1815," *Feminist Studies* 4 (June 1978): 101–126; Carl N. Degler, *At Odds: Women and the Family in America from the Revolution to the Present* (New York: Oxford University Press, 1980); Mary P. Ryan, *Cradle of the Middle Class: The Family in Oneida County, New York, 1790–1865* (Cambridge: Cambridge University Press, 1981); Steven Mintz, *A Prison of Expectations: The Family in Victorian Culture* (New York: New York University Press, 1983); Steven Mintz and Susan Kellogg, *Domestic Revolutions: A Social History of American Family Life* (New York: Free Press, 1988); Stephanie Coontz, *The Social Origins of Private Life: A History of American Families, 1600–1900* (London: Verso, 1988); Nancy M. Theriot, *The Biosocial Construction of Femininity: Mothers and Daughters in Nineteenth-Century America* (New York: Greenwood Press, 1988); Jan Lewis, "Mother's Love: The Construction of an Emotion in Nineteenth-Century America," in *Social History and Issues in Human Consciousness: Some Interdisciplinary Connections,* ed. Andrew E. Barnes and Peter N. Stearns (New York: New York University Press, 1989), 209–229; Robert L. Griswold, *Fatherhood in America: A History* (New York: Basic Books, 1993); Shawn Johansen, *Family Men: Middle-Class Fatherhood in Early Industrializing America* (New York: Routledge, 2001).

Along with the sources on youth and adolescence, historical studies of masculinity that include examinations of boys and boyhood in this period are Peter N. Stearns, *Be a Man! Males in Modern Society,* 2nd edition (New York: Holmes & Meier, 1990); Mark C. Carnes and Clyde Griffen, eds., *Meanings for Manhood: Constructions of Masculinity in Victorian America* (Chicago: University of Chicago Press, 1990); E. Anthony Rotundo, *American Manhood: Transformations in Masculinity from the Revolution to the Modern Era* (New York: Basic Books, 1993); Michael Kimmel, *Manhood in America: A Cultural History* (New York: Free Press, 1996); and Anne S. Lombard, *Making Manhood: Growing Up Male in Colonial New England* (Cambridge, MA: Harvard University Press, 2003). See also W. J. Rorabaugh, *The Craft Apprentice: From Franklin to the Machine Age in America* (New York: Oxford University Press, 1986).

In addition to the works from developmental psychology and cultural studies listed above, for exploration of the history of ideas about development and evolution, see Arthur O. Lovejoy, *The Great Chain of Being: A Study of the History of an Idea* (Cambridge, MA: Harvard University Press, 1936); Richard Hofstadter, *Social Darwinism in American Thought, 1860–1915* (Philadelphia: University of Pennsylvania Press, 1944); Stephen Jay Gould, *Ontogeny and Phylogeny* (Cambridge, MA: Belknap Press of Harvard University Press, 1977); Peter J. Bowler, *The Eclipse of Darwinism: Anti-Darwinian Evolution Theories in the Decades around 1900*

(Baltimore: Johns Hopkins University Press, 1983); Bowler, *Evolution: The History of An Idea* (Berkley: University of California Press, 1984); Bowler, *Theories of Human Evolution: A Century of Debate 1844–1944* (Baltimore: Johns Hopkins University Press, 1986); Bowler, *The Non-Darwinian Revolution: Reinterpreting a Historical Myth* (Baltimore: Johns Hopkins University Press, 1988); Thomas Laqueur, *Making Sex: Body and Gender from the Greeks to Freud* (Cambridge, MA: Harvard University Press, 1990); and Carolyn Steedman, *Strange Dislocations: Childhood and the Idea of Human Interiority, 1780–1930* (Cambridge, MA: Harvard University Press, 1995).

Scholars' readings of the works of Locke and Rousseau that were especially helpful for my purposes in this chapter are Jay Fliegelman, *Prodigals and Pilgrims: The American Revolution Patriarchal Authority, 1750–1800* (Cambridge: Cambridge University Press, 1982); Jane Roland Martin, *Reclaiming a Conversation: The Ideal of the Educated Woman* (New Haven, CT: Yale University Press, 1985); Julia Grant, *Raising Baby by the Book: The Education of American Mothers* (New Haven, CT: Yale University Press, 1998); Ernest Freeberg, *The Education of Laura Bridgman: First Deaf and Blind Person to Learn Language* (Cambridge, MA: Harvard University Press, 2001); Reinier, *From Virtue to Character*; Hemphill, *Bowing to Necessities*; and Hessinger, *Seduced, Abandoned, and Reborn*.

On antebellum health reform, see John Blake, "Health Reform," in *The Rise of Adventism: Religion and Society in Mid-Nineteenth-Century America*, ed. E. S. Gaustad (New York: Harper and Row, 1974), 30–49; Stephen Nissenbaum, *Sex, Diet, and Debility in Jacksonian America: Sylvester Graham and Health Reform* (Westport, CT: Greenwood Press, 1980); James C. Whorton, *Crusaders for Fitness: The History of American Health Reformers* (Princeton, NJ: Princeton University Press, 1982); Jayme A. Sokolow, *Eros and Modernization: Sylvester Graham, Health Reform, and the Origins of Victorian Sexuality* (Rutherford, NJ: Fairleigh Dickinson University Press, 1983); Regina Markell Morantz-Sanchez, *Sympathy and Science: Women Physicians in American Medicine* (New York: Oxford University Press, 1985); Martha H. Verbrugge, *Able-Bodied Womanhood: Personal Health and Social Change in Nineteenth-Century Boston* (New York: Oxford University Press, 1988); and Margaret Puskar-Pasewicz, " 'For the Good of the Whole': Vegetarianism in 19th-Century America," Indiana University Dissertation, 2003. On phrenology, see Stephen Tomlinson, *Head Masters: Phrenology, Secular Education, and Nineteenth-Century Social Thought* (Tuscaloosa: University of Alabama Press, 2005).

Chapter Two. "Persistence" versus "Periodicity"

Scholarship on the history of girls' and women's education and on the late-nineteenth-century debate incited by Clarke's work includes Mary Roth Walsh, *"Doctors Wanted: No Women Need Apply": Sexual Barriers in the Medical Profession, 1835–1975* (New Haven, CT: Yale University Press, 1977); Rosalind Rosenberg, *Beyond Separate Spheres: Intellectual Roots of Modern Feminism* (New Haven, CT: Yale University Press, 1982); Barbara Miller Solomon, *In the Company of Educated Women: A History of Women and Higher Education in America* (New Haven, CT: Yale University Press, 1985); Lynn D. Gordon, *Gender and Higher Education in the Progressive Era* (New Haven, Ct: Yale University Press, 1990); David Tyack and Elisabeth Hansot, *Learning Together: A History of Coeducation in American Public Schools* (New Haven,

CT: Yale University Press, 1990); Louise Michele Newman, *White Women's Rights: The Racial Origins of Feminism in the United States* (New York: Oxford University Press, 1999); Margaret A. Lowe, *Looking Good: College Women and Body Image, 1875–1930* (Baltimore: Johns Hopkins University Press, 2003); and Markell Morantz-Sanchez, *Sympathy and Science.*

Works on ideas about women and gender in late-Victorian science that helped with my exploration of conceptualizations of female adolescent development in the coeducation debate include Carroll Smith-Rosenberg and Charles Rosenberg, "The Female Animal: Medical and Biological Views of Woman and her Role in Nineteenth-Century America," *Journal of American History* 60, no. 2 (September 1973): 332–356; John S. Haller Jr. and Robin M. Haller, *The Physician and Sexuality in Victorian America* (Urbana: University of Illinois Press, 1974); Roland O. Valdisseri, "Menstruation and Medical Theory: An Historical Overview," *Journal of the American Medical Women's Association* 38, no. 3 (May/June 1983): 66–70; Louise Michele Newman, ed., *Men's Ideas/Women's Realities: Popular Science, 1870–1915* (New York: Pergamon Press, 1985); Cynthia Eagle Russett, *Sexual Science: The Victorian Construction of Womanhood* (Cambridge, MA: Harvard University Press, 1989); Ornella Moscucci, *The Science of Woman: Gynecology and Gender in England, 1800–1929* (Cambridge: Cambridge University Press, 1990); Ann E. Walker, *The Menstrual Cycle* (London: Routledge, 1997); Carroll Smith-Rosenberg, "Puberty to Menopause"; Rosenberg, *Beyond Separate Spheres;* Laqueur, *Making Sex;* Newman, *White Women's Rights;* Markell Morantz-Sanchez, *Sympathy and Science.* The most thorough treatment of Mary Putnam Jacobi's life and work is by Carla Jean Bittel, "The Science of Women's Rights: The Medical and Political Worlds of Mary Putnam Jacobi," Cornell University Dissertation, 2003.

Chapter Three. From "Budding Girl" to "Flapper Americana Novissima"

Indispensable to any exploration of the work of G. Stanley Hall is Dorothy Ross's biography, *G. Stanley Hall: The Psychologist as Prophet* (Chicago: University of Chicago Press, 1972). Other scholars' treatments of Hall and his work in the context of his times that are especially helpful include Robert E. Grinder and Charles Strickland, "G. Stanley Hall and the Social Significance of Adolescence," *Teacher's College Record* 64, no. 5 (February 1963): 390–399; Robert E. Grinder, "The Concept of Adolescence in the Genetic Psychology of G. Stanley Hall," *Child Development* 40, no. 2 (June 1969): 355–369; T. J. Jackson Lears, *No Place of Grace: Anti-Modernism and the Transformation of American Culture, 1880–1920* (New York, Pantheon, 1981); Ernest R. Hilgard, *Psychology in America: A Historical Survey* (San Diego, CA: Harcourt Brace Jovanovich, 1987), 529–534; Gail Bederman, *Manliness and Civilization: A Cultural History of Gender and Race in the United States, 1880–1917* (Chicago: University of Chicago Press, 1995); Heather Munro Prescott, *A Doctor of Their Own: The History of Adolescent Medicine* (Cambridge, MA: Harvard University Press, 1998); Jeffrey P. Moran, *Teaching Sex: The Shaping of Adolescence in the 20th Century* (Cambridge, MA: Harvard University Press, 2000); Kenneth B. Kidd, *Making American Boys: Boyology and the Feral Tale* (Minneapolis: University of Minnesota Press, 2004); Julia Grant, "A Real Boy and Not a Sissy: Gender, Childhood and Masculinity, 1890–1940," *Journal of Social History* 37, no. 4 (Summer 2004): 829–851; Kett, *Rites of Passage;* Gillis, *Youth and History;* Lesko, *Act Your Age!;* Sheldon H. White, "G. Stanley Hall: From Philosophy to Developmental Psychology," in *A Century of*

Developmental Psychology, 103–125; Driscoll, *Girls;* Dyehouse, *Girls Growing Up in Late Victorian and Edwardian England;* Rosenberg, *Beyond Separate Spheres;* Russett, *Sexual Science;* Rotundo, *American Manhood.*

In addition to this body of scholarship, my analysis in this chapter and Chapter 5 of the category of race in relation to Hall's thinking about adolescence is informed by John S. Haller Jr., *Outcasts from Evolution: Scientific Attitudes of Racial Inferiority, 1859–1900* (Urbana: University of Illinois Press, 1971); George W. Stocking Jr., *Victorian Anthropology* (New York: Free Press, 1987); Carl N. Degler, *In Search of Human Nature: The Decline and Revival of Darwinism in American Social Thought* (New York: Oxford University Press, 1991); Anne McClintock, *Imperial Leather: Race, Gender and Sexuality in the Colonial Conquest* (New York: Routledge, 1995); and Castañeda, *Figurations: Child, Bodies, Worlds.*

The following works helped with my discussions in this chapter and the next two on the meanings and experiences of adolescent sexuality during the late nineteenth and early twentieth centuries: Nathan G. Hale Jr., *Freud and the Americans: The Beginning of Psycho-analysis in the United States, 1876–1917* (New York: Oxford University Press, 1971); Carl N. Degler, "What Ought to Be and What Was: Women's Sexuality in the Nineteenth Century," *American Historical Review* 79 (1974): 1467–90; Michel Foucault, *The History of Sexuality*, trans. Robert Hurley, vol. 1, *An Introduction* (New York: Pantheon, 1978); Nancy F. Cott, "Passionlessness: An Interpretation of Victorian Sexual Ideology, 1790–1850," *Signs* 4, no. 2 (Winter 1978): 219–236; Peter Gay, *Education of the Senses,* vol. 1, *The Bourgeois Experience, Victoria to Freud* (New York: Oxford University Press, 1984); Carol Zisowitz Stearns and Peter N. Stearns, "Victorian Sexuality: Can Historians Do It Better?" *Journal of Social History* 18 (Summer 1985): 625–634; John D'Emilio and Estelle B. Freedman, *Intimate Matters: A History of Sexuality in America* (New York: Harper & Row, 1988), Steven Seidman, "Sexual Attitudes of Victorian and Post-Victorian Women: Another Look at the Mosher Survey," *Journal of American Studies* 23 (1989): 68–72; Paul Robinson, *The Modernization of Sex: Havelock Ellis, Alfred Kinsey, William Masters, and Virginia Johnson* (Ithaca, NY: Cornell University Press, 1989); Christina Simmons, "Modern Sexuality and the Myth of Victorian Repression," in *Passion and Power: Sexuality in History*, ed. Kathy Peiss and Christina Simmons (Philadelphia: Temple University Press, 1989), 157–177; Janice M. Irvine, *Disorders of Desire: Sex and Gender in Modern American Sexology* (Philadelphia: Temple University Press, 1990); Constance A. Nathanson, *Dangerous Passage: The Social Control of Sexuality in Women's Adolescence* (Philadelphia: Temple University Press, 1991); Pamela S. Haag, "In Search of 'the Real Thing': Ideologies of Love, Modern Romance, and Women's Sexual Subjectivity in the United States, 1920–1940," *Journal of the History of Sexuality* 2, no. 4 (1992): 547–577; Mary Jo Buhle, *Feminism and Its Discontents: A Century of Struggle with Psychoanalysis* (Cambridge, MA: Harvard University Press, 1998); Carol Groneman, *Nyphomania: A History* (New York: W. W. Norton), 2000; Jane F. Gerhard, *Desiring Revolution: Second-Wave Feminism and the Rewriting of American Sexual Thought 1920–1982* (New York: Columbia University Press, 2001); John Spurlock, "From Reassurance to Irrelevance: Adolescent Psychology and Homosexuality in America," *History of Psychology* 5, no. 1 (February 2002): 38–51; Laqueur, *Making Sex*; Moran, *Teaching Sex*; Peiss, *Cheap Amusements*; Odem, *Delinquent Daughters*; Brumberg, *The Body Project*; Schrum, *Some Wore Bobby Sox.*

Chapter Four. "New Girls for Old"

The most important works on female adolescence in the Progressive era and the 1920s are listed in the paragraph on girlhood and female adolescence under the sources for the "Introduction."

Helpful syntheses of the major developments in child welfare and the child sciences during this period include Hamilton Cravens, "Child-Saving in the Age of Professionalism, 1915–1930," in *American Childhood: A Research Guide and Historical Handbook,* ed. Joseph M. Hawes and N. Ray Hiner (Westport, CT: Greenwood Press, 1985); Joseph M. Hawes, *Children between the Wars: American Childhood, 1920–1940* (New York: Twayne Publishers, 1997); Hawes, *The Children's Rights Movement*; and Macleod, *The Age of the Child.* Also important for understanding the research about and treatment of children in this period is Hamilton Cravens, *Before Head Start: The Iowa Station and America's Children* (Chapel Hill: University of North Carolina Press, 1993). On the emergence and development of the child guidance movement from the Progressive era through the 1920s, see Margo Horn, *Before It's Too Late: The Child Guidance Movement in the United States* (Philadelphia: Temple University Press, 1989); Theresa Richardson, *The Century of the Child: The Mental Hygiene Movement and Social Policy in the United States and Canada* (Albany: State University of New York Press, 1989); and Kathleen W. Jones, *Taming the Troublesome Child: American Families, Child Guidance, and the Limits of Psychiatric Authority* (Cambridge, MA: Harvard University Press, 1999). On the range of expert advice produced about childhood in this period, see Hulbert, *Raising America.* For more on the dissemination of such advice and its reception by mothers, see Grant's book on parent education, *Raising Baby by the Book.*

On the history of eugenics, see Daniel J. Kevles, *In the Name of Eugenics: Genetics and the Uses of Human Heredity* (Cambridge, MA: Harvard University Press, 1995 [1985]); and Wendy Kline, *Building a Better Race: Gender, Sexuality, and Eugenics from the Turn of the Century to the Baby Boom* (Berkeley: University of California Press, 2001).

For the various roles of the "new woman" in American society in these periods, see Karen J. Blair, *The Clubwoman as Feminist: True Womanhood Redefined, 1868–1914* (New York: Holmes and Meier, 1980); Nancy F. Cott, *The Grounding of Modern Feminism* (New Haven, CT: Yale University Press, 1987); Robin Muncy, *Creating a Female Domain in Progressive Reform* (New York: Oxford University Press, 1991); Molly Ladd-Taylor, *Mother-Work: Women, Child Welfare, and the State, 1890–1930* (Urbana: University of Illinois Press, 1994); Lois Rudnick, *"The New Woman" in 1915: The Cultural Moment: The New Politics, the New Woman, the New Psychology, the New Art, and the New Theatre in America,* ed. Adele Heller and Lois Rudnick (New Brunswick, NJ: Rutgers University Press, 1991); Jean V. Matthews, *The Rise of the New Woman: The Women's Movement in America* (Chicago: Iran R. Dee, 2003); Newman, *White Women's Rights.* Christine Stansell's *American Moderns: Bohemian New York and the Creation of a New Century* (New York: Henry Holt, 2000) helped with situating the new woman in the context of the larger changes of the decades surrounding the turn of the century. Lynn Dumenil's *The Modern Temper: American Culture and Society in the 1920s* (New York: Hill and Wang, 1995) did the same for the period of the 1920s.

For the (gendered) history of the social and psychological sciences and women's contributions to them during the early twentieth century, especially the contributions of the three feminist psychologists explored in this chapter, see Harry L. Hollingworth, *Leta Stetter Hollingworth: A Biography* (Lincoln: University of Nebraska Press, 1943): Stephen J. Gould, *Mismeasure of Man* (New York: W. W. Norton, 1981); Lela B. Costing, *Two Sisters for Social Justice: A Biography of Grace and Edith Abbott* (Urbana: University of Illinois Press, 1983); Miriam Lewin, ed. *In the Shadow of the Past: Psychology Portrays the Sexes* (New York: Columbia University Press, 1984); Ellen Fitzpatrick, *Endless Crusade: Women Social Scientists and Progressive Reform* (New York: Oxford University Press, 1990); Mina Carson, *Settlement Folk: Social Thought and the American Settlement Movement, 1885–1930* (Chicago: University of Chicago Press, 1990); Mary Jo Deegan, *Jane Addams and the Men of the Chicago School, 1892–1918* (New Brunswick, NJ: Transaction Books, 1990); Linda Kreger Silverman, "Leta Stetter Hollingworth: Champion of the Psychology of Women and Gifted Children," *Journal of Educational Psychology* 84, no. 1 (1992): 20–27; Leonard Wilcox, *V. F. Calverton: Radical in the American Grain* (Philadelphia: Temple University Press, 1992); Elizabeth Lunbeck, *The Psychiatric Persuasion: Knowledge, Gender and Power in Modern America* (Princeton, NJ: Princeton University Press, 1994); Leila Zenderland, *Measuring Minds: Henry Herbert Goddard and the Origins of American Intelligence Testing* (New York: Cambridge University Press, 1997); Helen Silverberg, ed., *Gender and American Social Science: The Formative Years* (Princeton, NJ: Princeton University Press, 1998); Rosenberg, *Beyond Separate Spheres*; Degler, *In Search of Human Nature;* Russett, *Sexual Science;* Buhle, *Feminism and Its Discontents.*

On the history of endocrinology, see Nelly Oudshoorn, *Beyond the Natural Body: An Archeology of Sex Hormones* (London: Routledge, 1994); and Margaret Marsh and Wanda Ronner, *The Empty Cradle: Infertility in America from Colonial Times to the Present* (Baltimore: Johns Hopkins University Press, 1996).

Chapter Five. Adolescent Girlhood Comes of Age?

My understanding of the intellectual tenets of Victorian anthropology and of the meanings given and uses made of the concept of culture in early-twentieth-century anthropology is informed by George W. Stocking Jr., *Race, Culture, and Evolution: Essays in the History of Anthropology* (New York: Free Press, 1968); Stocking, *Victorian Anthropology;* Stocking, ed., *Romantic Motives: Essays on Anthropological Sensibility* (Madison: University of Wisconsin Press, 1989); Elizabeth Fee, "The Sexual Politics of Victorian Social Anthropology," in *Clio's Consciousness Raised: New Perspectives on the History of Women*, ed. Mary S. Hart and Lois Banner (New York: Harper & Row, 1974), 86–102; Hamilton Cravens, *The Triumph of Evolution: American Scientists and the Heredity-Environment Controversy* (Philadelphia: University of Pennsylvania Press, 1978); Joan Jacobs Brumberg, "Zenanas and Girlless Villages: The Ethnology of American Evangelical Women, 1870–1910," *Journal of American History* 69, no. 2 (1982): 347–371; Michael Cole, *Cultural Psychology: A Once and Future Discipline* (Cambridge, MA: Belknap Press of Harvard University Press, 1996); Susan Hegeman, *Patterns for America: Modernism and the Concept of Culture* (Princeton, NJ: Princeton University Press, 1999); Margaret D. Jacobs, *Engendered Encounters: Feminism and Pueblo Cultures 1879–1934* (Lincoln: University of Nebraska Press, 1999); Degler, *In Search of Human Nature;* Rosenberg, *Beyond Separate Spheres;* Silverberg, *Gender and American Social Science.*

On Margaret Mead's life and work, see Jane Howard, *Margaret Mead: A Life* (New York: Simon and Schuster, 1984); Lois W. Banner, *Intertwined Lives; Margaret Mead, Ruth Benedict, and Their Circle* (New York: Alfred A. Knopf, 2003); Newman, *White Women's Rights;* and Rosenberg, *Beyond Separate Spheres.* For the recent scholarly critique of Mead as a cultural determinist and of the limits of "Boasian Culturalism," see Derek Freeman, *Margaret Mead and Samoa: The Unmaking of an Anthropological Myth* (Cambridge, MA: Harvard University Press, 1983), especially Chapters 5–7; and Freeman, *The Fateful Hoaxing of Margaret Mead: A Historical Analysis of Her Samoan Research* (Boulder, CO: Westview Press, 1999), especially Chapter 14. For responses to Freeman and reassessments of Mead's work in light of his critique, see Ray A. Rappaport, "Desecrating the Holy Woman: Derek Freeman's Attack on Margaret Mead," *American Scholar* 55 (Summer 1986): 313–347; Lowell E. Holmes, *Quest for the Real Samoa: The Mead/Freeman Controversy and Beyond* (South Hadley, MA: Bergin & Garvey, 1987); Stephen O. Murray, "Problematic Aspects of Freeman's Account of Boasian Culture," *Current Anthropology* 31.4 (August–October 1990): 401–407; Murray, "On Boasians and Margaret Mead: Reply to Freeman," *Current Anthropology* 32, no. 4 (August–October 1991): 448–452; and Banner, 235–239.

Franz Boas's contributions to the emergence of cultural anthropology are explored in Walter Goldschmidt, ed., *The Anthropology of Franz Boas, Essays on the Centennial of His Birth* (Menasha, WI: American Anthropological Association, 1959); Stocking, *Race, Culture and Evolution;* Degler, *In Search of Human Nature;* and Hegeman, *Patterns for America.*

On Elsie Clews Parsons, see Peter H. Hare, *A Woman's Quest for Science: Portrait of Anthropologist Elsie Clews Parsons* (Buffalo, NY: Prometheus, 1985); Ute Gacs et al., eds., *Women Anthropologists: A Biographical Dictionary* (New York: Greenwood Press, 1988), 282–290; Louise Lamphere, "Feminist Anthropology: The Legacy of Elsie Clews Parsons," *American Ethnologist* 16 (August 1989): 519–520; Rosemary Lévy Zumwalt, *Wealth and Rebellion: Elsie Clews Parsons, Anthropologist and Folklorist* (Urbana: University of Illinois Press, 1992); Delsey Deacon, *Elsie Clews Parsons: Inventing Modern Life* (Chicago: University of Chicago Press, 1997); Rosenberg, *Beyond Separate Spheres;* Jacobs, *Engendered Encounters.*

On Miriam Van Waters, see Estelle B. Freedman, *Maternal Justice: Miriam Van Waters and the Female Reform Tradition* (Chicago: University of Chicago Press, 1996); and Odem, *Delinquent Daughters.*

The most important work on the history of scientific treatment of adolescence from the mid-twentieth century to the present is Heather Munro Prescott's *A Doctor of Their Own: The History of Adolescent Medicine* (Cambridge, MA: Harvard University Press, 1998). See also Prescott, " 'I was a Teenage Dwarf': The Social Construction of 'Normal' Adolescent Growth and Development in the United States," in *Formative Years: Children's Health in the United States, 1880–2000,* ed. Alexandra Minna Stern and Howard Markel (Ann Arbor: University of Michigan Press, 2005), 153–182.

Index

supplemental nutrition theory, 77–78
Sweetser, William, 33, 35, 37

Taft, Jessie, 149
Tarde, Gabriel, 210
Tarvis, Carol, 250
technological advances, 19
Terman, Lewis, 188, 191
Thomas, William I., 158, 174
Thompson, Michael, 251
Thorndike, Edward L., 152–153, 155
True, Ruth, 142

The Unadjusted Girl (Thomas), 174
universities. *See* colleges / universities
urban areas: African Americans and immigrants in, 146; opportunities for working-class girls in, 87; public schools in, 53

Van Gennep, Arnold, 219–220
Van Waters, Miriam, 196, 199, 216; background of, 216–218; cross-cultural comparisons used by, 220–221; functionalist approach of, 218–219; Parsons vs., 221–222; on puberty rites, 219–220, 232
von Baer, Karl Ernst, 59
von Bischoff, Theodor, 67
von Helmholtz, Hermann, 34
voting rights, for women, 151

wage system, 19, 20
Walkerdine, Valerie, 4, 252
Watson, John B., 148
Weissman, August, 96
Whorton, James, 29
Willard, Emma, 52
Williams, Frankwood E., 172
Wilson, J., 42
Wissler, Clark, 152–153
women. *See* females
Women and Leisure: A Study of Social Waste (Pruette), 184–185
Women's Bureau, 181
Women's Trade Union League, 181
Woodbridge, William Channing, 31
Woodward, Samuel, 44
Woodworth, Robert, 152–153
Woolley, Helen Thompson, 190, 191
working-class families: experiences of boys in, 20; institutions for, 136
working-class girls: experiences of, 15–16, 87; industrial employment for, 15–16; sexual exploitation of, 15, 16; sexual instinct and, 141–142
Wundt, Wilhelm, 93

Yerkes, Robert, 188
Youth: The Years from Ten to Sixteen (Gesell), 237, 238